37.95
68I

LOG ANALYSIS
OF SUBSURFACE GEOLOGY

"The Three Wise Monkeys" contemplate a log. From left to right: the geologist, the mathematician, the log analyst.

LOG ANALYSIS OF SUBSURFACE GEOLOGY
Concepts and Computer Methods

JOHN H. DOVETON

Kansas Geological Survey
University of Kansas
Lawrence, Kansas

A Wiley-Interscience Publication
JOHN WILEY & SONS
New York • Chichester • Brisbane • Toronto • Singapore

Copyright © 1986 by John Wiley & Sons, Inc.

All rights reserved. Published simultaneously in Canada.

Reproduction or translation of any part of this work beyond that permitted by Section 107 or 108 of the 1976 United States Copyright Act without the permission of the copyright owner is unlawful. Requests for permission or further information should be addressed to the Permissions Department, John Wiley & Sons, Inc.

Library of Congress Cataloging in Publication Data:

Doveton, John H., 1944-
  Log analysis of subsurface geology.

  "A Wiley-Interscience publication."
  Includes index.
  1. Geology—Data processing.   2. Borings—Data processing.   3. Geophysical well logging—Data processing.   I. Title.

QE33.2.B6D68   1986   550.1'3'028   85-17790
ISBN 0-471-89368-4

Printed in the United States of America

10 9 8 7 6 5 4 3 2 1

To my mother, Valerie

ACKNOWLEDGMENTS

Acknowledgment is made for illustrations used or adapted from the following sources: American Association of Petroleum Geologists (Figs. 2.6–2.8, 3.3, 3.6, 3.15, 5.15); Society of Economic Paleontologists and Mineralogists (Figs. 2.9, 2.23, 5.21); Society of Petroleum Engineers (Figs. 4.5, 4.9, 4.12); Schlumberger (Figs. Intro-1, 1.11, 1.13, 5.20); Dresser Atlas (Figs. 1.12, 1.14, 2.2, 3.1, 3.14, 4.3); *American Scientist* (Fig. 7.13); *Mathematical Geology* (Fig. 3.10); Canadian Society of Petroleum Geologists (Fig. 1.15); Clay Minerals Society (Fig. 3.13); Gulf Coast Association of Geological Societies (Fig. 3.5); Cornell University Press and Chapman and Hall Ltd. (Fig. 2.10); Van Nostrand Reinhold Inc. (Fig. 1.3). The terms laterolog, LL3, LL7, microlaterolog, MLL, proximity log, BHC, GNT, SNP, CNL, lithodensity log, *M–N* plot, MID plot, Z plot are registered marks of Schlumberger Ltd.

J. H. D.

PREFACE

It does not seem so long ago that I was a young exploration geologist who was learning the rudiments of log analysis. It was clear that the log traces recorded variations in physical properties of stratigraphic sequences whose length often surpassed the column exposed in the walls of the Grand Canyon. But if log analysis was intrinsically geological, it was certainly very different from field studies of outcrop that were the focus of my student days. As we worked on the problems with our slide rules, our goal was the location of reservoir horizons with favorable porosities and hydrocarbon "shows" that indicated commercial production. The logs gave useful insights on subsurface geology, but this knowledge was subordinated to the task of improving quantitative estimations of porosities and hydrocarbon saturations.

In recent years, there has been a steady evolution within log analysis from methods grounded firmly in classical reservoir engineering theory to a more integrated approach which incorporates geological concepts and interpretations. As with most changes, this development has reflected the influence of a variety of factors. The subtleties of the stratigraphic trap and the complexities of tertiary oil recovery projects have spurred the design and application of appropriate geological models. The revolutionary advances in seismic data processing which led to seismic stratigraphy often require simulation models based on logs to extract their full geological meaning. At the same time, a new generation of logging tools has been introduced which measure properties that reveal patterns of deposition and diagenesis to those who care to read them.

In a parallel development, the mechanics of log interpretation have moved beyond the slide rule, through the era of the programmable calculator, and finally to the microcomputers of today. The use of the computer has a tremendous potential to transform the measurements of modern logging tools into continuous profiles of geological variability which express lithofacies, mineralogy, and internal geometry. This multivariate data base is the mathematical geologist's dream: it is both numerical and numerous.

However, the full potential of modern logs as a medium for enhanced interpretations of subsurface geology draws ultimately on the fusion of an intelligent understanding of several areas. These include a familiarity with the principles of logging measurements, a knowledge of reservoir engineering fundamentals, an informed appreciation of simple mathematics and computer processing, and last, but definitely not least, a sound background in geology. I have symbolized the multidisciplinary approach as the "Three Wise Monkeys" in the frontispiece illustration, which, while whimsical, attempts to make the point in deadly earnest.

The aim of this book is to present a concise introduction to log analysis which is keyed to geological implications and mathematical methods for geological interpretation. The first four chapters provide a preliminary review of the most common logging tools. Exhaustive descriptions of the tools and the physics of their measurements are readily available in manuals published by various logging companies. Chapter 5 describes the common graphical methods of multiple log interpretation which are used widely by log analysts. Chapter 6 introduces basic concepts of matrix algebra and illustrates their use in the geological analysis of well logs. Chapter 7 outlines examples of statistical analysis and signal processing techniques as methods to reveal patterns and trends of geological variation. Chapter 8 reviews remedial methods directed to the suppression of measurement error associated with logs. Finally, in Chapter 9, the concepts developed in the analysis of individual well profiles are extended to the mapping of mineralogies and lithofacies across entire areas.

As is the case with almost every book, the author cannot assume sole credit (but must assume all the blame). Many of the studies cited here were team efforts, and I am grateful to Ted Bornemann, Harold Cable, Curt Conley, John Davis, Roberto Peveraro, Pat St. Clair, Lynn Watney, and Zhu Ke-an for their various contributions. Cora Cowan typed the various versions of this manuscript and Jennifer Sims drafted many of the illustrations. Finally, I owe a special thanks to the Kansas Geological Survey as a stimulating institution for research and to its director, William W. Hambleton, for his encouragement and support.

JOHN H. DOVETON

Lawrence, Kansas
November 1985

CONTENTS

INTRODUCTION		1
1	**RESISTIVITY**	8

    Basics, 8
    Resistivities of Rocks, 9
    Quantitative Relationships Between Resistivity and Porosity, 11
    Rock Texture Properties and Resistivity–Porosity Relationships, 14
    Advanced Models of Textural–Resistivity Relationships, 18
    Facies Interpretation from Resistivity and Porosity Logs, 19
    Textural–Resistivity Relationships in Carbonates, 21
    The Resistivity of Hydrocarbon-Bearing Rocks, 24
    The Measurement of Resistivity by Logging Tools, 27
        The Normal Device, 28
        The Induction Device, 29
        The Laterolog, 30
        Microresistivity Tools, 30
        Vertical Resolution and Radius of Investigation, 32

2	**THE DIPMETER**	35

    Structural Geologic Interpretations from Dipmeter Patterns, 40
        Structural Dip and Unconformities, 40
        Faults, 41
        Compactional Drape, 43

Dipmeter Interpretation of Sedimentary Structures and
Depositional Environments, 45
    Eolian Dunes, 46
    Braided Stream Deposits, 47
    Meandering Stream Point-Bars, 48
    Delta Distributary Channels, 51
    Distributary Mouth Bars, 52
    Estuarine and Tidal Channels, 53
    Barrier Bars, 54
    Marine Shelf Sands, 56
Analysis of Dipmeter Vectors, 57
Fabric Orientation Analysis from the Dipmeter, 64

3 THE SPONTANEOUS POTENTIAL AND GAMMA-RAY LOGS   70

The Spontaneous Potential Log, 71
    Factors Influencing the SP Log, 74
    Use of SP Shapes in the Recognition of Sandstone Depositional
    Environments, 75
The Gamma-Ray Log, 83
    Application of the Gamma-Ray Log to Shale Content
    Evaluation, 84
    Simple Sedimentary Environmental Interpretation from
    Gamma-Ray Logs, 88
    General Applications of the Gamma-Ray Log, 89
    Geologic Radioactive Sources of the Gamma-Ray Log, 92
    Applications of the Spectral Gamma-Ray Log, 95

4 POROSITY LOGS: SONIC, DENSITY, NEUTRON   100

The Sonic Log, 101
The Density Log, 106
The Neutron Log, 109
Analysis of Shale Compactional Trends, 113
Geologic Implications of Porosity Trends and Patterns, 118

5 GRAPHICAL METHODS OF LITHOLOGY DETERMINATION   123

Overlay of Porosity Logs on a Common Reference Scale, 124
Porosity Log Frequency Crossplots, 128
A Third Dimension: Z-Plots, 132
Condensed Projection Plots, 133

Contents   xiii

ROMA–UMA Crossplots, 138
Graphical Methods for Shaly Sandstones, 140
Crossplots as Tools of Geometry, 147

6 NUMERICAL METHODS FOR LITHOLOGY ESTIMATION FROM WELL LOGS   153

The Basics of Compositional Analysis Using Matrix Algebra, 154
Matrix Algebra Compositional Analysis of Logs, 157
Case Study 1: Analysis of a Cambro-Ordovician Dolomite, 161
Case Study 2: Analysis of a Pennsylvanian Sandstone–Shale Sequence, 164
   Volumetric Solutions of Shales, 164
Case Study 3: Permian Evaporites, 167
Advanced Matrix Algebra Methods for Component Estimation, 170
   The Underdetermined System, 170
   The Uniquely Determined System, 173
   The Overdetermined System, 175
   A Coordinated Solution Procedure, 177
Normative Log Solutions and Modal Petrography, 180

7 MATHEMATICAL ANALYSIS OF LOG TRENDS AND PATTERNS   184

Patterns in Compositional Associations, 185
Time-Series Analysis of Log Data, 190
   Polynomial Analysis of Long-Term Trends in Log Traces, 190
   Polynomial Filtering of Short-Term Trends, 194
   Fourier Analysis of Cyclic Components, 196
   Short-Term Spectral Analysis, 199
Pattern Recognition Methods, 203
   Principal Component Analysis, 204
   Discriminant Function Analysis, 207

8 REMEDIAL CORRECTION OF LOGS   213

Single-Well Methods for Normalization, 214
   Use of Calibration Units, 214
   Application of Crossplots, 214
   Comparison with Core Analysis, 215
Dual-Well Methods, 216

Overlays, 216
Histograms, 217
Multiwell Methods for Normalization, 217
Simple Mapping, 217
Trend Surface Analysis, 217

9 LITHOFACIES MAPPING FROM LOGS — 225

Simple Lithofacies Mapping from a Single Log, 226
Moments and Polynomial Regression, 234
Three-Dimensional Slice Mapping, 238
Further Applications, 245
Complex Lithofacies Mapping from Multiple Logs, 246
Interpretation and Validation of the Wireline
Lithofacies Map, 250

APPENDIX 1 KIWI—A Computer Program for Compositional Analysis from Logs — 262

APPENDIX 2 Physical Properties of Common Geological Components — 267

INDEX — 269

INTRODUCTION

On a fine September day in 1927, a battered station wagon drove through the countryside of Alsace to the Péchelbronn oilfield. Assisted by Scheibli and Jost, Henri Doll unloaded an experimental resistivity sonde at the well Diefenbach 2905, Tower No. 7, and attached it by wire to a makeshift winch. The sonde was lowered to the bottom of the hole and resistivity measurements were made at 1-meter intervals in a time-consuming sequence of discrete recordings. On returning to Paris, Doll plotted the resistivity readings against depth on a piece of graph paper and, by linking successive observations with lines, drew the first electrical well log. An extract from this historic record is shown in Figure 1.

The Schlumberger company had been engaged in surface electrical studies for a number of years and had been successful in establishing relationships between bedrock geology and resistivity profiles. The Péchelbronn company had contracted with Schlumberger for some surface work and became interested in the possibility that borehole resistivity measurements might prove useful in helping their geologists with subsurface correlation problems. Conrad Schlumberger had independently considered the theoretical aspects of such a procedure and entrusted the design and operation of an experimental resistivity device for this purpose to his son-in-law, Doll.

In the Péchelbronn area, the stratigraphic problem was the recognition of the top of a bed of Hydrobiae marls which was sometimes missed in drilling. Doll's first electric log clearly indicated the Hydrobiae marls as a zone of uniformly low resistivity which was capped by a mixed sequence of conglomerates, sandstones, and hard marls, marked by more resistive peaks and troughs. In further logging runs in neighboring wells, it was found that many of these resistivity features could be correlated laterally, and the method of "electrical coring" was established as a useful aid in correlation. Details concerning these early events are described by Allaud and Martin (1977).

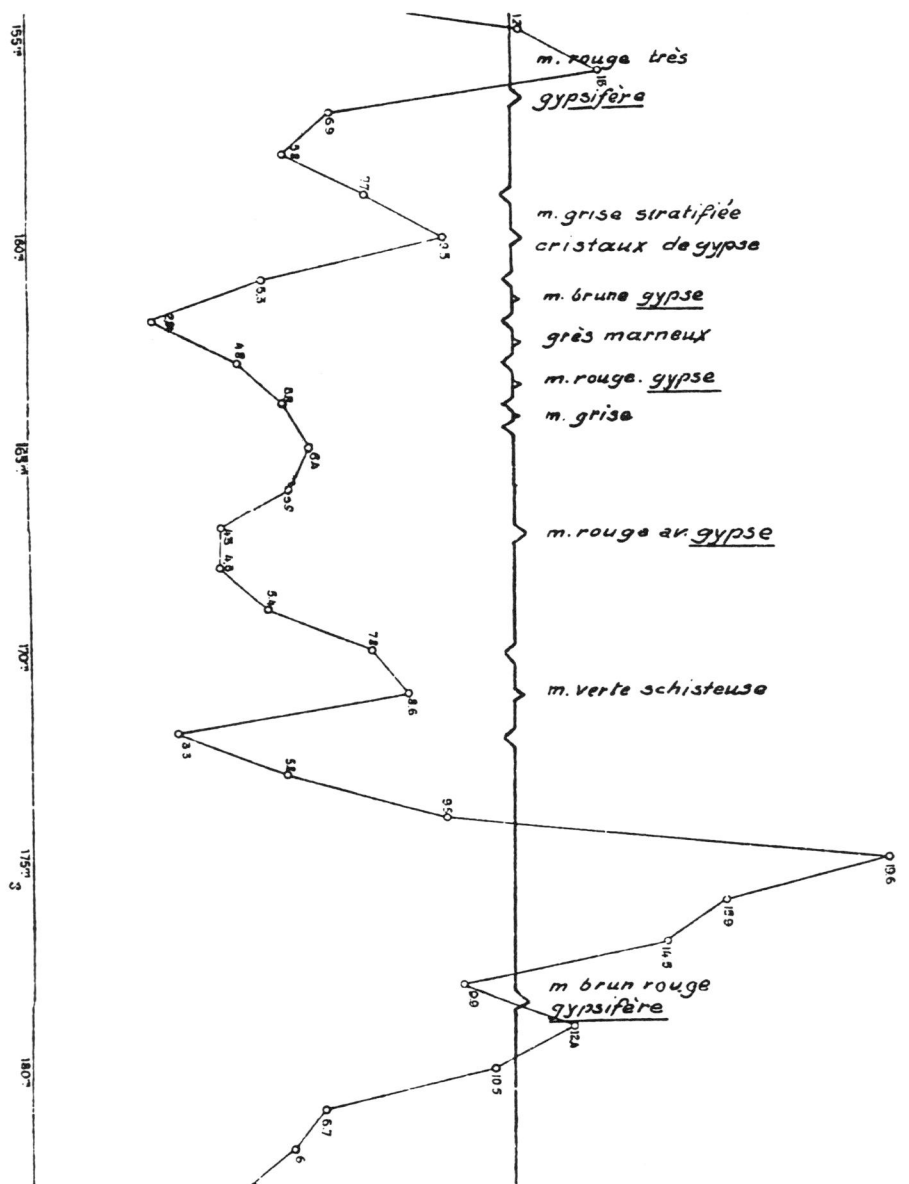

FIGURE 1. Extract from the first Schlumberger log, Alsace, 1927.

Introduction 3

There is an interesting point in this story. The first log was run to resolve a specific *geological* problem. The experiment succeeded in showing distinctive relationships between borehole resistivity measurements and lithology which were useful for stratigraphic correlation. Since that time, the major emphasis of log interpretation has shifted to the location and evaluation of porous, permeable hydrocarbon zones and a reservoir engineering perspective. More recently, there has been a growth of interest in the broader implications of logs as aids in the exploration for stratigraphic traps. Subtle changes in depositional facies and diagenetic patterns are the key to many modern plays. Clearly, there is a potential for logs to be used in lithofacies studies since, in the final analysis, *logs are physical measurements of rocks*.

The range of logging devices has expanded dramatically since the early days to include the measurement of a host of electrical, thermal, nuclear, and acoustic properties. Consequently, the logging program of a typical modern well records a wealth of numerical information from subsurface formations. The results are presented as analog traces of variation, drawn as functions of depth with scaling conventions which are observed by all logging service companies. These conventions have been chosen both to maintain a common standard and to allow the simultaneous presentation of a number of log traces in a meaningful manner (e.g., Fig. 2).

Some authors, such as Jeffries (1965), have objected to the modern trend in multiple trace presentation on the grounds that "too many curves cause chaos." The neophyte log analyst would certainly agree with this perception, but would generally find with practice that the multiple presentation format is a useful device in the determination of trends, patterns, and discriminations.

A useful metaphor can be found in music. A melody consists of a sequence of notes which are marked off in time intervals and bars. This aspect is commonly termed a horizontal pattern and is supplemented by a vertical pattern of chords and counterpoint which carry harmony. Written as sheet music, a tune therefore has two dimensions of horizontal–melodic and vertical–harmonic. A musical score may have limited meaning for nonmusicians, but its mastery is a matter of applied practice rather than an intellectual accomplishment. An experienced log analyst can read and interpret a multiple set of log traces with much the same ease that a musician can play a tune on a piano from a reading of sheet music. The responses of logs to a variety of idealized lithologies are themes that are the basis of repeated variations along the record of a stratigraphic sequence. The major composite variations in fluctuation of the log suite carry the "melody" of the interbedded lithologies and their gross reservoir characteristics. Differences between the logs are "harmonic" elements which convey useful information concerning changes in mineralogy and fabric.

There are, of course, some limitations. The visual interpretation of an extensive log suite recorded from a complex lithological sequence can be as challenging as the reading of a musical score written for a full orchestra. At this point, the use of a computer for log analysis is the obvious means to condense the large information base to transformations which are meaningful to both geologists and reservoir engineers. Because log measurements are numerical and numerous, the computer

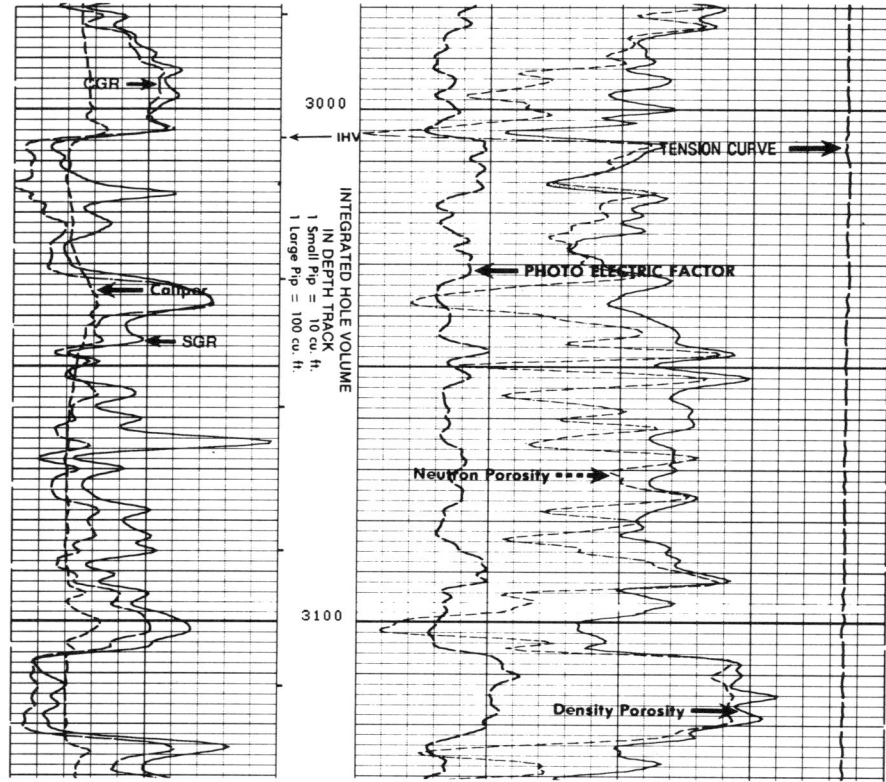

FIGURE 2. A modern wireline log combination. Caliper, gamma-ray, photoelectric factor, density, neutron, and tension logs from a Permian section in southern Kansas.

is a powerful device both as a workhorse in bulk processing of data, and as a medium for the mathematical analysis of relationships. Perhaps the major obstacle to work along these lines is the fact that useful results require a skillful integration of geology, mathematics, and log analysis. The viewpoints of narrow specialists whose understanding of logs are restrictive are shown symbolically by the "Three Wise Monkeys" illustration used as the frontispiece of this book.

HEAR NO EVIL: THE GEOLOGIST

Although logs have been used as graphic frameworks for subsurface stratigraphy for decades, concepts in their application by many geologists have barely advanced beyond those used in the analysis of the first log. Fundamentally, log traces are used in the delineation of boundaries of subsurface units for the production of structure and isopach maps. Secondarily, the "shapes" of log profiles within units (particularly the SP, gamma ray, and resistivity) are qualitative aids in the interpretation of depositional origin (mostly in sandstone–shale sequences). The numerical and physical implications of the log responses have not received anything like the attention they deserve until fairly recently. This is regrettable because orthodox sedimentary petrographic data tend to be either qualitative or poorly quantitative. Most observations of sedimentary rocks are purely descriptive, with assignations to broad categories of mineral composition and texture. Quantitative aspects are generally both expensive and time-consuming measurements. In contrast, the measurements of logging tools are in immediate numerical form and so are suitable material for data analysis. The major current drawback to interpretations of geology from the quantitative analysis of logs is the limited documented research that can be cited as supporting evidence. However, this is a result of sins of omission, rather than an intrinsic failing in the information content of logs. As stated before, logs are records of the physical properties of rocks.

SEE NO EVIL: THE MATHEMATICIAN

The process of logging a hole can be considered as remote sensing of the subsurface, particularly as some of the physical measurements are the same as those made by orbiting satellites. Computer processing of both satellite and seismic data is crucial to their successful analysis and suggests that mathematical methods should be valuable in novel log interpretation techniques. However, the most common geological application to date has been in research on methods of automated correlation. The depth axis of each log is some function of geological time. Unfortunately, the treatment of logging data by conventional time-series methods is complicated by the vagaries of depositional histories as a result of varying rates of sedimentation and periods of erosion and nondeposition. These and other problems are peculiar to geological data. Consequently, reasonable geological models must be the source of hypotheses which dictate both the choice of logs pertinent to the problem and the types of mathematical methods to be used in their analysis.

SPEAK NO EVIL: THE LOG ANALYST

Most log analysts are employed by oil companies to locate economic oil and gas reservoirs and estimate engineering properties. They are interested in pores: their size, shape, and fluid content. Mineralogy is of secondary importance, although important to the degree that its evaluation is necessary for accurate porosity estimates. Consequently, the log analyst is less aware of some significant geological implications in his analyses. This situation has changed in recent years owing to major efforts in the location and definition of stratigraphic traps and the demands of efficient engineering practice in complex reservoirs and tertiary oil recovery projects. The effective use of logs is common to all these applications, and their geological information content is a crucial aspect. Quantitative analyses are preferable to qualitative interpretations, since these will form the basis of important decisions concerning exploration drilling programs and choice of methods for reservoir exploitation.

Physics, mathematics, and geology are the fundamental facets of log interpretation. Each is a science, honored by time and the names of great men. Log interpretation is a newcomer and a mongrel which draws on an integrated knowledge of these three areas and is sometimes dignified by the name "petrophysics" (proposed by Archie in 1942). The trend toward increasing specialization within the sciences during modern times has caused the creation of artificial demarcation lines. These barriers are often evident in the limited success of engineers, geologists, and log analysts to communicate their perceptions of log properties to one another. Their common ground of log analysis has a great potential for enriched interpretations of geology that would be useful to all.

Both the problem and the promise are echoed by Wiener (1948) in his recollections concerning the development of cybernetics. His remarks are worth extensive quotation. He stated (pp. 2–3) that he "shared the conviction that the most fruitful areas for the growth of the sciences were those which had been neglected as a no-man's land between the various established fields. . . . Today there are few scholars who can call themselves mathematicians or physicists or biologists without restriction. A man may be a topologist or an acoustician or a coleopterist. He will be filled with the jargon of his field, and will know all its literature and all its ramifications, but, more frequently than not, he will regard the next subject as something belonging to his colleague three doors down the corridor, and will consider any interest in it on his own part as an unwarrantable breach of privacy. It is these boundary regions of science which offer the richest opportunities to the qualified investigator. They are at the same time the most refractory to the accepted techniques of mass attack and the division of labor. If the difficulty of a physiological problem is mathematical in essence, ten physiologists ignorant of mathematics will get precisely as far as one physiologist ignorant of mathematics and no further. If a physiologist who knows no mathematics works together with a mathematician who knows no physiology, the one will be unable to state his problem in terms that the other can manipulate, and the second will be unable to put the answers in any form that the first can understand."

The preceding remarks have shaped the philosophy behind this book. The fundamental aim is to describe approaches and methods whose outcome is information with meaning in the context of conventional geological models. The theory of tool design, rock properties, petrophysical relationships, and mathematical algorithms are described as essential prerequisites. In the earlier chapters, the principles of measurement and application of the most commonly used logs are described individually. In a later chapter, the analysis of multiple logs are shown which use graphic methods of log overlays and crossplots. These geometrical solutions are a natural preliminary to the mathematical development that follows. The more complex problems of analysis are suitable subjects for the use of a computer. Finally, the methods used in the analysis of rock sequences from a single well are extended to applications in the areal mapping of lateral variation of lithofacies in both two and three dimensions.

REFERENCES

Allaud, L., and Martin, M., 1977, *Schlumberger, the History of a Technique*, John Wiley & Sons, Inc., New York, 333 pp.

Jeffries, F., 1965, Wireline logs—the infant oil finding technique, *Canadian Petroleum Geology Bull.*, Vol. 13, No. 2, pp. 291–302.

Wiener, N., 1948, *Cybernetics, or Control and Communication in the Animal and the Machine*, The M.I.T. Press, Cambridge, Mass., 212 pp.

CHAPTER ONE

RESISTIVITY

BASICS

In 1827 Georg Ohm published a pamphlet in Berlin, in which he proposed that the current in a metallic wire is proportional to the potential difference between its ends. This idea was ridiculed widely as being overly simplistic and Ohm resigned from his mathematics professorship at Cologne as a consequence. After some years in obscure poverty, Ohm was finally vindicated. His thesis is now known as "Ohm's law" and is expressed by the equation

$$E = Ir$$

where the potential difference (E) is related to the current (I) by a constant called resistance, which is a characteristic of the conducting medium and whose units are named "ohms" in his honor.

In keeping with the simplicity of Ohm's law, the algebra that governs combinations of resistance is extremely straightforward. The total resistance of resistors in series is given by

$$r = r_1 + r_2 + \cdots + r_n$$

but differs from that of a set arranged in parallel when it is

$$\frac{1}{r} = \frac{1}{r_1} + \frac{1}{r_2} + \cdots + \frac{1}{r_n}$$

The resistance of any material will vary with both its size and shape. In order to have some common basis for comparison of electrical measurements in rocks, we need a parameter that characterizes each specific component, in much the same way that density, rather than weight, functions as a diagnostic property. This parameter is resistivity, given by

$$R = \frac{rA}{L}$$

where A is the cross-sectional area of the resistor and L is its length. The relationship matches intuition: long, narrow resistors are shown to have higher resistances than short, broad forms of the equivalent volume of the same material. The units of resistivity follow from dimensional analysis. Since units of resistance are ohms, units of area are meters squared, and units of length are meters, resistivity is measured in ohm-meters (also written as ohm-m^2/m, ohm-m, or Ωm).

RESISTIVITIES OF ROCKS

A resistivity log from a borehole represents a composite measurement of resistivities of rock-forming minerals and fluid and gaseous phases within the pore spaces. Although the world's museums are filled with exotic mineral collections, the bulk mineral compositions encountered by the drill bit in penetrating a section of a typical sedimentary basin are extremely restrictive in variety. Most sedimentary sequences consist almost entirely of sandstones, limestones, dolomites, shales, and cherts, with sporadic occurrences of evaporites and coal beds. The resistivities of their component minerals are all in excess of a million ohm-meters. Similar high resistivities also characterize hydrocarbon oils and gases. All of these materials can be considered as insulators in practice, particularly since the effective measurement scale of resistivity logging tools is generally limited to a range between 0.1 and 2000 ohm-meters.

The resistivity log is obviously controlled by factors other than the resistivities of mineral components and these have simple explanations. Shales have low resistivities which are caused by clay mineral properties. Dry clay minerals have forbiddingly high resistivities. However, when wet and subjected to an electrical field, ions associated with clay mineral cleavage surfaces migrate and carry electrical current. The magnitude of this conduction effect in any shale will vary according to its clay mineral composition, species, and their morphologies.

In sandstone, limestone, and dolomite beds which are shale free, electrical current is almost entirely carried by the aqueous phase of the rock pore space. Although pure water is virtually an insulator, all subsurface water contains dissolved minerals whose ionic dissociation provides the medium for electrical conductance. The resistivity of a subsurface brine varies inversely with the concentration of dissolved solids subject to the relative ionic mobility of the constituent ions. In a first approximation, pore water can generally be considered as a simple salt solution since sodium chloride is the dominant mineral. As a secondary control, brine resistivity decreases with increasing temperature.

The cumulative effect of these two factors results in the observation that resistivities of formation waters tend to decrease with depth in the subsurface as a result of

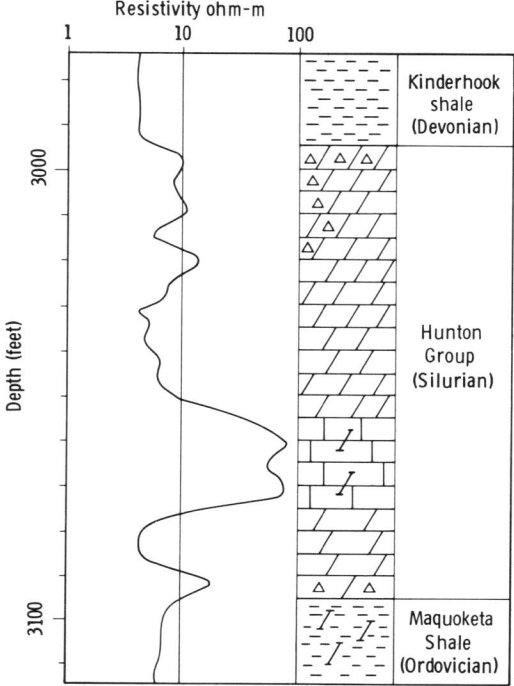

FIGURE 1. Induction resistivity log of the Hunton Group from a well in central Kansas. Lithology key: dashed, shale; orthogonal brick, limestone; oblique brick, dolomite; triangles, chert.

increasing concentrations of dissolved solids and increasing temperatures. However, this general trend will be perturbed in sequences with anomalous geothermal gradients or hydrodynamic infiltrations of fresher waters.

Since electrical current is conducted primarily by the brine contained in the pore space of a shale-free rock, the profile of its resistivity log trace is controlled largely by the volume of contained brine. In a unit that contains no oil or gas, this volume corresponds to the porosity. High resistivities match low porosities; low resistivities indicate higher porosities. This association can be applied in the interpretation of a resistivity log of a subsurface unit to implications that reach beyond simple assessments of porosity variation. Figure 1 shows an induction resistivity log of the Silurian Hunton Group in a well in central Kansas which is referenced with a lithology description from drill cuttings. The overlying Kinderhook Shale and underlying Maquoketa Shale are marked by low resistivities caused by the electrical conductance carried by wet clay minerals. The resistivity trace within the Hunton Group shows a distinctive sequence of zones of varying resistivity. A simple pattern is discernible which orders dolomitic limestone, cherty dolomite, and dolomite in terms of decreasing resistivity and, therefore, increasing porosity. Dolomites in this sequence correspond with dolomitized oolitic limestones in contrast with undolomitized lithographic limestones (Lee et al., 1948). Consequently, the resistivity trace reflects indirectly the results of dolomitization and chertification through its response to volume. In turn, diagenesis is controlled by fabric selectivity of the original sediment in the differential ability of separate carbonate depositional facies to transmit and react with pore fluids.

QUANTITATIVE RELATIONSHIPS BETWEEN RESISTIVITY AND POROSITY

The systematic pattern of dependence of resistivity on porosity in shale-free reservoir rocks was understood from the inception of well logging and used in qualitative interpretation. However, the pioneer paper by Archie (1942) marked the first attempt to pin down practical mathematical functions which could be used in quantitative analysis. Archie derived relationships from extensive laboratory measurements on sandstone samples but his basic conclusions can be drawn from a simple theoretical model.

Consider a meter cube of a hypothetical sandstone (Fig. 2). At one extreme, the lower limiting case is a sandstone with zero porosity, which is a quartz cube with both resistance and resistivity effectively infinite for any simple measuring device. At the other extreme is a totally hypothetical sandstone with 100% porosity, which is a tank of brine whose resistivity is denoted as R_w. As an intermediate case, we could construct an artificial sandstone by sketching a 100 × 100 square grid on the face of a quartz meter cube, drilling a number of channels through the cube at selected grid cells, and filling these with brine whose resistivity is R_w. The resistance

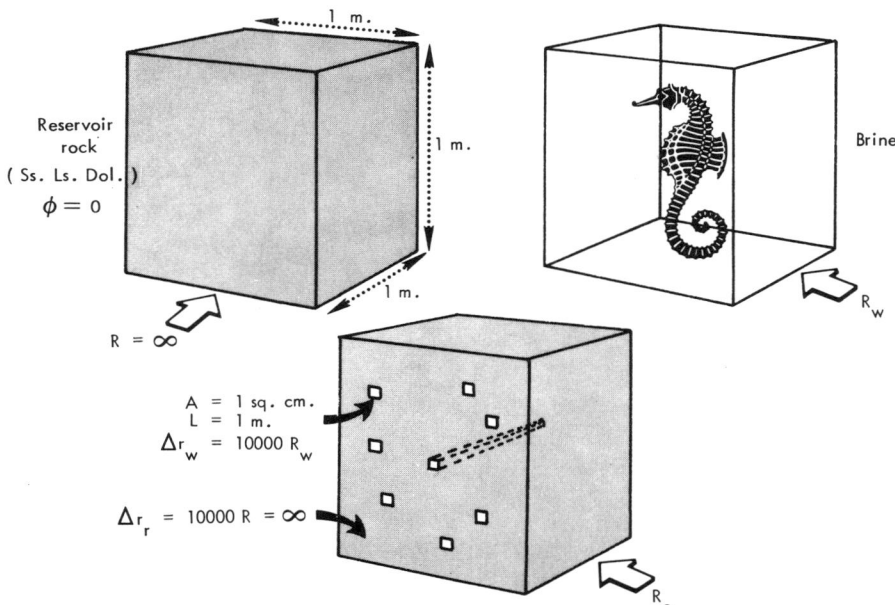

FIGURE 2. Relationship between formation factor and porosity in a simple cube model.

of this "sandstone" cube is r_o and its resistivity is R_o. [The meter cube is a convenient model, since both its cross-sectional area (A) and its length (L) are equal to one.] In log analysis, R_o is the standard designation for the resistivity of a rock which is completely saturated with a brine, whose resistivity is designated R_w. The resistance of the "sandstone" block is easily calculated, since the block can be considered as a stack of quartz prisms and brine channels which are arranged as resistances in parallel. The number of brine channels is equal to the "porosity" of the "sandstone" times 10,000 or

$$n = 10,000\, \phi$$

Now,

$$\frac{1}{r_o} = \frac{n}{\Delta r_w} + \frac{(10,000 - n)}{\Delta r_r}$$

where Δr_w is the resistance of one of the brine channels and Δr_r is the resistance of one of the quartz rods. Then,

$$\frac{1}{R_o} = \frac{10,000\, \phi}{\Delta r_w} + \frac{10,000(1 - \phi)}{\Delta r_r}$$

Since the resistance of a quartz rod approaches infinity and,

$$\Delta r_w = 10,000\, R_w$$

it follows that

$$\frac{1}{R_o} = \frac{\phi}{R_w}$$

Rearranging terms, this becomes

$$\frac{R_o}{R_w} = \frac{1}{\phi}$$

In other words, the ratio between the resistivity of this "sandstone" and the resistivity of its brine is a constant, and controlled by its porosity. The ratio R_o/R_w is termed the "formation factor" and is conventionally denoted by F.

The "sandstone" model used here is obviously grossly oversimplified in its representation of pore space by straight channels. In reality, the brine medium is a three-dimensional maze of void spaces connected by pore throats as a complex branching network. Both pore casts and the scanning electron microscope have given a direct insight of the true nature of the problem in topology (Fig. 3) which

FIGURE 3. Metallic pore cast from a consolidated sandstone X100. From Collins (1961).

is difficult to appreciate from the two-dimensional view provided by the conventional thin section (Fig. 4). Electrical current paths will twist and turn in a meter cube of real sandstone and take a longer route than 1 meter. The average ratio of the path length to the straight line distance between the initial and ending points is known as the "tortuosity" of the pore network. Consequently, the extended length of the brine channels will increase their resistance, and the resistivity of a real sandstone meter cube will be greater than that predicted by the simple model.

However, the equation $F = 1/\phi$ has the desirable properties that the formation factor is infinite at zero porosity and is unity at a theoretical porosity of 100%. (Note that porosity is used as a proportional quantity in log analysis equations.) A modified equation that preserves these endpoints but allows the intermediate range to fit higher formation factors is the relationship

$$F = \frac{1}{\phi^m}$$

where m is some value greater than unity. This function was proposed by Archie (1942) as the most reasonable empirical relationship to satisfy measurements from sandstone cores. It is now known as the "Archie equation" in his honor. He noted that the quantity m appeared to have a value of about 1.3 for unconsolidated sands and a range between 1.8 and 2.0 for many consolidated sandstones.

Subsequent workers have given the unfortunate name of "cementation factor" to the exponent m, while the following hypothetical scale is printed in many log analysis texts:

m	
1.3	Unconsolidated sandstones
1.4–1.5	Very slightly cemented
1.5–1.7	Slightly cemented
1.8–1.9	Moderately cemented
2.0–2.2	Highly cemented

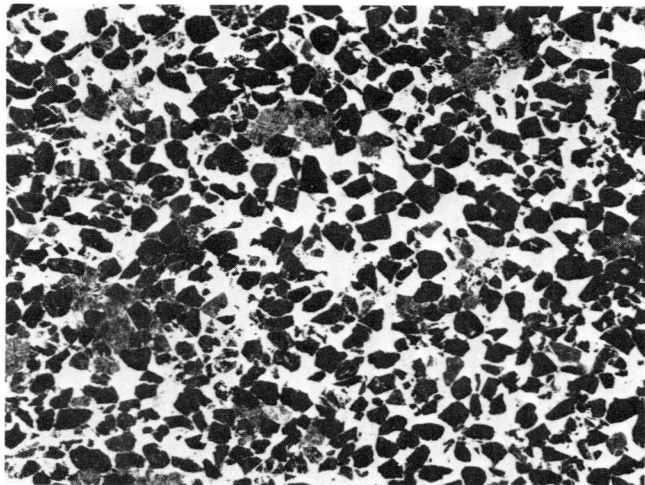

FIGURE 4. Two-dimensional view of pore network as seen in thin section. Bartlesville Sandstone (Pennsylvanian), Kansas. Magnification approximately ×20.

In fact, m is regulated by the tortuosity of the pore network, but this is the complement of the matrix geometry, which is in turn controlled by the rock texture. The "cementation factor" is therefore a loosely generic term that reflects the composite effect of a host of textural properties.

ROCK TEXTURE PROPERTIES AND RESISTIVITY–POROSITY RELATIONSHIPS

The practical application of the Archie equation in reservoir evaluation was hindered by the necessity to define a quantity for the cementation factor, m. This in turn required some poorly defined notion of sandstone texture. Winsauer et al. (1952) made detailed measurements of a variety of core properties in an attempt to integrate measures of pore geometry and sandstone textures in a more specific resistivity–porosity equation. After reviewing several alternatives, they generalized Archie's equation to

$$F = \frac{a}{\phi^m}$$

and proposed the now famous "Humble equation"

$$F = \frac{0.62}{\phi^{2.15}}$$

as a best-fit relationship to measurements from 29 North American sandstones (Fig. 5).

At first sight, the equation appears implausible because it predicts that the projected formation factor at 100% porosity will be less than one. However, the extreme diversity of source localities and ages of the fitted sandstone data necessarily implies a variation in cementation factor and the need for a correction term, a, if m is held constant. The quantity a has never been dignified with a name, but its function can be understood from comparison of the trace of the Humble equation with Archie equation lines defined by different values of m (Fig. 5). The Humble equation line cuts across the radiating fan of Archie equation lines, intersecting $m = 2$ at a porosity of about 10% and $m = 1.8$ at a porosity of approximately 30%. Since low-porosity sandstones are more highly consolidated than higher-porosity sands, the constant a functions as a slippage element which automatically incorporates the tortuosity changes generally associated with sandstones of differing porosities. In the final analysis, the modified Archie equation is a recognition that the tortuosity

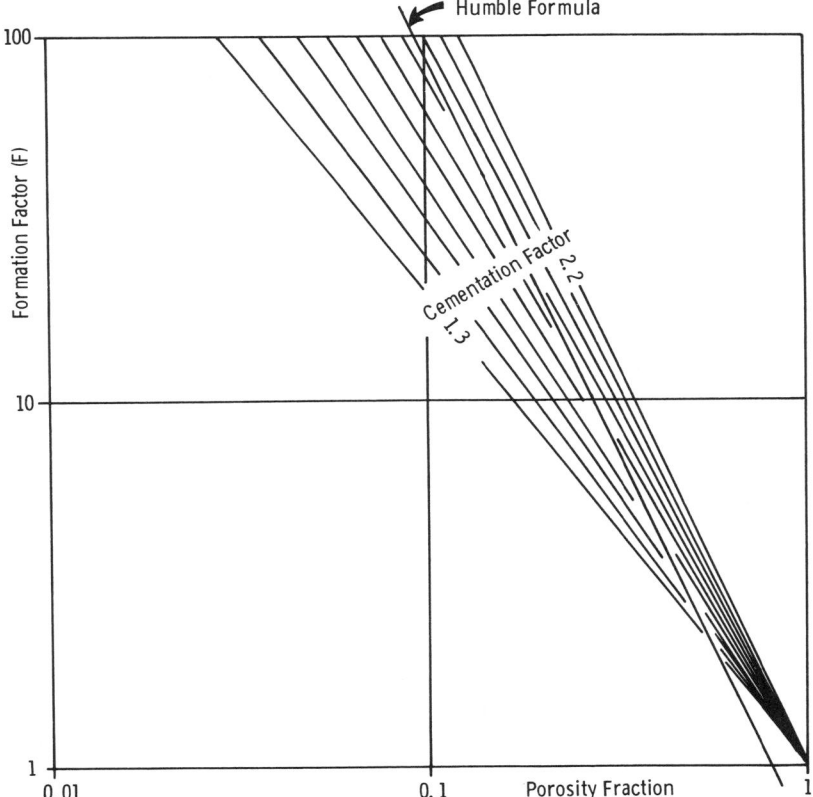

FIGURE 5. Geometric relationship between the Humble equation and the Archie equation with a set at unity and m ranging between 1.3 and 2.2.

registered in the "cementation factor" is not a constant but varies in a sympathetic manner with rock textural properties.

It is unfortunate that Winsauer et al. (1952) did not pursue a more detailed analysis of the interrelationships between pore geometry characterizations and textural measurements from their suite of sandstone samples. They limited their attention to properties linked with the formation factor and concluded that the only obvious relationship beyond a primary function with porosity was a positive correlation with the packing index of the sand grains. However, individual sample estimates of the "cementation factor," m, may be calculated directly using the basic Archie equation applied either to core values or logging measurements. Any systematic relationships with rock properties would prove valuable in facies interpretations from field studies.

As part of their study, Winsauer et al. (1952) made direct measurement of tortuosity through the determination of the transit times of ions moved through core samples under a potential gradient. The potential gradient was made sufficiently high to cause dominance of the rate of ionic migration over random velocity components caused by diffusion. A Spearman rank correlation coefficient of 0.72 between their measures of tortuosity and sample estimates of m demonstrates that the cementation factor is directly dependent on tortuosity.

Spearman rank correlation coefficients were also computed to search for significant relationships with sandstone textural properties. The choice of a ranked measurement of similarity was made in preference to the more commonly used Pearson product–moment coefficient, since it reveals systematic monotonic trends but is not restricted to linear functions. The equation for the Spearman rank correlation coefficient is given by

$$r_s = 1 - \frac{6\Sigma d^2}{n(n^2 - 1)}$$

where n is the number of sample observations and d is the difference in rank of each sample when ordered with respect to each of the variables. The coefficient can take any value between -1 and $+1$. Positive values indicate agreement between rankings on the two variables; negative values suggest a diametrically opposite trend. The significance of any coefficient can be tested through the computation of a Student t value using the formula

$$t = r_s \sqrt{\frac{n-2}{1-r_s^2}}$$

If this quantity exceeds the tabulated critical value of t at $(n-2)$ degrees of freedom and an appropriate level of significance (generally 5%), the null hypothesis of no association can be rejected in favor of a significant correlation.

The Spearman rank correlation coefficient between permeability and cementation factor was computed as -0.64. A systematic trend in decreasing permeabilities with increasing cementation factor (and increasing tortuosity) is consistent with the notion of a common geometrical control on the transmission of fluids and electrical current. With regard to textural properties, the cementation factor shows significant associations with grain size standard deviation (0.34) and roundness (-0.34). Weaker associations with phi-scale skewness (-0.22), phi-scale mean diameter (0.20), and packing index (0.19) are judged not significant, but some allowance should be made for the conservatism of the significance level and the small sample size (29 cores). The cumulative results of the tests match intuition. Better sorted sands with more rounded grains generate lower values of the cementation factor. If the weaker correlations reflect poorly defined, but systematic, trends, then loosely packed grain frameworks with a bias in size distribution to the coarser range are also linked with low m values.

Studies by many sedimentary petrographers reviewed in such texts as Pettijohn, Potter, and Siever (1972) have demonstrated that textural properties are by no means independent but are a composite association controlled by cycles of weathering and the hydraulic physics of transport and deposition. The covariation of this ensemble is crystallized in the concept of "textural maturity." Using the classification scheme of Folk (1951), the core samples were typed according to grade of maturity and average cementation factors computed for each class. The values of m show a generalized decline with increasing textural maturity, ranging from 1.87 (immature), through 1.78 (submature) to 1.76 (mature). A failure to detect any coherent patterns would imply that the primary matrix geometry of the original sediment had been obscured effectively by mechanisms of mechanical compaction and diagenesis. The important role of these factors in sandstone lithification are well documented by Schmidt and McDonald (1979) and are probably a cause in the deterioration of trends between textural properties and the cementation factor.

The preceding remarks are based on relationships observed in the highly varied suite of sandstone samples used in the pioneer study of Winsauer et al. (1952). As such, they express trends that range across a broad spectrum of sandstone types. We must therefore be cautious in extending these concepts to the interpretation of resistivity and porosity logs, which will often be focused on more restrictive variation within a single genetic unit.

Doveton, Ebanks, and Conley (1978) analyzed petrophysical and petrographic data measured in core samples from the Skinner Sandstone (Pennsylvanian) in southeast Kansas. Both the stratigraphy and sedimentology of the cored section indicated an alluvial channel fill as the depositional model. Correlations between the cementation factor and average grain size, sorting, cement proportion, and mica content were uniformly weak and not statistically significant. However, a comparison of the cementation factor profile with the core description of bedding structure (Fig. 6) shows a distinctive pattern. Ripple-bedded zones have a higher average cementation factor of 2.20 than cross-bedded zones, whose corresponding value is 2.05. The difference is statistically significant as judged by a Student t-test. The distinction is matched by a contrast in average permeability of 25 md for rippled zones and 40 md for cross-bedded units.

The observations suggest that grain textural controls of cementation factor may be relatively muted within a genetic unit and instead are dominated by changes in fabric. This finding conforms with other reservoir studies where the pronounced

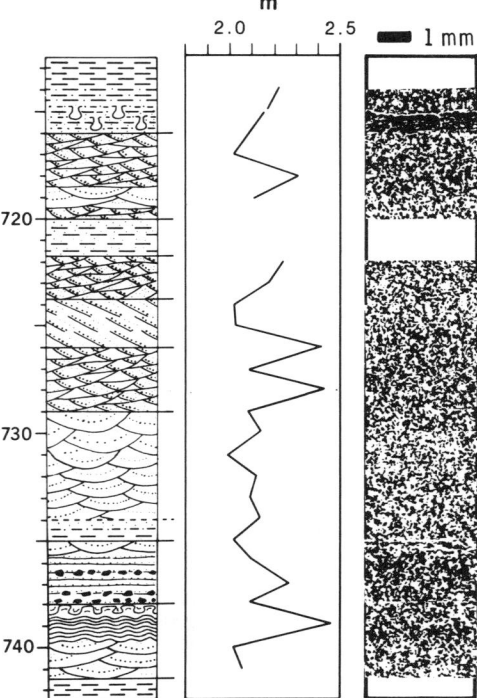

FIGURE 6. Bedding structures, cementation factor trace, and photomicrograph grain-size profile of a Skinner Sandstone (Pennsylvanian) section in southeast Kansas.

effect of fabric on permeability variation has been widely documented (e.g., Hewitt and Morgan, 1965). In particular, the strong distinction in permeability between ripple-bedded and cross-bedded units has been stressed, although Hrabar and Potter (1969) note that this can be attributed to both fabric and differences in grain size, sorting, and cement between the two facies.

ADVANCED MODELS OF TEXTURAL–RESISTIVITY RELATIONSHIPS

A frustrating aspect of empirical methods is the intimate linkage between rock properties which makes it difficult to isolate the major cause(s) of cementation factor variation. More recent theoretical and experimental studies have attempted to go beyond the empiricism of Archie and other workers. Madden (1976) pointed out that there is no rigorous theoretical explanation for the Archie equation. Its main justification is as a "law" of observation, which has been confirmed many times by laboratory measurements of numerous core samples. The capillary tube model used earlier in this chapter is a feeble representation of pore geometry, particularly as it ignores the intrinsic branching structure. Even a more sophisticated network model of interconnected tubes requires an arbitrary adjustment of tubular cross section in order to reproduce the Archie equation.

Recent applications of percolation theory (Kirkpatrick, 1973; Straley, 1978) and a self-similar model (Sen et al., 1981) are probably more realistic approximations. However, all authors concede that the Archie equation is an adequate description of porosity–resistivity relationships and that the cementation factor is a valid measure of pore geometry and framework texture. Advanced models coupled with laboratory studies of synthetic grain packs appear to have succeeded in the isolation of textural properties which are the direct cause of variations in the cementation factor.

Measurements of resistivity–porosity–particle-shape interactions in marine sands by Jackson et al. (1978) demonstrated that the cementation factor was sensitive to particle shape. The factor increased as the particles became less spherical and was highest in sediments where platy materials became a significant component. This empirical association was supported by Sen et al. (1981) in a mathematical derivation. Jackson et al. (1978) also noted that variations in size and sorting of sizes appeared to have little effect on the cementation factor in sands where the shape distribution was unchanged.

Mendelson and Cohen (1982) examined the effect of particulate anisotropy on resistivity variation and arrived at the following conclusions. The cementation factor has a minimum value for rocks in which the grains have the same shape and orientation. If the grains are aligned with a distribution of shapes, or if they are randomly arranged with or without a distribution of shapes, the cementation factor is increased to a higher value. The cementation factor is suggested therefore to be a composite function of shape, shape distribution, and degree of orientation.

At first sight, the synthesis of results from this more recent work may appear to conflict with some of the earlier conclusions concerning cementation factor–textural relationships. However, the ensemble covariation of textural properties is a collective response to sand transport history and the hydraulics of the depositional site. Mechanisms of sorting will control distributions of both size and shape. Fabric geometry is the internal structure of bedding forms which are themselves products of the local flow regime. Consequently, although the cementation factor is sensitive primarily to shape and fabric, it will show secondary associations with other textural properties. The major contribution of these new data is that they pinpoint the causal textural controls as grain shape, shape sorting, and degree of fabric anisotropy. As a result, more specific interpretations of cementation factor profiles calculated from logs can be extended beyond vague notions of changes in textural maturity.

FACIES INTERPRETATION FROM RESISTIVITY AND POROSITY LOGS

The textural interpretation of the cementation factor can be drawn from the analysis of a subsurface sandstone logged by a resistivity and a porosity device. In a water-saturated unit with known water resistivity (deduced either from log measurements or analysis of a drill-stem sample), the resistivity of any zone is an estimate of the quantity R_o.

The formation factor is given by

$$F = \frac{R_o}{R_w}$$

Substitution of the zone porosity into the Archie equation,

$$F = \frac{1}{\Phi^m}$$

yields an estimate of the cementation factor, m.

As a demonstration example, the procedure was applied to a logged section of the St. Peter Sandstone (Middle Ordovician) in a well in northern Kansas. Resistivity and porosity log readings were combined in the production of estimates of m as a function of depth. The "cementation factor log" is shown as a profile together with the trace of the gamma-ray log in Figure 7. The average value of m is 1.72 which is consistent with core and cuttings descriptions of friable and moderately cemented sandstones. The markedly high values (in excess of 4) that characterize the thin upper sand unit were excluded from the calculation. They are caused by oil saturations which would decrease the proportion of conductive brine and increase the tortuosity of the electrical current path. The abrupt decrease in values in the middle of the unit coincides with the intervening shale bed (note the increase in gamma-ray response), whose conductivity distorts the estimate. The two "anomalous" zones demonstrate that the method is adversely sensitive to significant variations in shale content and oil or gas saturations.

In the lower part of the sandstone, there is a striking trend in decreasing cementation factor moving upward through the unit. Drawing on earlier conclusions, we can be reasonably confident that this reflects a parallel trend in declining tortuosity of the pore network. As a more tentative interpretation, the pattern indicates an upward transition in sandstone texture to better sorted, more equant grains and/or a stronger

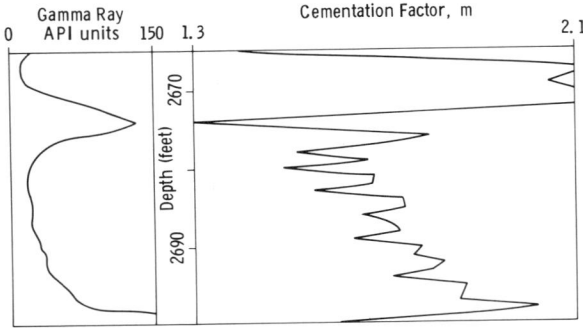

FIGURE 7. Gamma-ray log and cementation factor profile of the St. Peter Sandstone (Ordovician) in a well in northern Kansas.

preferred fabric orientation marked by a change in depositional bedding form. Alternatively, it is the product of selective diagenesis which has modified the framework geometry with effective obliteration of primary textural character. However, the poor degree of cementation makes this second possibility unlikely.

The St. Peter Sandstone was interpreted to be a littoral deposit by Dapples (1955). Later authors have all considered the St. Peter to represent a complex of submarine sand waves and dunes on a shallow marine shelf (Potter and Pryor, 1961; Pryor and Amaral, 1971; Dott and Roshardt, 1972). The literature on ancient sand bodies attributed to a shallow marine origin (e.g., Asquith, 1970; Brenner and Davies, 1974; Hobday and Reading, 1972) shows a concensus on a basic pattern of vertical variation. Ripple-bedded and thin-bedded siltstones and fine sandstones grade upward into trough and planar cross-stratified coarse sandstone. The cementation factor profile is fully consistent with this model and most of its variation can be attributed to fabric changes associated with the change in bedding form.

In fact, this case study is so satisfactory that it can be considered to be a textbook example, which is exactly what it is! In practice, grain textural and fabric variations are often observed to show complex fluctuations. Their use in the typing of depositional environments must be considered as thematic rather than precision criteria. Amaral and Pryor (1979) recorded a variety of trends in their detailed textural analyses of the St. Peter from outcrop samples in Wisconsin. In particular, they noticed apparent differences in vertical variability of thin units contrasted with thicker equivalents. Such differences are to be expected as a natural consequence of the range of local hydraulic regions present on a shallow marine shelf. These ambiguities and subtleties will inevitably be reflected in profiles of cementation factors.

The methodology offers some promise as an interpretation tool in the analysis of textural facies patterns in subsurface sandstone units. Its realistic application is subject to the following considerations. The formulation of the Archie equation presumes the sandstone to be water-saturated and relatively shale-free. Significant oil or gas saturation, or moderate clay contents will distort the cementation factor to anomalous values. The cementation factor is linked directly with pore network tortuosity which, in turn, is the complement of the matrix geometry. Interpretation must take account of both the textural properties of the original sediment and the degree to which the primary characters may have been obscured in compaction and diagenesis. There are a multiplicity of alternative explanations for the same cementation factor value. Consequently, field studies should be coordinated with explicit petrographic and sedimentological information from available core and cuttings for genetic interpretations.

TEXTURAL–RESISTIVITY RELATIONSHIPS IN CARBONATES

The preceding sections have considered the resistivity characteristics of both theoretical and laboratory models composed of packed grains. These models are adequate descriptions of sandstone fabrics but are a limited subset of a more complex range of possibilities for carbonate frameworks. However, Archie (1952) concluded that the formation factor and porosity variation of most limestones were fitted by a linear function when plotted on logarithmic scales. He noted a tendency for greater scatter about the line than was the case for sandstones, but this is easily explained by the wider range of limestone pore structures. The Archie equation is therefore a useful empirical description of resistivity–porosity variation in carbonates as well as sandstones. However, the "meaning" of the cementation factor is more difficult to interpret. The most widely used form of the Archie equation for both limestones and dolomites is the basic

$$F = \frac{1}{\Phi^2}$$

The apparent simplicity of this equation represents an average relationship that best typifies carbonates with intergranular and intercrystalline porosities.

Carbonate sediments at the time of deposition commonly have porosities which exceed 40%. By contrast, most ancient carbonates contain only a few percent of pore space as the cumulative result of multiple processes of diagenesis. The heterogeneity of carbonate sequences is shown by rapid changes in porosity and the occurrence of thin porous horizons between low-porosity layers. The variation in pore volume is matched by a great diversity of possible pore geometries. In their genetic classification, Choquette and Pray (1970) recognized 15 basic porosity types of which the most common are interparticle, intraparticle, intercrystalline, moldic, fenestral, fracture, and vug. Many carbonate rocks contain several porosity types in compound pore systems.

Cementation factors measured from carbonate core samples show generalized associations with textural character. Chombart (1960) reported that cementation factors are generally between 1.8 and 2.0 for crystalline and granular carbonates, 1.7 and 1.9 for chalky limestones, and 2.1 and 2.6 in carbonates with vugs. The presence of fractures causes a reduction in the cementation factor to values in the neighborhood of 1.4 (Suau and Gartner, 1980). These numbers are a reflection of the relative tortuosity of the aggregate pore network in each case. As such, they can be used as indicators of pore morphology, but any genetic implications are a matter for geological interpretation.

It is difficult to design theoretical resistivity models which are realistic simulations of carbonate pore geometries. An exception is provided by fracture porosity which can be represented adequately by planar channels with minimal tortuosity. Aguilera (1976) considered a double-porosity model made up of two pore systems linked in parallel. The model can be imagined as a "Rubik's cube" arrangement in which cells of limestone with matrix porosity are separated by fracture channels. If the cube is saturated with conductive water, the resistivity relationships of the various components lead to the following equation,

$$\Phi_t^M = (1 - \Phi_f) \Phi_x^m + \Phi_f$$

where Φ_f, Φ_x, Φ_t are the fracture, matrix, and total porosities and m and M are the cementation factors of the matrix and the total rock, respectively. The relative proportions of the matrix and fracture porosities control the variation of the rock cementation factor between a theoretical lower limit of unity (all fractures) to an upper limit in the neighborhood of two (no fractures).

In practice, the influence of fractures on resistivity logs is highly variable and is determined by their size, orientation, and the nature of their contained fluids (mud filtrate, formation water, or hydrocarbons). Induction resistivity tools are influenced by horizontal resistivity, laterologs by vertical resistivity. Consequently, the induction log is generally insensitive to vertical fractures, but can detect horizontal fractures when they are filled with conductive water. By contrast, vertical fractures may be shown by a pronounced resistivity reduction on the laterolog when compared with an induction log run in combination. Microresistivity tools are most sensitive to fractures, since they respond to only a small volume of formation. Fractures entirely filled with mud filtrate will register as highly conductive anomalies.

The estimation of carbonate cementation factors in water-saturated sections may be made from resistivity and porosity logs using the procedure described earlier. As a simple example, a cementation factor profile is shown for a section of Arbuckle Limestone (Cambro-Ordovician) in northeast Kansas (Fig. 8). The sequence consists of medium-crystalline dolomite with some chert and has localized occurrences of vuggy and fracture porosity. The cementation factor tracks closely with the generalized expectation of a value of two, but shows minor oscillations to higher values (possible vugs) and lower values (possible fractures).

By contrast, a second example (Fig. 9) shows a highly variable cementation factor profile for a Pennsylvanian limestone in southern Kansas. The upper part of the section is an oolitic zone. Here, ooids have been selectively dissolved in the production of oomoldic porosity, whose high tortuosity is reflected in extreme

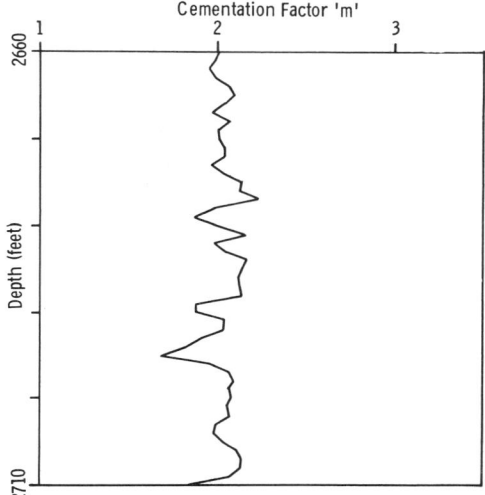

FIGURE 8. Cementation factor profile of an Arbuckle Limestone section from a well in northeast Kansas.

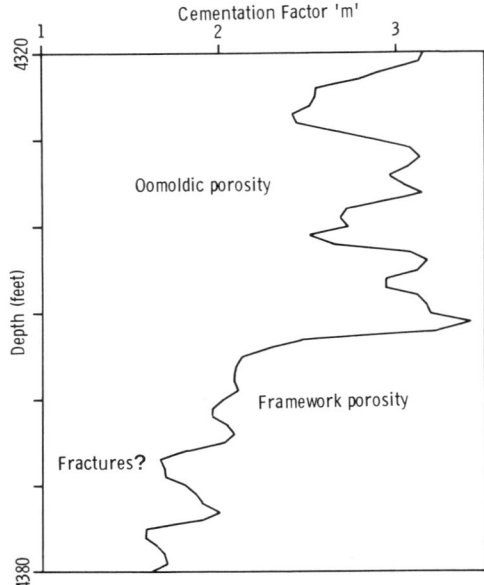

FIGURE 9. Cementation factor profile of a Pennsylvanian limestone from a well in southern Kansas.

cementation factor values. Beneath the oolite, more typical carbonate cementation factors prevail and suggest simple matrix porosity in these wackestones. However, local excursions of the factor to lower values are indications of possible fracture occurrence.

More definitive relationships could (and probably will) be developed as a result of closer cooperation between carbonate petrologists and petrophysicists. The use of cementation factor profiles as a means to characterize carbonate framework properties is more speculative than is the case for sandstones. In practice, meaningful results would require particularly close integration of core data on an individual case study basis to relate pore morphology suggested by resistivity with more conventional genetic descriptions.

THE RESISTIVITY OF HYDROCARBON-BEARING ROCKS

The resistivity of a rock which is completely saturated with formation water is specified by the symbol R_o. The more general term, R_t, is used to signify the resistivity of a rock whose fluid content is either unknown, or which is a mixture of water and hydrocarbons. In a reservoir rock with pore space completely filled with brine, R_t is equivalent to R_o. At the other extreme, if all the water could be replaced by oil or gas, the entire rock (mineral framework and contained fluids) would be an insulator with extremely high resistivity. This situation is hypothetical because all oil-bearing rocks are characterized by an "irreducible water saturation" caused by a film of water which clings tenaciously to the matrix surface. In practice, the resistivity R_t will be higher than the lower limiting value of R_o as a function of water saturation. The oil or gas phase reduces the volume of conductive brine in the pore space. It also increases the tortuosity of current paths around the oil globules or gas bubbles which develop in the central part of the void spaces. The "resistivity index" is defined by the equation

$$I = \frac{R_t}{R_o}$$

This ratio functions as a comparison test between the observed resistivity of a porous rock and the expectation of its resistivity if the pore space contained nothing but formation water.

Archie (1942) found an empirical relationship between the resistivity index and water saturation that took the form

$$I = \frac{1}{S_w^n}$$

where S_w is the fractional water saturation and n is termed the "saturation exponent." The quantity n appears to be controlled by pore geometry and, consequently, by rock texture. Laboratory measurements have indicated that the saturation exponent ranges between about 1.8 and 2.5. In most log analysis computations, a value of 2 is selected as both simple and adequate for exploration applications.

The use of these functions with the Archie equation gives a quantitative procedure to predict water saturation in a subsurface zone, based on its porosity, resistivity, and the resistivity of the brine in its pore space. Since

$$I = \frac{1}{S_w^n}$$

then

$$S_w = \sqrt[n]{\frac{1}{I}}$$

$$= \sqrt[n]{\frac{R_o}{R_t}}$$

$$= \sqrt[n]{\frac{FR_w}{R_t}}$$

The formation factor is estimated by insertion of the fractional porosity into the Archie equation with the choice of a cementation factor appropriate for the rock framework. The quantity R_t is drawn directly from a resistivity log reading. The formation water resistivity may be computed from the spontaneous potential log (see Chapter 3) or deduced from a resistivity–porosity crossplot of neighboring zones which include water-saturated horizons. Alternatively, the formation water resistivity can be measured directly from a sample taken from a drill-stem test or production unit. In this latter case, the water resistivity must be corrected to its value at the formation temperature. An approximate transformation is given by Arp's empirical formula

$$R_{w2} = \frac{R_{w1}(T_1 + 7)}{T_2 + 7}$$

where R_{w1} and R_{w2} are water resistivities at Fahrenheit temperatures T_1 and T_2, matched by laboratory measurement and subsurface formation conditions, respectively. The temperature of the subsurface zone can be estimated easily from interpolation between the mean annual temperature at the surface and the maximum temperature recorded at the bottom of the hole.

This procedure is the fundamental approach used by log analysts in the search for hydrocarbons using logs. The basic model is appropriate for shale-free reservoir rocks and will generally give acceptable results for units with minor contents of shale. A significant shale component in sandstones require expansion to more complex models which accommodate the conductivity caused by cation-exchange mechanisms. The pioneer laboratory and theoretical work of Hill and Milburn (1956) and Waxman and Smits (1968) has modified the Archie model to include this additional factor.

The log characterization of textural properties in oil and gas reservoirs is complicated by the hydrocarbon phase as a new source of variability. However, systematic changes in matrix geometry are reflected in the fluctuations of irreducible water saturation with porosity. Plots of irreducible water saturation versus porosity in relatively homogeneous reservoirs generally form a distinctive trend which is approximated by a hyperbolic curve (Fig. 10). The empirical equation for this relationship is

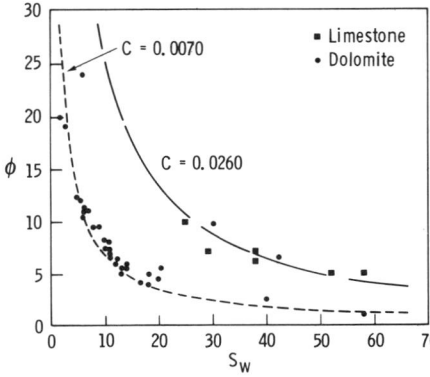

FIGURE 10. Hyperbolic curves of porosity–irreducible water saturation relationships in Canadian carbonates. After Dorin (1968).

$$\Phi S_{wi} = c$$

where c is a quantity controlled by pore geometry and, consequently, framework texture. Petroleum engineers have studied the associations from the standpoint of capillary pressure models and their implications regarding permeability. However, geologists have not exploited the interplay of irreducible water saturations and porosity as diagnostic indications of textural variations in reservoir rocks.

THE MEASUREMENT OF RESISTIVITY BY LOGGING TOOLS

When resistivity logging was first introduced, it was generally thought that the tools then in use were accurately recording the actual resistivities (R_t) of the formations penetrated by the borehole. It was soon realized that a variety of complicating factors were involved which required corrections to be made to the recorded values in order to obtain good estimates of the true resistivities. All resistivity tools are to some extent "averaging" devices that record resistivities of zones rather than resistivities of discrete points. So, for example, the resistivity of a thin resistive horizon will generally be underestimated by most tools since its reading will be partly reduced by contributions of more conductive adjacent beds.

A clearer understanding of the role of drilling in the modification of the resistivities of formations in the vicinity of the borehole was an important step in the development of models that related measured apparent resistivities to virgin formation values. In addition to its other functions, the drilling mud forms a mudcake seal on the borehole wall of permeable formations. In doing this, mud filtrate penetrates the formation, displacing formation water and oil or gas. In a zone immediately adjacent to the borehole the mud filtrate displaces all the formation water and any "movable oil saturation" (the "flushed zone"). Beyond this, the mud filtrate displaces part of the formation water in a "transition zone" which ultimately peters out at a contact edge with virgin formation. This process is termed "invasion" and the relative depth of invasion is broadly a function of formation porosity/permeability properties, so that less porous formations (e.g., typical carbonates) are often more highly invaded than moderately porous units (e.g., typical sandstones).

The replacement of formation water by mud filtrate involves a change of pore water resistivity from R_w to R_{mf}. In a typical logging operation, the mud is "freshwater" as contrasted with the formation waters encountered. The result of invasion is generally to create a more highly resistive annulus surrounding the borehole. The cumulative result of this process is a complex structure of resistivity shells centered on the borehole (Fig. 11). When the objective of most commercial logging is to evaluate the oil or gas potential of stratigraphic units, a resistivity tool is selected that will best estimate the resistivity of the virgin formation by taking into account borehole characteristics, drilling mud properties, formation lithologies, and degrees of invasion. Four classes of the most common resistivity devices are reviewed briefly.

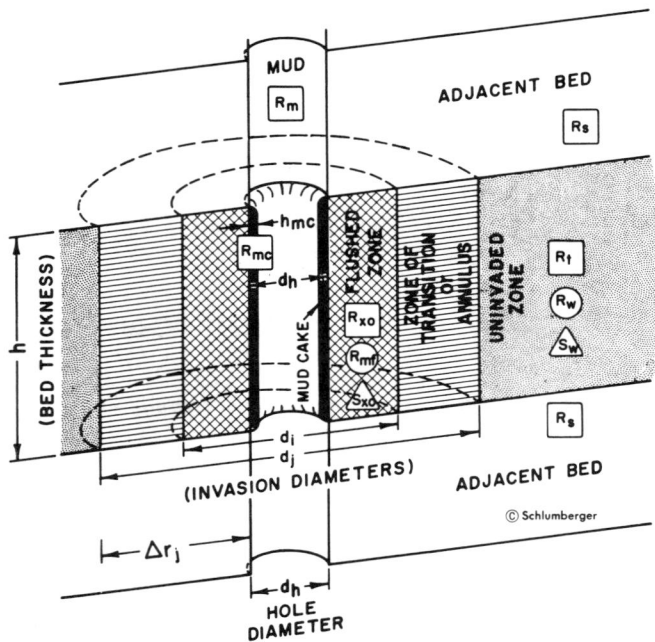

FIGURE 11. Model representation of resistivity variation in the vicinity of a borehole. From Schlumberger publications.

The Normal Device

The design of the normal device tool is fairly simple and consists of two electrodes separated by a fixed spacing on an insulating mandrel (Fig. 12). A current is passed between the lower electrode and another electrode at ground level and the potential measured at the upper electrode and an electrode intermediate between it and ground level. The current and measured potentials allow the computation of the resistivity of a zone whose reference point is midway between the two tool electrodes. The resistivity corresponds to that of a volume surrounding the tool whose radius is approximately twice the tool electrode spacing under favorable conditions. As a result the measurement tends to be a composite of resistivities from the mud, invaded zone, and virgin formation.

Since the "radius of investigation" is largely dictated by the electrode spacing, two common variants of the device are the "short normal" (16-inch spacing) and the "long normal" (64-inch spacing). The normal device has been entirely superseded by induction and laterolog tools, which give much better estimates of true formation resistivity. However, the short normal is still run with the induction resistivity tools and comparison of its trace with the induction curve gives indications of relative degrees of invasion.

The Induction Device

The focused induction tool was developed to measure conductivities deep within the formation with minimal disturbance by the invaded zone. The tool contains transmitter coils with a high-frequency alternating current which induce eddy currents in the adjacent section (Fig. 13). Most of these eddy currents are focused beyond the diameter of the typical flushed zone and their magnitude is an approximation of the conductivity of the virgin formation. In turn, they induce voltages in the receiver coil which are transformed to estimates of formation conductivity and, as a reciprocal, resistivity.

Since the induction tool actually measures conductivity directly, rather than resistivity, more reliable readings tend to be made within lower-resistivity sections. As a result, the induction tool is ideally suited for sandstone sections, which typically have high porosities, but may not be a satisfactory first choice in highly resistive sequences such as low-porosity carbonates. Unlike other resistivity tools, the induction tool can be run in holes drilled with air or with oil-base muds since it does not require electrical contact with the mud or formation. The tool operates well in

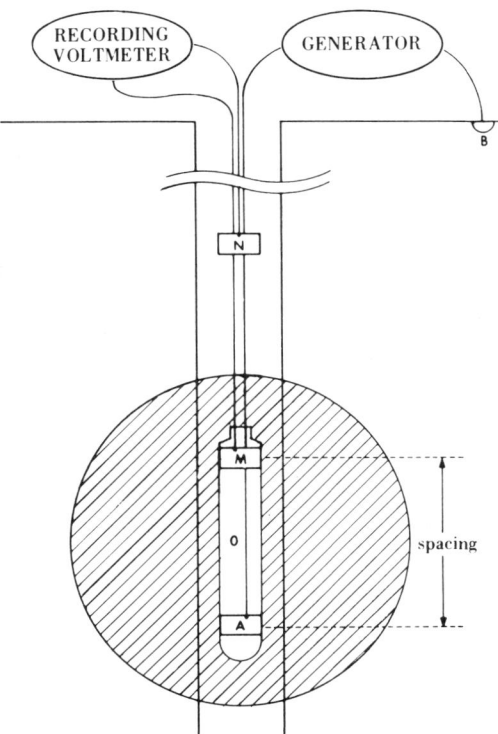

FIGURE 12. Schematic illustration of the normal device electrode arrangement and resistivity measurement. From Dresser-Atlas publications.

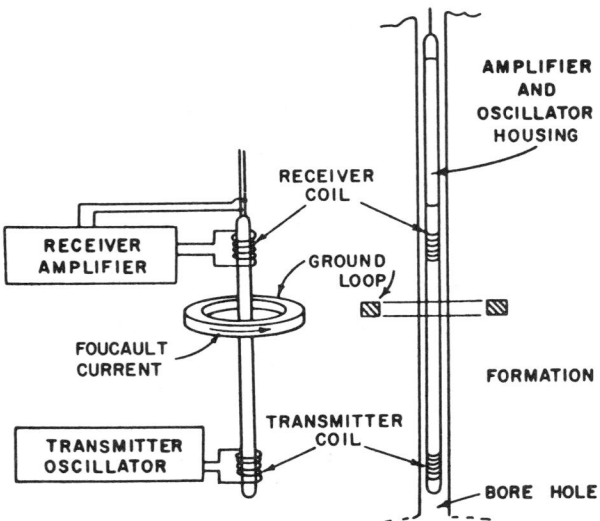

FIGURE 13. Schematic diagram of the induction tool design. From Schlumberger publications.

"fresh muds" but readings become strongly degraded in "salt muds" due to the greatly increased contribution of the borehole to the total conductivity reading.

The Laterolog

The laterolog (or guard log) was developed to provide accurate readings of formation resistivity in holes drilled with saltwater muds. There are various designs of laterolog tools, but the central principle of operation is a three-electrode arrangement in which a current supply of constant intensity is supplied to the central electrode (Fig. 14). A variable current intensity is transmitted to the two surrounding ("guard") electrodes whose magnitude is adjusted so that there is a zero potential with the central electrode. As a result, current in the central electrode is constrained to flow radially outward as a "current disc" into the surrounding formation. The thickness of the disc is determined by the spacing of the guard electrodes while the current density at any lateral distance from the central electrode is inversely proportional to this distance times the spacing. The drop in potential of the current disc radiating into the formation is monitored by a remote return electrode. As a result, an apparent resistivity is deduced which is the summation of resistivity contributions by the mud, invaded zone, and virgin formation. In situations where the mud is relatively conductive, degree of invasion restricted and resistivity of the formation is fairly high, this apparent reading is a close approximation of the true virgin formation resistivity.

Microresistivity Tools

A variety of microresistivity devices are available (e.g., microlaterolog, proximity log, etc.) whose primary purpose is to measure the resistivity of the flushed zone,

R_{xo}. In design, they are scaled-down versions of their larger counterparts, but the electrodes are mounted on a pad which is pressed directly against the formation wall. Their radius of investigation is limited essentially to the flushed zone, but they have a much finer vertical resolution than the conventional resistivity tools. As a result, they are especially useful in the definition of thin beds, which tend to be lost in the "averaging" process of the larger tools.

Measurements of the resistivity of the flushed zone in productive sections are used to calculate residual hydrocarbon saturations using the standard formulas of reservoir evaluation with a substitution of the mud filtrate resistivity, R_{mf} for the formation water resistivity. R_{mf} does not require independent evaluation since it is measured at the well site and recorded on the log heading together with the temperature of its measurement. If estimations are made of the hydrocarbon saturation of the virgin formation from resistivity readings of an induction or laterolog tool, the difference between these figures and the corresponding residual saturations is the "movable hydrocarbon saturation." Calculations of this type utilize the flushing action of drilling as an ersatz production test which gives general indications of such properties as oil gravity and permeability.

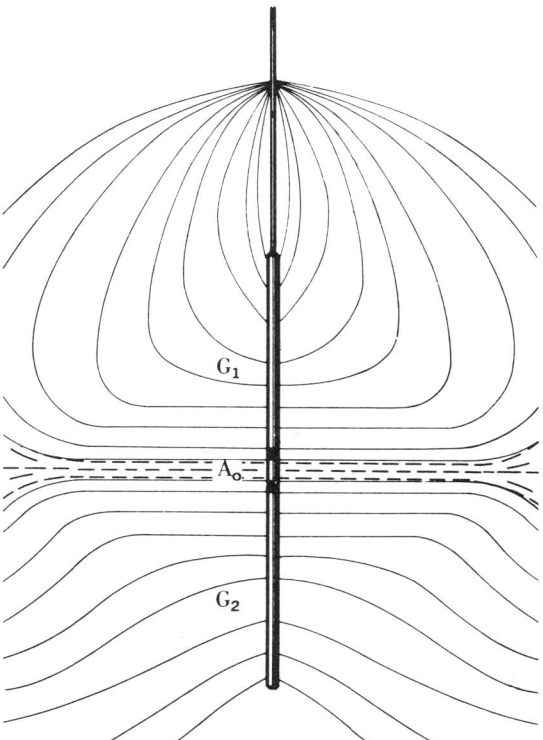

FIGURE 14. Schematic illustration of the guard or focused system laterolog tool and the paths of current in the vicinity of the borehole. From Dresser-Atlas publications.

Vertical Resolution and Radius of Investigation

The ideal resistivity tool would obviously be one that had extremely good vertical resolution (thereby defining very thin beds) and a large radius of investigation (giving reliable readings deep in the formation). These two criteria are almost impossible to meet in the design of a practical resistivity device, since it requires the vertical component to be extremely small (effects of adjacent beds) and the lateral component to be very large. As a result, different resistivity tools are run for different purposes and several types may be run in the same hole.

A comparison of three different laterologs run through the same section of a Mississippian carbonate in Saskatchewan (Jeffries, 1965) is shown on Figure 15. Moving from left to right (across log types) there is a progressive increase in vertical resolution but, at the same time, a reduction in the radius of investigation, with increasing resistivity contributions from the flushed zone. If a stratigrapher wished to use one of these logs for bed correlation purposes, the microlaterolog (MLL) would be his choice. If a petroleum geologist required a reservoir analysis of the formation, he would use the laterolog LL7, since the readings are the best estimate of the virgin formation resistivity.

The petroleum geologist would recognize that the LL7 trace was a "moving average" of the real resistivity variation whose "averaging window" corresponded

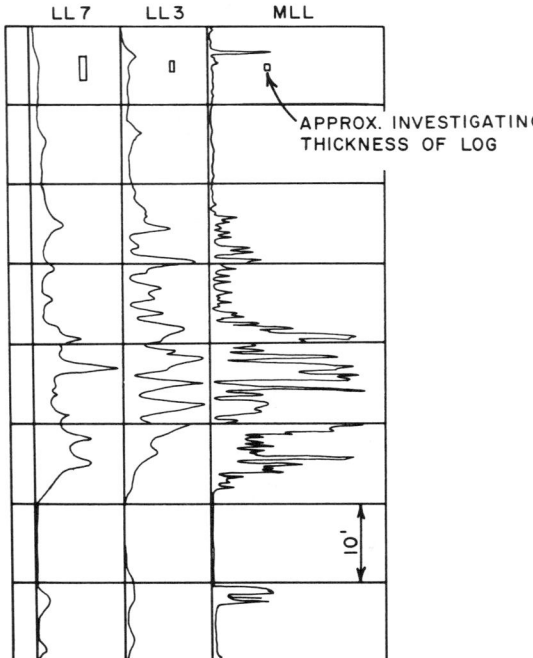

FIGURE 15. Comparison of resistivity logs (laterolog 7, laterolog 3, microlaterolog) run through a Mississippian carbonate section in Saskatchewan. After Jeffries (1965).

to the vertical definition of the tool (in this case about 3 feet). This averaging effect can be compensated to some degree by subdividing the trace into "zones" of depth increments. Each zone would be identified with a peak, trough, or shoulder and its boundaries located at curve inflection points between these features. The best estimate resistivity value ascribed to each zone corresponds to the peak, shoulder, or trough, because the averaging process dictates that a peak will be an underestimate, a trough an overestimate of the true value in the zone. The zoning procedure considerably expedites the task of the log analyst in practice, as it reduces the infinity of resistivity values of the analog trace to a manageable number of zones for reservoir evaluation. Computer reservoir evaluations sidestep this zonation process as they perform calculations on log data digitized at fixed intervals (normally 6 inches). However, the computer evaluations are not invalidated so long as it is realized that the computed water saturation profiles are moving averages of the real variation, whose averaging window is set by the vertical resolutions of the resistivity and porosity tools which were used.

The preceding remarks on vertical resolution apply to all logging tools so that analysis by hand is done most rationally after zoning the logs as an initial step. As a general rule of thumb, the vertical resolution of the induction and larger laterologs ranges between 3 and 5 feet; the vertical resolution of the porosity tools is on the order of 2 to 3 feet. Variations of these figures are introduced by different design features in any tool series and the magnitude of contrast between adjacent beds.

REFERENCES

Aguilera, R., 1976, Analysis of naturally fractured reservoirs from conventional well logs, *J. Petrol. Tech.*, Vol. 28, pp. 764–772.

Amaral, E. J., and Pryor, W. A., 1977, Depositional environment of the St. Peter Sandstone deduced by textural analysis, *J. Sed. Pet.*, Vol. 47, No. 1, pp. 32–52.

Archie, G. E., 1942, The electrical resistivity log as an aid in determining some reservoir characteristics, *Trans. Amer. Inst., Mech. Eng.*, Vol. 146, pp. 54–62.

Archie, G. E., 1952, Classification of carbonate reservoir rocks and petrophysical considerations, *Am. Assoc. Petrol. Geolog. Bull.*, Vol. 36, No. 2, pp. 278–298.

Asquith, D. O., 1970, Depositional topography and major marine environments, Late Cretaceous, Wyoming, *Am. Assoc. Petrol. Geolog. Bull.*, Vol. 54, No. 7, pp. 1184–1224.

Brenner, R. L., and Davies, D. K., 1974, Oxfordian sedimentation in Western Interior United States, *Am. Assoc. Petrol. Geolog. Bull.*, Vol. 58, No. 3, pp. 407–428.

Choquette, P. W., and Pray, L. C., 1970, Geologic nomenclature and classification of porosity in sedimentary carbonates, *Am. Assoc. Petrol. Geolog. Bull.*, Vol. 54, No. 2, pp. 207–250.

Chombart, L. G., 1960, Well logs in carbonate reservoirs, *Geophysics*, Vol. 25, No. 4, pp. 779–853.

Dapples, E. C., 1955, General lithofacies relationships of St. Peter Sandstone and Simpson Group, *Am. Assoc. Petrol. Geolog. Bull.*, Vol. 39, No. 4, pp. 444–467.

Dorin, A. H., Chase, E. A., and Link, W., 1968, A guide to improved formation evaluation in carbonates in northwestern Alberta, *Trans. 2nd Formation Evaluation Symposium of The Canadian Well Logging Society*.

Doveton, J. H., Ebanks, W. J., Jr., and Conley, C. D., 1978, Integrated petrophysical and petrographic studies of sedimentary facies in some Cherokee Sandstones of Southeast Kansas, *Geol. Soc. Amer.*, South-Central Section, Abstracts with Programs, Vol. 11, No. 1, pp. 4–5.

Dott, R. H., Jr., and Roshardt, M. A., 1972, Analysis of cross-stratification orientation in the St. Peter Sandstone in Southwestern Wisconsin, *Geol. Soc. Amer. Bull.*, Vol. 83, No. 9, pp. 2589–2596.

Folk, R. L., 1951, Stages of textural maturity in sedimentary rocks, *J. Sed. Pet.*, Vol. 21, pp. 127–130.

Hewitt, C. H., and Morgan, J. T., 1965, The Fry in situ combustion test—reservoir characteristics, *J. Petrol. Tech.*, Vol. 17, pp. 337–353.

Hill, H. J., and Milburn, J. D., 1956, Effect of clay and water salinity on electrical behavior of reservoir rocks, *Trans. Amer. Inst. Mech. Eng.*, Vol. 207, pp. 65–72.

Hobday, D. K., and Reading, H. G., 1972, Fair weather versus storm processes in shallow marine sandbar sequences in the Late Precambrian of Finmark, North Norway, *J. Sed. Pet.*, Vol. 42, No. 2, pp. 318–324.

Hrabar, S. V., and Potter, P. E., 1969, Lower West Baden (Mississippian) Sandstone body of Owen and Greene counties, Indiana, *Am. Assoc. Petrol. Geolog. Bull.*, Vol. 53, No. 10, pp. 2150–2160.

Jackson, P. D., Taylor-Smith, D., and Stanford, P. N., 1978, Resistivity-porosity-particle shape relationships for marine sands, *Geophysics*, Vol. 43, No. 6, pp. 1250–1262.

Jeffries, F., 1965, Wireline logs—the infant oil finding technique, *Can. Petrol. Geol. Bull.*, Vol. 13, No. 2, pp. 291–302.

Kirkpatrick, S., 1973, Percolation and conduction, *Rev. Mod. Phys.*, Vol. 45, pp. 574–588.

Lee, W., Leatherock, L., and Botinelly, T., 1948, The stratigraphy and structural development of the Salina Basin in Kansas, *Kansas Geol. Survey Bull.*, No. 74, 155 pp.

Madden, T. R., 1976, Random network and mixing laws, *Geophysics*, Vol. 41, No. 6A, pp. 1104–1124.

Mendelson, K. S., and Cohen, M. H., 1982, The effect of grain anisotropy on the electrical properties of sedimentary rocks, *Geophysics*, Vol. 17, No. 2, pp. 257–263.

Pettijohn, F. J., Potter, P. E., and Siever, R., 1972, *Sand and Sandstone*, Springer-Verlag, New York, 618 pp.

Potter, P. E., and Pryor, W. A., 1961, Dispersal centers of Paleozoic and later clastics of the upper Mississippi Valley, *Geol Soc. Amer. Bull.*, Vol. 72, No. 8, pp. 1155–1250.

Pryor, W. A., and Amaral, E. J., 1971, Large-scale cross stratification in the St. Peter Sandstone, *Geol. Soc. Amer. Bull.*, Vol. 82, No. 1, pp. 229–244.

Schmidt, V., and McDonald, D. A., 1979, The role of secondary porosity in the course of sandstone diagenesis in Aspects of Diagenesis (Scholle and Schluger, Eds.), *Soc. Econ. Pal. Min. Spec. Publ.* No. 26, pp. 125–207.

Sen, P. N., Scala, C., and Cohen, M. H., 1981, A self-similar model for sedimentary rocks with application to the dielectric constant of fused glass beads, *Geophysics*, Vol. 46, No. 5, pp. 781–796.

Straley, J. P., 1978, Critical phenomena in resistor networks, *J. Phys.*, Vol. C9, pp. 783–795.

Suau, J., and Gartner, J., 1980, Fracture detection from well logs, *The Log Analyst*, Vol. XXI, No. 2, pp. 3–13.

Waxman, M. H., and Smits, L. J. M., 1968, Electrical Conductivities in oil-bearing shaly sands, *Trans. Amer. Inst. Mech. Eng.*, Vol. 243, pp. 107–122.

Winsauer, W. O., Shearin, H. M., Jr., Masson, P. H., and Williams, M., 1952, Resistivity of brine-saturated sands in relation to pore geometry, *Am. Assoc. Petrol. Geolog. Bull.*, Vol. 36, No. 2, pp. 253–277.

CHAPTER TWO

THE DIPMETER

The dipmeter was introduced in 1941 as a logging device to measure the orientation of bedding planes and other features within a borehole. Up to that time, the mapping of subsurface geometry had been based exclusively on the stratigraphic correlation of marker horizons in wells which were often miles apart. The function of the dipmeter is to imitate the procedure whereby a geologist establishes the dip and strike of beds in surface outcrops. Within a well any planar feature will intersect the borehole wall as an ellipse whose relative vertical displacement around the hole is controlled by the dip and strike of the plane and the orientation of the borehole axis (Fig. 1). Because the majority of exploration holes have a diameter of about 8 inches, moderately dipping features will be matched by short maximum vertical displacements and will be often less than a foot. Consequently, a practical dipmeter must make measurements which have a fine vertical resolution. The initial use of an SP dipmeter was soon replaced by the deployment of a microresistivity device.

The modern dipmeter tool consists of either a three-arm or four-arm device, in which the arms are arranged radially around the central tool housing, and are linked by a spring mechanism which presses the arms against the borehole wall as the tool is run through the hole. On each arm there is a rubber pad in which are embedded small electrodes which make microresistivity measurements of the adjacent formation. The tool housing includes a compass and weighted pendulum which continuously monitor the geographic orientation of the arms and the deviation of the tool from the vertical.

The raw results of a dipmeter run are three or four microresistivity traces, together with the azimuth of one of the arms used as a reference and the borehole deviation angle (Fig. 2). Since lithological changes are very minor when traced across the width of the borehole, the form of the resistivity traces is very similar. Major differences between them are expressed in a relative depth shift between correlative features, which is a function of their strike and dip. In the earlier analyses of dipmeter results, dips and azimuths were computed within the logged sequence by

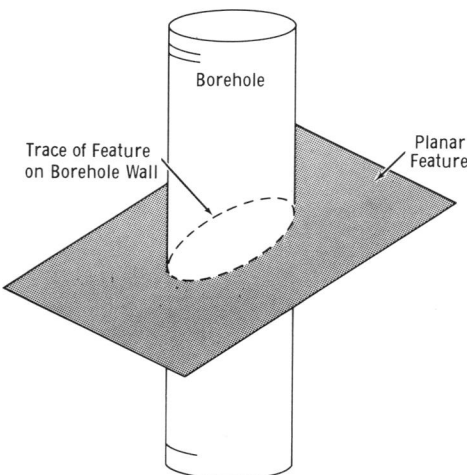

FIGURE 1. Borehole wall registration of dipping planar geologic feature as an inclined elliptical trace.

an extraordinarily time-consuming procedure of correlating common peaks and troughs, and computing the azimuth and dip of their orientation planes from simple geometrical considerations. However, in most modern dipmeter runs, this task is performed by a computer program which searches for a best match between resistivity segments on adjacent arms and computes the corresponding azimuth and dip.

The basic algorithm for the most widely used dipmeter processing program is straightforward. It is applied to the microresistivity traces which are recorded on magnetic tape as data strings digitized at an extremely fine frequency. Three fundamental parameters control the operation of the program: the window length, search length (or search angle), and the step length. The window length determines the number of feet of the reference resistivity trace which will be used as a match with corresponding (but displaced) lengths of traces from the other arms. The search length is the number of inches of maximum allowable displacement in the program's hunt for the best match on adjacent traces. This parameter clearly controls the maximum dip angle that the program can "see." Since all boreholes deviate from the bit size, a search angle is often specified and the corresponding search length computed from the caliper measurement of the diameter of the hole. The third parameter is the step length which determines the vertical distance in feet between successive correlation attempts. A graphic representation of the interrelationships between these three parameters is shown in Figure 3.

In matching a segment of the reference trace with the resistivity trace from another arm, a statistical correlation coefficient is computed as a simple measure of match. When the coefficient is calculated at all possible match positions within the search length range, the maximum positive correlation is selected as the position of best match. If this best value fails to exceed some lowest acceptable limit, the computation at this position may be abandoned and new computations initiated at the next position determined by the step length. In the case of a set of satisfactory matches, the relative displacements between the trace defines the orientation of a plane relative to the borehole axis. Since three points uniquely define a plane, the modern useage of the four-arm dipmeter provides a further diagnostic check on whether the four computed displacements are sensibly consistent with a planar feature. The incorporation of the dipmeter readings of the borehole deviation and azimuth allows correction of the orientation of the feature with respect to a true horizontal plane.

The choice of appropriate values for the controlling parameters clearly requires careful consideration of the overall geometry of the subsurface section and problems introduced by marked deviations of the borehole from the vertical. In gently dipping

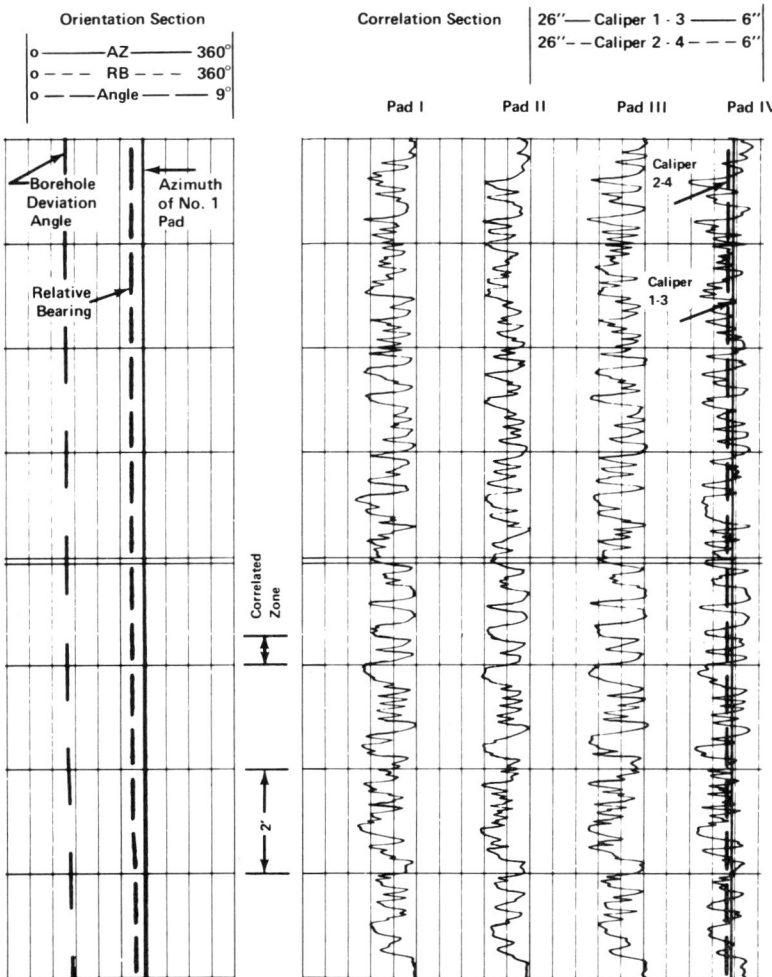

FIGURE 2. Four-arm dipmeter orientation traces and microresistivity curves. From Dresser-Atlas publications.

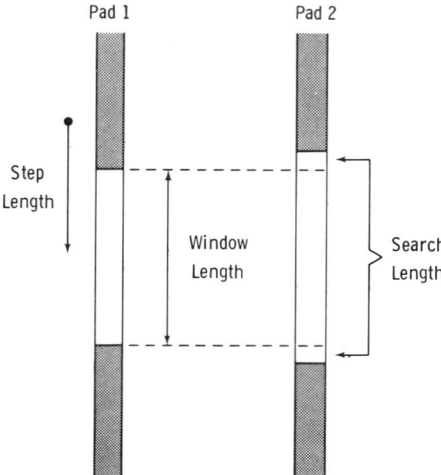

FIGURE 3. Schematic illustration of the correlation matching procedure used to match a segment of the microresistivity trace on the reference electrode (pad 1) with that of an adjacent electrode.

sequences, search angles can be held within low ranges and keep computation costs at acceptable levels. When penetrating high-angle beds, the appropriate search angle will often dictate a greatly exaggerated search length. Similar problems will be caused in the analysis of highly deviated holes drilled from offshore platforms. The size of window length and step length should not be chosen independently. A large window length with respect to step length will sometimes generate spurious trends of constant dip with depth when a pronounced peak or trough is caught within the window at several successive steps (Robertson, 1972).

The window length obviously prescribes the lower limiting resolution of the maximum size of features that can be detected by the processing. From conventional statistical reasoning, a large sample size (and so window length) is preferable. This will tend to subdue the contribution of laterally impersistent features and tool errors. The result is an averaging of dip information within the window. This is desirable in the extraction of structural dips since the effect is to filter out anomalous variation in favor of repetitive parallel features. However, in the case of stratigraphic analysis, current bedding features will generally be lost both as a consequence of their thinness and dip variability. In dipmeter studies that are aimed specifically at the study of internal structures, the window length must be shrunk to a finer size. At this level, the attempted correlations will become increasingly sensitive to "noise," much of which is caused by microrugosity variations of the hole. The increasing importance of stratigraphic studies in oil exploration has encouraged research into the improvement of dipmeter analysis to fine detailed resolution. The crude correlation algorithm described previously has been expanded in a more complex pattern recognition approach keyed to the characterization of peak and trough features on the microresistivity traces. By this means, fine-scaled but systematic variation can be differentiated from most spurious features. Examples of these more advanced algorithms are described by Vincent et al. (1977) and Kerzner (1982).

Regardless of the method of computation, the usual form or presentation of the computed results is as a "vector plot" or "tadpole plot" (Fig. 4). The vertical axis is depth and the horizontal axis measures the magnitude of dip on either an arithmetic or logarithmic scale. The location of each "tadpole" registers its depth of computation and calculated dip. The tail of the tadpole indicates the azimuth of the associated dip as related to a conventional compass circle. Many of the tadpoles will correspond to minor features or will be the result of errors introduced by tool measurements or computer processing. Others will show a tendency to be organized in coherent patterns which reflect distinctive variations in the dip of the units logged. A color convention is widely used by log analysts as a screening device to discriminate tadpole clusters which appear to represent significant common trends (see Fig. 4). The colors and their meanings are:

GREEN—dips with a common azimuth and essentially constant dip.
RED—dips with a common azimuth and increasing dip with depth.
BLUE—dips with a common azimuth and decreasing dip with depth.

As a supplementary motif, we shall also refer to a "white" pattern for intervals in which dips appear to be random, both in orientation and magnitude. This corresponds broadly to the "white noise" term used in signal processing. In many cases, white patterns are generated by faulty tools, poor borehole conditions, or computer processing problems. In other cases, the patterns are a legitimate expression of fabric heterogeneity and are of diagnostic significance.

The interpretation of the meaning of the dip patterns is rooted in orthodox geological theory of the structural geometry of subsurface units. Dip patterns can be variously attributed to tectonic dips, fault assocations, compactional drapes, and features internal to lithologies, such as cross-bedding. The dipmeter results are never considered in isolation, but are coordinated with other logs which are used to define stratigraphic markers, boundaries of units, and their internal mineralogies.

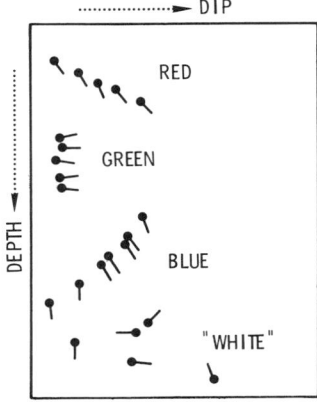

FIGURE 4. Tadpole or vector plot of computed dips and azimuths from a section in an offshore Canadian well. Examples of color assignation of dip trends are also indicated.

STRUCTURAL GEOLOGIC INTERPRETATIONS FROM DIPMETER PATTERNS

In this section, we consider the recognition and analysis of geometrical patterns from dipmeter data that reflect structural features on all scales. Under this loose rubric are included regional dip, local structure, faults, unconformities, and the results of draping caused by differential compaction of stratigraphic units within a sedimentary sequence. All these features are characterized by sets of parallel (or subparallel) surfaces or as discontinuities which separate such sets. Since they are generally large in scale, they are most easily extracted from computer processing of dipmeter logs and less equivocal in interpretation than current depositional structures.

Structural Dip and Unconformities

A tadpole plot is shown of a section of a dipmeter log run in a Canadian offshore exploration well, together with a portion of a seismic line (Fig. 5). This well was

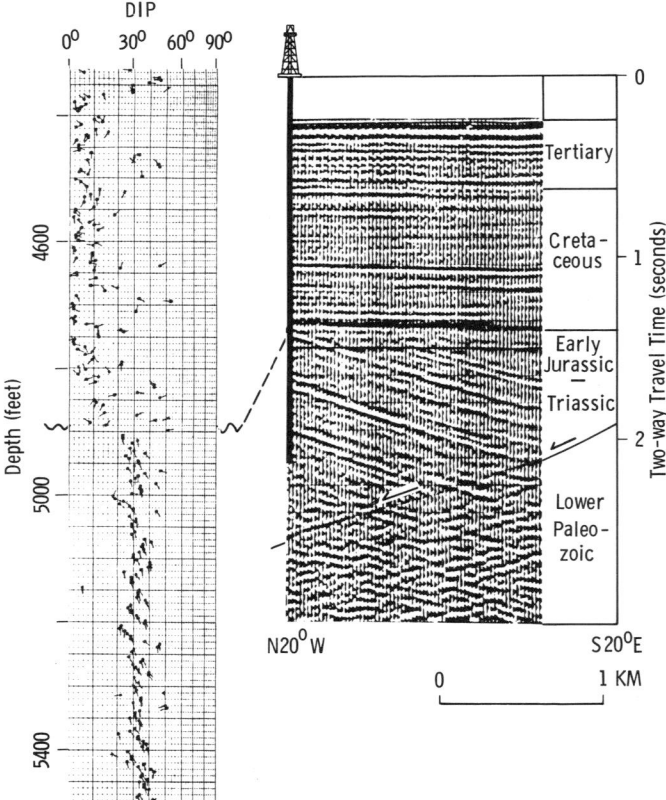

FIGURE 5. Dip vector plot and seismic line from a well on the LaHave Platform adjacent to Nova Scotia. Courtesy John Wade.

drilled on the LaHave Platform adjacent to Nova Scotia. The dip vectors show a dramatic shift in the middle of the section. Below this level, there is a strong pattern of dips which trend approximately 30 degrees in a southeasterly orientation. This lower sequence consists of thick deposits of terrestrial red sandstones and shales of probable Late Triassic to Early Jurassic age. They accumulated in one of the numerous grabens and half-grabens formed in the early stages of seafloor spreading which initiated the Atlantic Ocean (Wade, 1978). Taken over a large interval, the basic pattern can be characterized as "green" and reflects structural dip. At a smaller scale, repetitive sequences of red and blue patterns indicate local drape and sedimentological features.

The LaHave Platform became emergent during the Early Jurassic and was subjected to erosion. The platform was slowly transgressed during the later part of the Jurassic and the Early Cretaceous. In this well, the transgression is marked by a major unconformity at a depth of 4886 feet. The immediately overlying sediments are suggested to be of Early Cretaceous age, based on palynological evidence (G. L. Williams, personal communication). This transgressive sequence consists of gray shales and sandstones with a long-term trend of very shallow dips in contrast with the underlying red beds.

As a general rule, structural dip will show as monotonous but easily discernible green patterns which collectively form a long-term trend. Other patterns usually occur as localized motifs which match the limited vertical scale of the geological features that control them. The recognition of unconformities on the dipmeter log is mainly contingent on the geometrical relationship between beds above and below the erosional or nondepositional surface. Discrimination is obviously easiest in an instance of a pronounced angular unconformity. However, similar breaks in structural dip can be caused by faulting. Clearly, some knowledge of local geology is crucial to intelligent interpretation as is the case with all dipmeter studies.

Faults

In order to be detected on a dipmeter log, faulting must cause bedding plane distortions about the fracture zone or result in differential tilting of the fault blocks. The nature and degree of bedding distortion will be determined by the relative competence of the sequence. The widespread application of the dipmeter in sediments of the Gulf Coast region has provided many examples of fault-induced patterns whose interpretation has been confirmed by detailed subsurface mapping. "Growth" faults are normal faults which are formed contemporaneously with deposition and often show greater formation thicknesses on the downthrown side. They commonly develop "rollover" features which provide local anticlinal traps for hydrocarbons. The resulting distortion of bedding planes in the hanging wall is easily detected by the dipmeter (Fig. 6a) and is generally contained within a 10-foot interval above the fault plane (Gilreath and Maricelli, 1964). "Drag" faults are also normal faults that occur more frequently in the north of the coastal plain. These are generally formed after deposition, but at a stage before complete compaction of the sediments. The poor degree of mechanical competence results in a relative local warping of

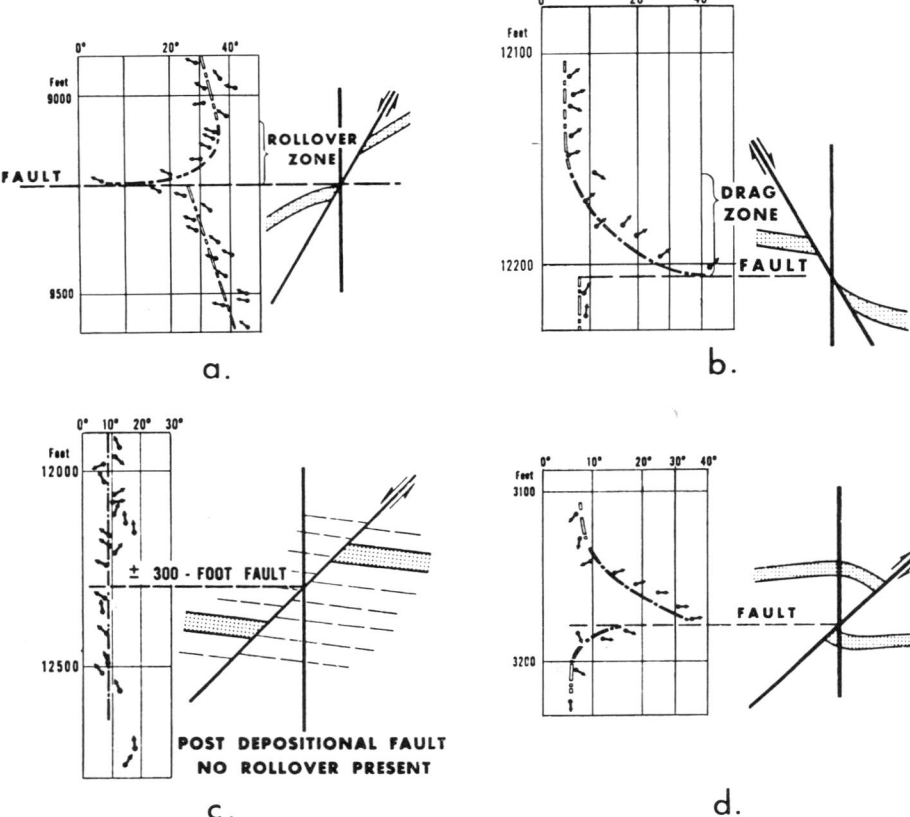

FIGURE 6. Dip patterns associated with a growth fault (*a*), a drag fault (*b*), a normal fault (*c*), and a thrust fault (*d*). From Gilreath and Maricelli (1964).

the hanging wall into the fault plane with effects registered by the dipmeter (Fig. 6*b*).

In more consolidated, and therefore competent, sequences the effect of a normal fault may go undetected in the dipmeter log (Fig. 6*c*) if the result is one purely of fault block displacement with no distortion. However, normal fault associations similar to "rollover" were first noted in the Grand Canyon region by Powell (1875) in his exploration down the Colorado River. Subsequent studies of faulting in the Colorado Plateau have confirmed the widespread existence of anticlinal flexures in the hanging wall which curve into the plane of the normal fault. Hamblin (1965) hypothesized that the "reverse drag" phenomenon (equivalent to rollover) was associated with normal faults having curved planes that shallowed at depth. He concluded that the curvature caused the fault blocks to be pulled apart as well as displaced vertically. The incipient gap would then be filled either by rupture as a set of antithetic faults or by flexure to generate reverse drag.

Similar remarks are true for thrust or reverse faults, although the stress mechanics of thrust faulting commonly result in bed rotations in both hanging and foot walls (Fig. 6d). Overturned beds are commonly observed in thrust-fault zones in the Rocky Mountains of the northern United States and southern Canada. Their geometry can be analyzed through application of the dipmeter, although the wide range of dip angles present significant problems in dipmeter processing and interpretation.

Compactional Drape

Stratigraphic sequences typically consist of beds with varying degrees of mechanical competence. When subjected to burial and the associated pressures of lithostatic loading, a sediment pile exhibits differential degrees of compaction, both vertically and laterally. If relatively rigid lensoid bodies are surrounded by less competent units, originally planar surfaces will be distorted in local structural features of compactional draping. The dipmeter is sensitive to the results of this phenomenon and can be used to pinpoint the location, shape, and orientation of nearby barrier sands, channels, and carbonate reefs. As used in these applications, the dipmeter has proved especially useful in the detection of potential reservoir units which are not actually encountered within the well bore.

Barrier bar sands generally parallel ancient shorelines and have a convex upper surface. Differential compaction of bounding transgressive shales results in a draping of beds over the crest of these features. In wells drilled at flank positions, the drape is shown in the shales overlying the sandstones as red patterns, in which dips diminish upward to overlying green patterns representing local structural dip (Fig. 7a). The crest of the bar is therefore located updip from the pattern trend and the bar axis parallels their strike.

Channel sandstones form elongate bodies which are often highly sinuous at a regional scale, but can be approximated as a local linear segment for purposes of dipmeter analysis from a nearby well. Their cross-sectional shape varies according to the geological context, although almost all have a convex-downward erosional base. When the channel has cut down into compacted lithologies, the upper surface may be broadly concave-upward in form. The draping of overlying shales is effectively a synclinal feature which dips inward to the sandstone axis. The resulting dip sequences in the shales of flank positions are red patterns which, again, specify the direction to the axis and its local orientation (Fig. 7b).

When the channel is cut into soft beds such as shales, the compaction of the sequence generally results in a biconvex cross section as the profile of the sandstone body. This situation causes draping characteristics in the shales overlying the channel sand similar to those of a barrier bar sand. This equivalence makes the important point that dip interpretation of draping is specific to the shape and orientation of the sandstone bodies, rather than diagnostic of depositional environment. However, dip patterns measured within the sandstone units often discriminate current bedding surfaces which successfully discriminate these alternative models. Their interpretation will be reviewed in later sections of this chapter.

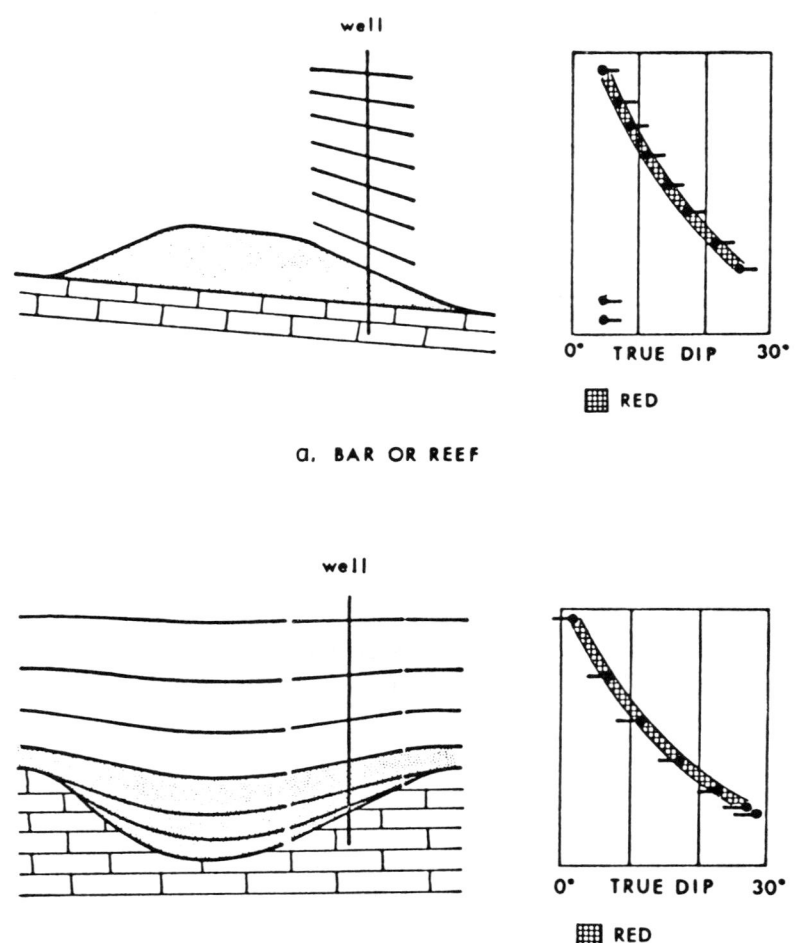

FIGURE 7. Dip patterns associated with draping over a barrier bar sand (*a*) and a channel sandstone (*b*). From Campbell (1968).

Reefs of limestone and dolomite form subsurface features with a relative relief that may reach hundreds of feet. When their adjacent and overlying facies are marine shales, the compactional draping effect over the reef flanks is highly pronounced. Examples of the situation include the Leduc Reef–Ireton Shale association in the Upper Devonian of Alberta and the Mississippian reefs of northcentral Texas. The drape structure can be discriminated with ease on dipmeter logs run in wells located less than a quarter of a mile from the reef. The dip sequences in the overlying shales at flank positions form red patterns, and interpretation of geometry follows the same logic as for barrier bar sands. Within the reef core itself, the dips are generally highly inclined but chaotically oriented (white patterns), and are caused by fractures, vugs, and other diagenetic features of no regional significance (Fig. 8). Below and seaward of the reef core, a sequence of forereef talus ideally shows a blue pattern of downward-decreasing dips sloping away from the reef. In detailed studies from several wells, the forereef and backreef sides of the structure can often be determined. The forereef dips are usually greater (up to 35°) than most of the more gently sloping backreef flanks (up to 25°).

Even reefs overlain and flanked laterally by relatively competent facies show similar but more subdued draping structures which can be detected by dipmeter logs. Dipmeter studies of reef geometries are documented from the Middle Devonian Rainbow pinnacle reefs of northern Alberta (Cox, 1968) and the Silurian reefs of the Michigan Basin (Bigelow, 1973). In both of these cases, the reefs are overlain by repetitive sequences of carbonates and evaporites rather than shales. These authors also report in detail on the proven success of the dipmeter as a means of locating the position of a reef core from an exploration well which narrowly missed its target.

DIPMETER INTERPRETATION OF SEDIMENTARY STRUCTURES AND DEPOSITIONAL ENVIRONMENTS

The scale of sedimentary structures contained within stratigraphic units is usually drastically smaller than forms associated with tectonic folding, faulting, and elements of significant compactional draping. With the notable exception of eolian and some submarine dunes, most cross beds are less than 1 foot thick. In addition, there is the common tendency for cross-bedded surfaces to show complex polymodal patterns of variable dip and orientation.

Although the computer correlation algorithm described earlier is adequate for the task of discriminating major or repetitive dip patterns, the fine resolution demanded

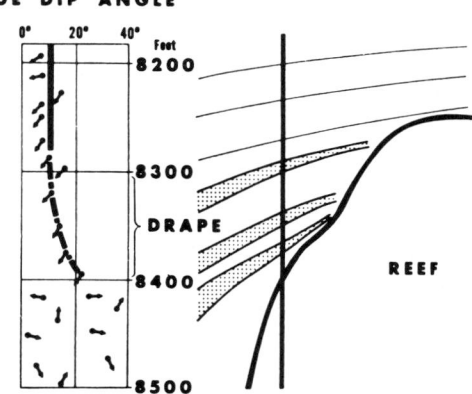

FIGURE 8. Dip patterns associated with draping over a carbonate reef. From Gilreath and Maricelli (1964).

by more detailed analysis introduces severe problems in processing. By necessity, the window length used for correlation must be shrunk to a size smaller than crossbed thickness. However, the frequency of spurious correlations and statistical noise increases as the window length is decreased. As a remedial operation, the dip correlations from a computer program can be edited by a process of cluster analysis (Hepp and Dumestre, 1975) to screen out aberrant dips. Schoonover (1974) also describes a pattern recognition algorithm applied to dip vector sets which discriminates systematic patterns by type, classified by a mathematical decision function based on a training set.

Vincent et al. (1977) extended the notion of pattern recognition to analysis of the raw microresistivity traces. "Feature extraction" is first applied to locate and identify peaks, troughs, and other curve components. Similar features are then correlated through comparison of pattern vectors under the constraint of noncrossing correlations. The purpose of the pattern recognition approach is to mimic the human aptitude to discriminate individual systematic features. These must correlate across the borehole, be consistent with a single dipping surface, and match lithologic variation rather than the geometry of the borehole. Because the dipmeter is a "contact" device, it is sensitive to the rugosity of the borehole wall. The continuous measurement of borehole diameter by the caliper is particularly helpful in the distinction of curve events which reflect borehole characteristics. The sophisticated nature of this algorithm introduces elements which are more consistent with a traditional geological approach to the problem than the statistical rationale of the simpler correlation program. By this means, specific curve features which are "anomalous" but systematic, will not be eliminated by statistical criteria as atypical. At the same time, safeguards must be included within the program which inhibit acceptance of anomalous but incoherent features caused by tool errors or microrugosity variations.

When the results of a dipmeter analysis are interpreted in terms of a sedimentological model, it must always be remembered that dips are measured with respect to a present-day horizontal datum. Consequently, measured angles need to be rotated by a factor that corrects for tectonic tilting and restores the dip vectors in their orientation at the time of deposition. It is normally a simple operation to determine the structural dip of a sequence by locating persistent and reasonable green patterns of common dips. The corrective operation is then a rotation opposite and equal to the vector of structural dip.

Eolian Dunes

Eolian cross beds cause extremely distinctive and coherent patterns to be registered on the dip vector plot. They are easily discriminated in computer processing because of the great thicknesses of many of the cross-bed sets. Blue patterns characterize the toeset, with low angles close to structural dip at the base. These patterns merge upward into high-angle (around 30°) green patterns in the dune foreset which match the angle of repose. The angular relationships are shown for a hypothetical borehole through a transverse dune in Figure 9. Here, an "outcrop" view of cross-bedding normal to transport direction is matched with the corresponding dip vector patterns.

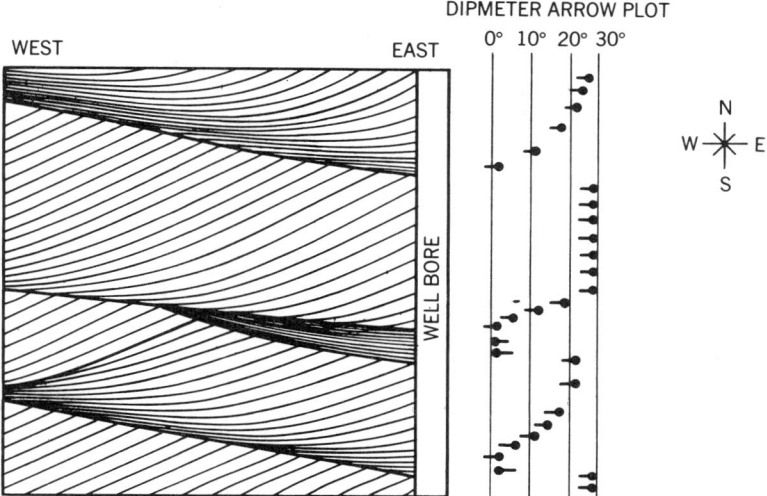

FIGURE 9. Dipmeter patterns from a hypothetical borehole penetrating eolian cross-bed sets. From Nurmi (1978).

The internal structure of modern eolian dunes has been described by authors such as McKee (1966). The application of this knowledge in conjunction with dipmeter patterns allows some interpretation of the dominant dune types within a logged section. Idealized relationships between different dune morphologies and the orientation spread of their cross beds are shown in Figure 10. Transverse dune cross beds are generally considered to occur most commonly in desert eolian depositional systems. Cross-bedding patterns of this type have been interpreted by both Glennie (1972) and Van Veen (1975) to typify dipmeter patterns from the Permian Rotliegendes Group in the southern North Sea region.

A typical field example of a dipmeter profile through a set of dunes (Fig. 11) shows the diagnostic arrangement of green pattern foreset and blue pattern toeset dips. The high consistency of dip orientation within each dune set is characteristic of eolian deposits, but there is often a marked variation between sets. The change in orientation reflects both fluctuations in wind direction and the location of the borehole with respect to its relative position on each successive dune.

Braided Stream Deposits

Braided stream deposits are produced by complex networks of channels which carry high loads of gravels and coarse sands. They are usually found in areas undergoing significant rates of erosion and form thick and extensive sheetlike bodies which often overlie an irregular surface of unconformity. In plan view, the channel system is irregular with numerous bifurcations, but exhibits a relatively low degree of sinuosity.

A dipmeter profile of a braided stream sequence generally shows a highly variable set of patterns which are impersistent and erratic. The small diameter of the borehole

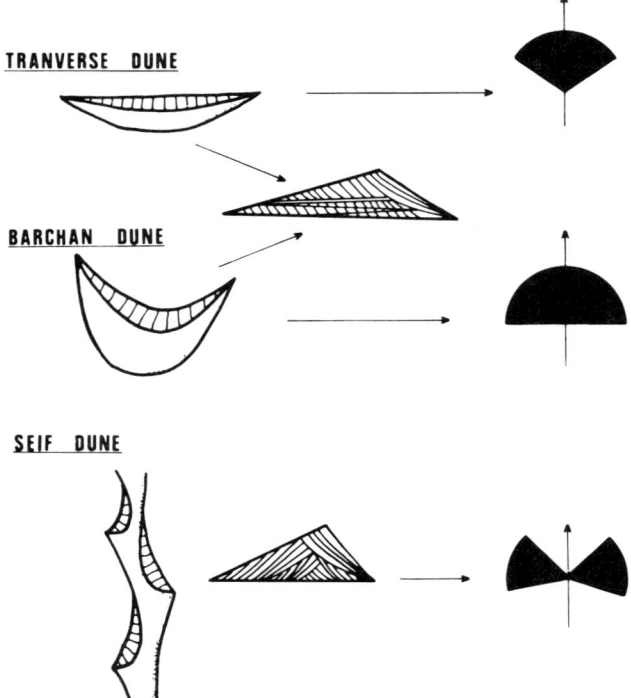

FIGURE 10. Relationship between dune morphology and cross-bedding orientation. From Richard C. Selley: *Ancient Sedimentary Environments, Second Edition*. Copyright © 1970, 1978. Used by permission of the publishers, Cornell University Press and Chapman and Hall.

results in a dipmeter perception of nonplanar surfaces and localized lenses of internally coherent dips. White patterns are commonly observed, while limited red and blue patterns pick out trough cross beds and erosional surfaces. Shales formed in abandoned channels may register green patterns which indicate the general direction of structural dip. Within the coarse clastics, dips can vary from horizontal to angles as high as 35°. The broad expectation of a range of between 20° and 25° as dips for foreset beds (Selley, 1978) is an aid in their identification. The foresets show variability in bearing, but are approximately constrained within a 90° arc whose mean orientation gives an indication of the downstream direction (Fig. 12). This trend should roughly parallel the axis of elongation of the sandstone body.

Meandering Stream Point Bars

Fluvial sediments are finer grained than braided channel deposits, with high proportions of both sand and shale. Both outcrop and cored intervals show a dominant motif of upward-fining grain size sequences. These range upward from a channel floor erosion surface, which may have an intraformational conglomerate, through a succession of cross-bedded point-bar sands into cross-laminated fine sands and overbank shales. The meander of the channel causes the stream bed to migrate laterally, with accretion at point-bar surfaces. Within the point bar, trough cross beds of relatively coarse sands grade upward into thinner tabular sets with finer sands. However, frequent reworkings result in sequence interruptions and repetitions of the basic pattern as variations about the general trend. Abandoned channel loops are filled with deposits of silts and clays. The meander belt deposit forms a sinuous body which may be up to 20 times the width of the stream (Goetz et al., 1977).

The sedimentology of both ancient and modern alluvial systems has been studied extensively (well summarized in Reading, 1978) and the geological literature provides models which are guides in the interpretation of dipmeter profiles (Fig. 13). Dips are generally observed to be highly inclined and somewhat erratic at the base of

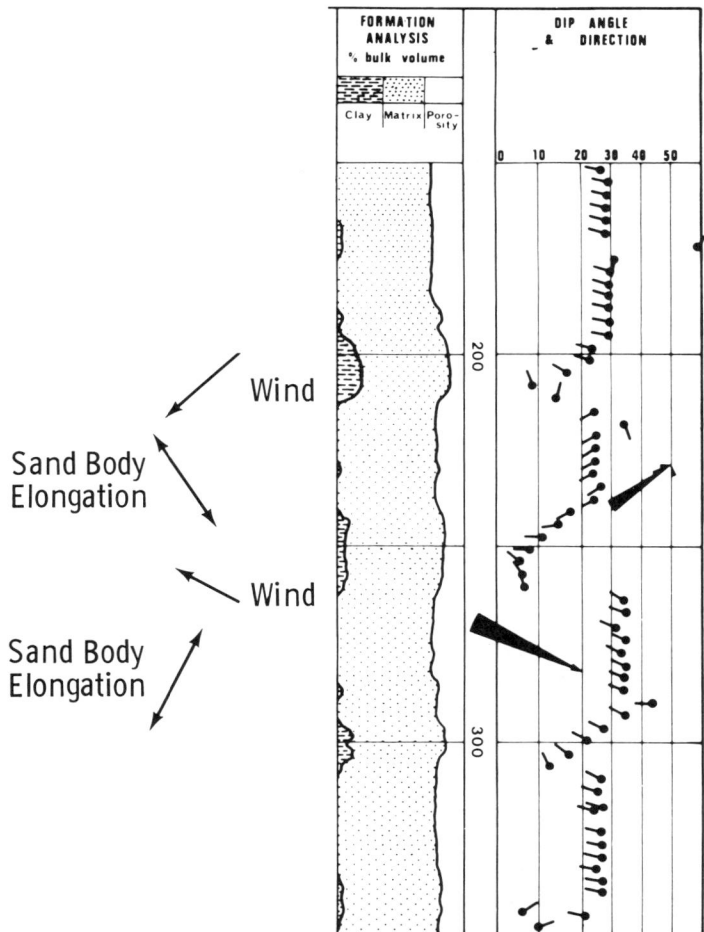

FIGURE 11. Dip patterns in eolian sand dunes. From Goetz et al. (1977).

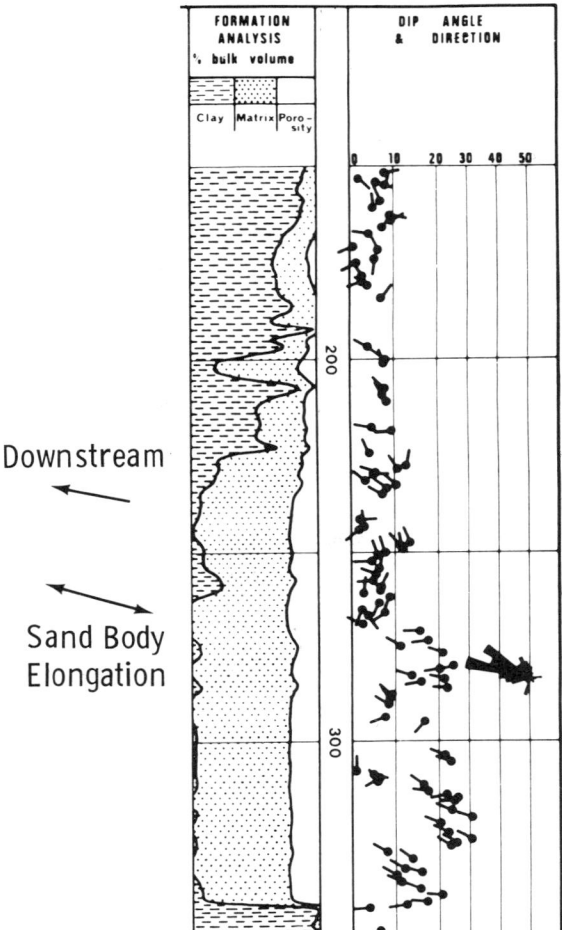

FIGURE 12. Dip patterns on a braided stream alluvial deposit. From Goetz et al. (1977).

trough cross-bedded units, but become more coherent and shallower in the overlying tabular cross beds. As a consequence, downward-increasing red dip patterns typify channel sections. Detailed analysis often reveals a bimodal pattern of orientation. One mode is directed toward the channel axis and is related to large-scale point-bar bedding surfaces; the other points down the channel axis and reflects cross-bedding surfaces (Selley, 1978). Spreads of up to 180° can be expected in the orientations of cross-bed dips as a response to trough shapes and the variability of current flow. However, their statistical average indicates the overall downstream direction of the meander belt in the immediate vicinity of the borehole. Red patterns in the overlying shales suggest compactional draping features as discussed earlier. Green patterns within the shales discriminate the local structural dip.

Delta Distributary Channels

The meandering streams of the upper alluvial plain pass into distributary channels of the delta. These channels are less sinuous in plan and form natural levees of clays and silts. The width of the channel fill is generally much less than those of alluvial meander belts, so that the deposits form long narrow bodies. Delta distributary channels can show complex variations in dip pattern. However, the basic motifs follow similar patterns to those found in meandering stream point bars. An example of a dipmeter profile through a delta distributary channel is shown in Figure 14.

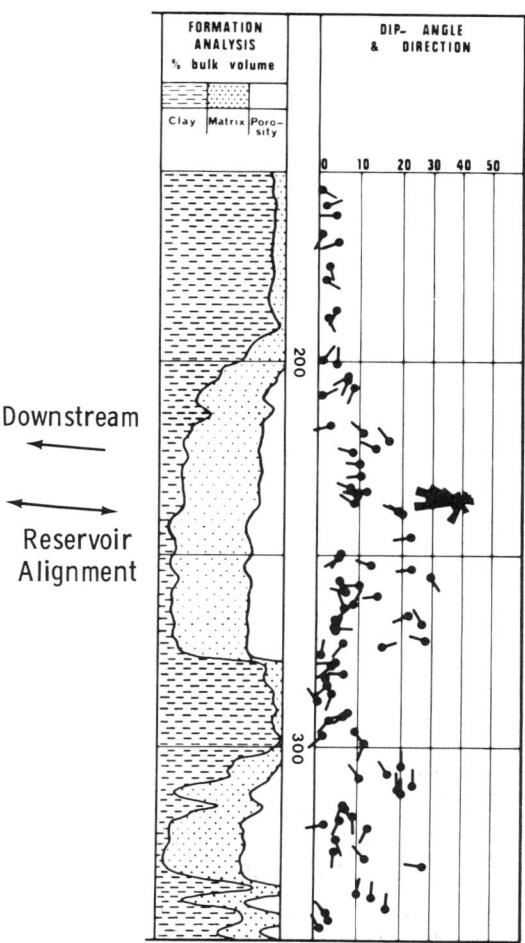

FIGURE 13. Dip patterns on a meandering stream point-bar accumulation. From Goetz et al. (1977).

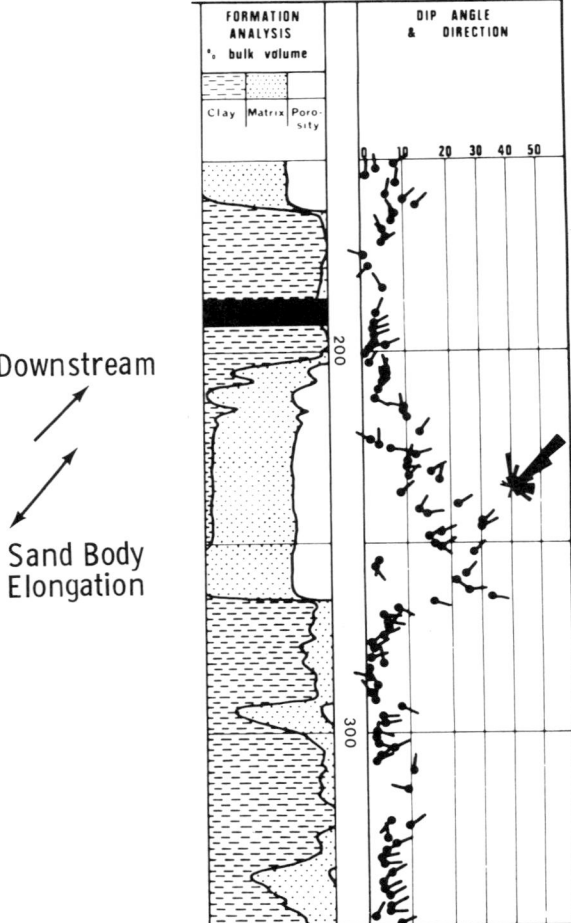

FIGURE 14. Dip patterns on a delta distributary channel fill. From Goetz et al. (1977).

Distributary Mouth Bars

The morphology of deltas is controlled primarily by the sediment load and the balance between fluvial and marine regimes. A spectrum of delta shapes are possible according to the strength of destructive wave processes relative to constructive fluvial deposition. These range from birdsfoot, through lobate, arcuate, and cuspate forms. Highly constructive deltas, such as that of the Mississippi, are particularly favorable environments for the production and preservation of distributary mouth bars. These form lobate sand bodies at the mouths of distributaries and extend seaward.

While relatively fine grained at the base, there is a general trend of a coarsening-upward profile. At the same time, the cross-bedding angles are generally steepest at the top and shallow downward to cause distinctive blue patterns on the dipmeter vector plot (Fig. 15). The cross beds are normally tabular in form and inclined seaward, unless reworked or influenced by long-shore currents. As the delta progrades seaward, the feeder channel may cut down through the upper part of the bar. Gilreath and Stephens (1971) document several examples of mouth bar dipmeter vector patterns from both the Tertiary Mississippi Delta and a Cretaceous delta in Wyoming and North Dakota.

Estuarine and Tidal Channels

In cases where deltas are dominated by strong tides there is oscillating interplay of fluvial and marine processes. At flood tide, seawater moves upstream into estuaries

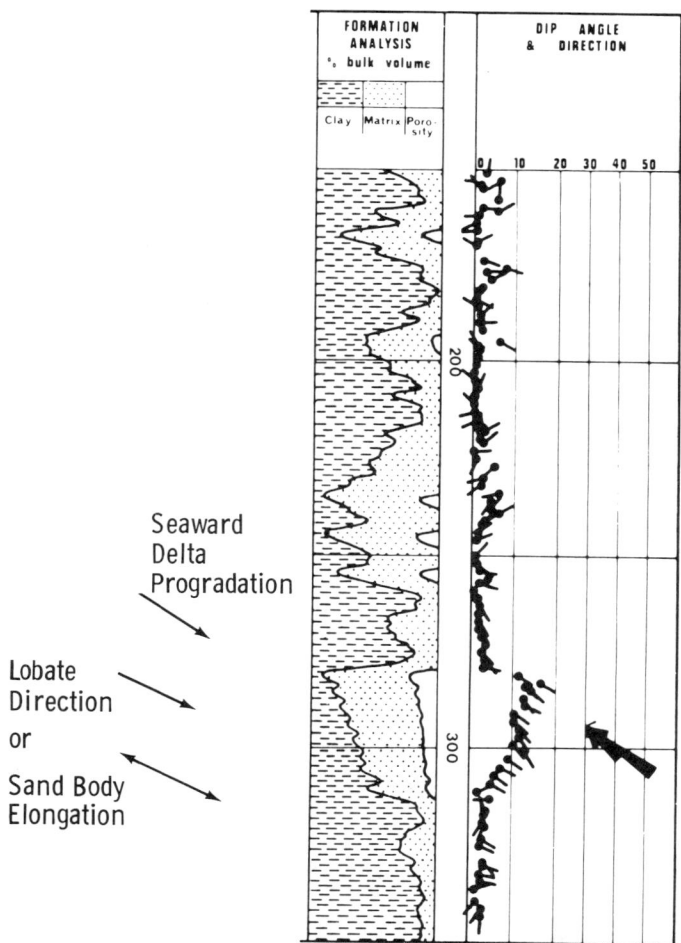

FIGURE 15. Dip patterns on a distributary mouth bar. From Goetz et al. (1977).

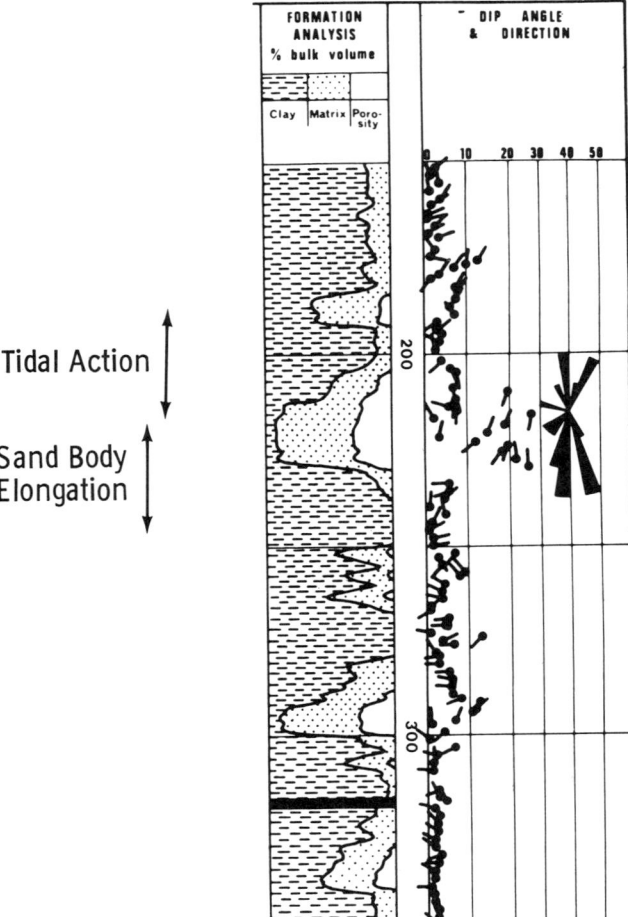

FIGURE 16. Dip patterns on tidal channel or tidal flat deposits. From Goetz et al. (1977).

and tidal channels, but recedes in time of ebb tibe. The net effect often causes a bimodal pattern of diametrically opposed dip directions which are picked up by the dipmeter (Fig. 16). Otherwise, the internal characteristics of these sands resembles those of distributary channel fills when deposited in moderately narrow estuaries. Wider estuaries generally develop extensive tidal flats as silt and mud complexes cut by subparallel sand bars. The grain-size profile typically shows a ragged fining-upward gradation from coarse sands at the base to shales at the top.

Barrier Bars

On coastlines with strong marine currents and a source of sediment, linear shoreline elements are formed parallel to the coast. These include both beaches and offshore bars and islands found at or close to wave base. Most beaches show upward coarsening in grain size and tabular cross beds which dip toward the sea. Submerged offshore bars show characteristics similar to beaches on their seaward side. Dips on the landward side slope in the opposite direction and are often greater than those on the seaward flank. With sufficient sediment supply, the bars may emerge and form a barrier complex with lagoons or swamps behind them.

Dipmeter vector profiles through a bar sand show some variation in form according to the location of the borehole on the bar. As a general motif, dips are characterized by downward-decreasing blue patterns (Fig. 17). At flank positions, the dips are generally unimodal in orientation and reflect either shallow seaward-dipping beds or the steeper foresets on the landward side. Close to the bar axis, the patterns may be bimodal and diametrically opposed dips (Campbell, 1968). In all these cases,

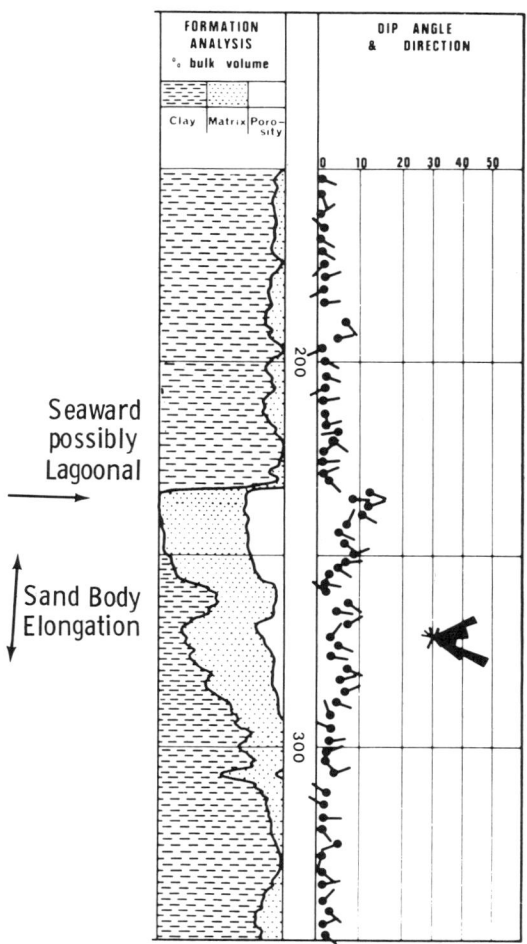

FIGURE 17. Dip patterns on a bar-type sand. From Goetz et al. (1977).

the dips are arranged at right angles to the bar axis. The upper surface of the bar is generally convex-upward and causes compactional draping in the overlying shales. The draping effect is most pronounced at flank positions and has the same orientation as dips within the sandstone. This arrangement contrasts with channel sands in which the dips of the shale drape are at right angles to the sandstone cross-bed dips.

Marine Shelf Sands

Considerably less is known about the detailed external morphology and internal structures of sand bodies on modern marine shelves than is the case for continental and coastline deposits. A major reason for this discrepancy is that this environment is less accessible for detailed observation. However, it appears that the variability

FIGURE 18. Dip patterns on marine shelf blanket sands. From Goetz et al. (1977).

of marine shelf hydraulics makes a standard model of dip profile an unrealistic goal. Many marine sands are sheetlike in form with repetitive regressive–transgressive units. Current bedding is generally shallow in inclination. Dips may form localized blue or red patterns but are often polymodal or random in orientation as broad white patterns (Fig. 18).

ANALYSIS OF DIPMETER VECTORS

Up to this point, vector results from a dipmeter run have been considered as patterns in a preliminary reconnaissance of structural elements and sedimentary environments. This step is necessary since it both dictates the interpretation model and identifies the geologic meaning of specific vector patterns. No features are uniquely diagnostic, because they are responses to geometry rather than geologic process. Consequently, all sources of information must be integrated with the dipmeter log, including the analysis of other logs (described in other chapters), cuttings and core descriptions, and context within regional geology. However, once the appropriate model has been identified, dipmeter vectors can be analyzed to extract detailed measures of the three-dimensional configuration of both structural and sedimentological elements. The most commonly used methods are reviewed in the following example.

The Cherokee Group sandstones of Kansas often take the form of elongate, "shoe-string" bodies, whose location and orientation is often difficult to determine without recourse to an extensive drilling program. The example to be described is one of several dipmeter surveys made in a series of neighboring wells through a Cherokee Group section which contained a productive development of the Bartlesville Sandstone. The purpose of the dipmeter work was to deduce the long-axis orientation of the sandstone and to determine the geographic location of the central axis (the thickest part of the sandstone) as related to the existing well control. A successful analysis would enable a more efficient siting of further development wells.

Gamma-ray, SP, and resistivity logs are shown for part of the Cherokee Group in a well in southeast Kansas (Fig. 19). The log traces show a section that is primarily shale with a development of the Bartlesville Sandstone toward the base. The results of the dipmeter survey are presented in a variety of forms: first, a listing of the dips and azimuths at various depths (Table 1), which are the output of computer processing of the raw microresistivity traces; second, a vector or tadpole plot of the same information (Fig. 20); and third, a less widely used transformation of the same data, a cylindrical plot (Fig. 21). If the cylindrical plot is rolled on itself about a vertical axis, the sinusoidal lines form ellipses which represent the intersection of dipping planes with a hypothetical borehole wall and give a scale model representation. All three forms present alternative views of the same dipmeter information.

More detailed graphic analysis may be made of subsections of the whole interval through the use of polar plots. The most commonly used variants are the Schmidt net (the Lambert equal area projection) or the Wulff net (the stereographic projection). On either projection, the normals to the dipping planes are plotted with respect to

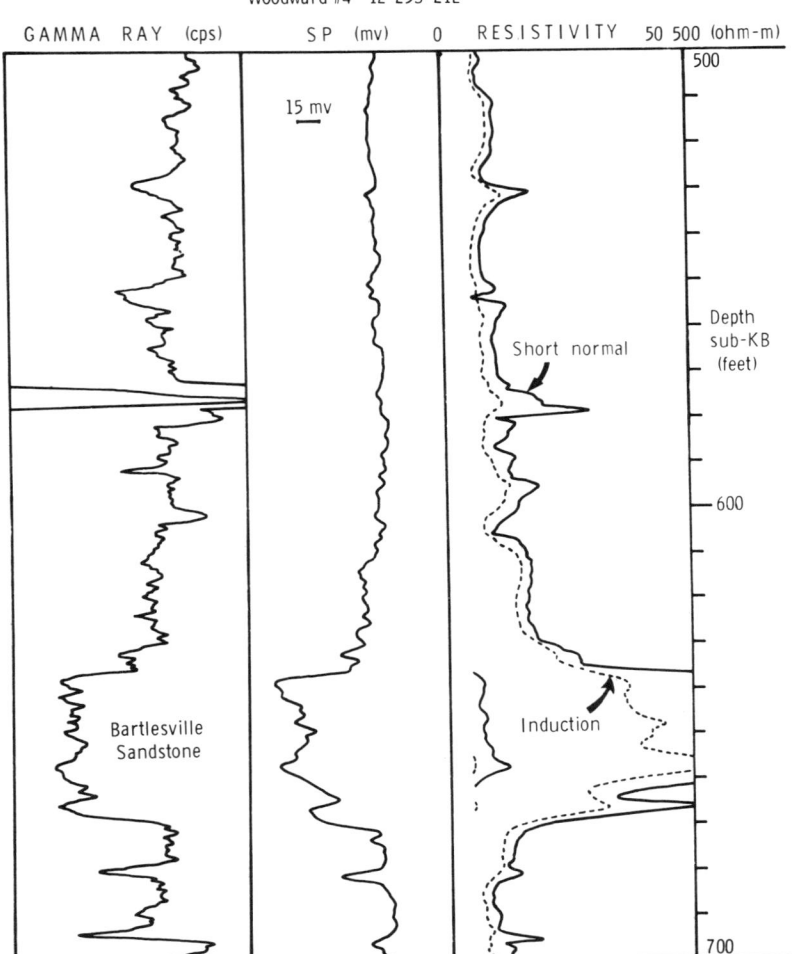

FIGURE 19. Gamma-ray, SP, and resistivity logs of a Cherokee (Pennsylvanian) section from a well in southeastern Kansas.

a hypothetical lower hemisphere (the conventional choice) or the upper hemisphere (used in this example). The Schmidt net presentation has the advantage that equal areas on the hemisphere surface will be preserved as equal areas on the projection; the Wulff net preserves great circles on the hemisphere as great circles on the projection. (Both these techniques are widely used in structural geology for the analysis of tectonic elements.)

Schmidt nets are shown for subdivisions of the sequence (Fig. 22). High in the sequence, the shales show very minor dips with no discernible preferred orientation, which grade downward to a zone with small dips in an eastern orientation. These sets probably reflect local structural dip. In the zone immediately above the sandstone, there is a pronounced increase in dip, with an overall western orientation; this probably represents structural drape of the shale over the sandstone. The data suggest that the long axis of the sandstone is oriented north–south, and, if the upper surface of the sandstone is convex, that the central axis lies to the east of the well. Since the sandstone overlies shales, the theoretical considerations discussed earlier suggest that the upper surface should be broadly convex-upward. This hypothesis is confirmed

TABLE 1
Depths, Dips, and Azimuths from a Cherokee Group Section in a Southeast Kansas Well

501	4.7	121	587	2.5	62
503	3.7	100	589	1.7	82
505	4.5	156	593	4.8	31
507	2.6	132	595	2.6	84
509	3.7	122	596	7.8	84
511	0.0	0	605	9.9	232
513	0.8	107	611	7.8	316
515	1.9	27	615	7.6	306
517	1.2	158	617	9.7	308
518	1.8	125	619	7.4	310
519	1.4	161	625	5.3	176
521	1.4	154	634	8.5	272
523	0.7	210	635	11.6	262
525	0.7	97	637	6.2	248
527	1.4	156	638	5.2	277
529	2.8	72	646	1.6	64
531	1.7	65	648	1.5	101
533	1.4	143	651	1.7	200
535	0.5	299	653	4.8	172
537	2.1	358	655	2.9	193
539	1.1	9	657	3.6	168
541	1.5	50	662	2.8	176
543	1.1	5	665	4.6	173
545	1.5	48	666	6.4	177
552	1.6	291	667	6.7	168
553	2.8	217	671	3.5	21
555	2.9	211	673	2.7	158
559	1.5	42	675	1.6	75
563	2.5	321	681	4.0	279
565	2.2	249	683	4.7	359
567	1.5	313	685	1.4	278
569	3.4	286	689	3.3	278
574	2.7	86	691	1.4	285
575	3.2	16	693	0.9	149
577	1.1	357	659	0.9	331
579	1.3	134	696	1.4	94
581	6.1	154	697	1.7	62

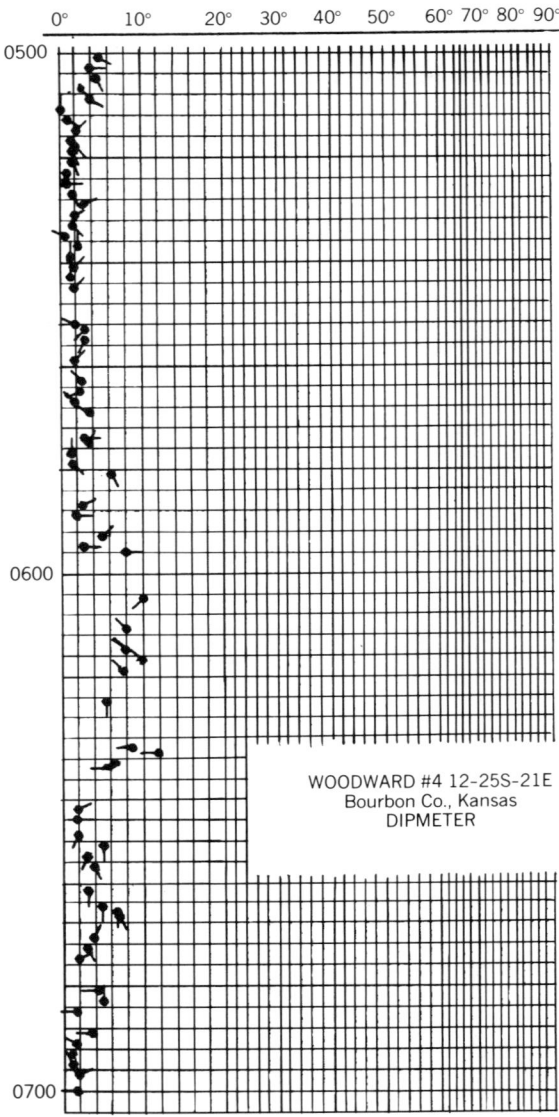

FIGURE 20. Dip vector plot of the Cherokee section.

from mapping by Van Dyke (1975), who demonstrated local structural uplifts of limestones higher in the section over similar sand bodies in the area.

The dips within the sandstone show an abrupt contrast, with orientations strongly aligned to the south. If these dips discriminate the surfaces of cross-bedding sets, then the cross-bedding direction is aligned with the axis of the sandstone body. This interpretation would suggest a current flow from north to south, and the

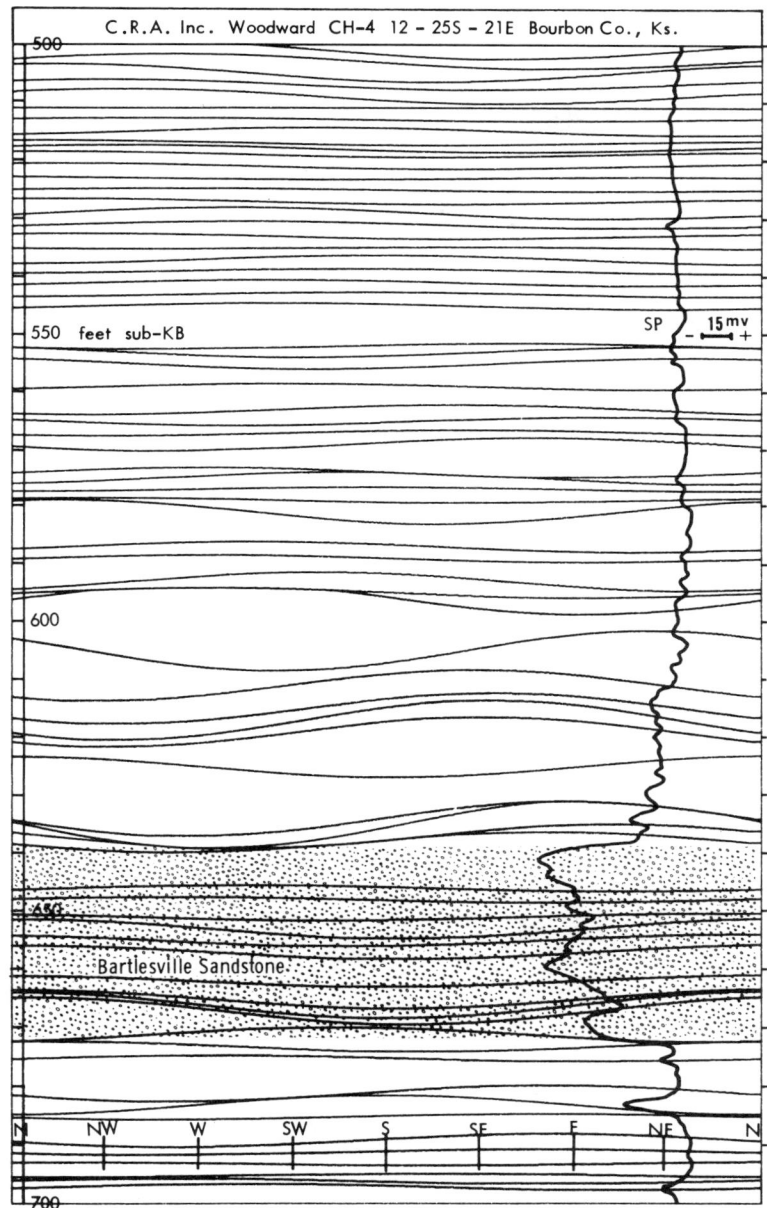

FIGURE 21. A cylindrical plot presentation of the Cherokee section dips.

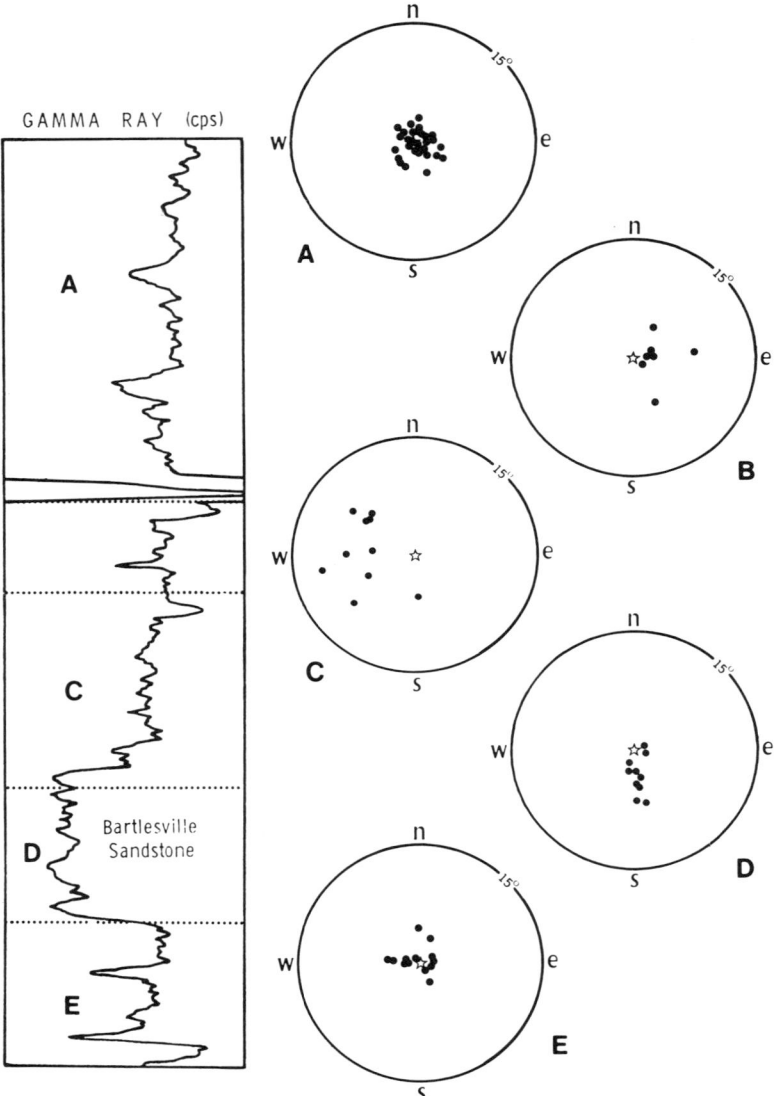

FIGURE 22. Schmidt net polar plots of dips from subdivisions of the Cherokee section.

depositional origin of the sandstone body as a channel. (The most likely alternatives of a marine bar or a strandline deposit would be expected to have cross-bed orientations at right angles to the long axis.) The SP and gamma-ray profiles of the sandstone are also consistent with a channel interpretation (see Chapter 3). An anomalous feature of these dips is their low angle which is significantly more shallow than those that typify alluvial channel cross beds. However, this apparent angle is an artifice of the crude computer algorithm used in microresistivity trace correlation. As mentioned previously, the relatively coarse window used in the traditional form of crossing results in an averaging effect. This tends to suppress the fine sedimentological detail revealed by more refined algorithms.

The dip vectors of this sequence can also be analyzed through the use of azimuth-frequency diagrams ("rose diagrams") in a similar manner to that widely used by geologists in the study of paleocurrent directions (Potter and Pettijohn, 1963). In common with dip vector plots, and polar plots, this method is purely graphical in the interpretation of trends and relative dispersions. Since the data are numerical, mathematical methods of analysis are clearly appropriate.

Scheidegger (1965) demonstrated that the correct approach to the analysis of directions in three-dimensional space lay in statistical analysis of vectors summed in the form of an "orientation tensor." The principal eigenvector of this orientation tensor specifies the mean direction of the component vector. Further work by Woodcock (1977) detailed the role of eigenvalues on the characterization of the overall fabric shape which characterizes the set of vectors taken as a group pattern of orientation. The major concepts are reviewed in the context of the Cherokee example.

The preferred orientation of the dip plane normals in the Schmidt net plots may be numerically analyzed in terms of their distribution on the reference sphere. If azimuth is denoted by A, dip by D, direction cosines, L, m, and n are computed with respect to the east–west, north–south and vertical axes:

$$L = \sin A \sin D$$

$$m = \cos A \sin D$$

$$n = \cos D$$

An orientation tensor, $\mathbf{A}$, is computed for the N observations of the analytical interval:

$$\mathbf{A} = \frac{1}{N} \begin{bmatrix} \Sigma L^2 & \Sigma Lm & \Sigma Ln \\ \Sigma Lm & \Sigma m^2 & \Sigma mn \\ \Sigma Ln & \Sigma mn & \Sigma n^2 \end{bmatrix}$$

The eigenvectors of the orientation tensor specify the major, intermediate, and minor axes of the distribution of normals, and are identified by the size of their respective eigenvalues. Each eigenvector contains the direction cosines that specify its orientation. Each eigenvalue corresponds to the proportion of the total variance which is accounted for by its associated eigenvector. A "scattering angle" of the normals about the principal axis may be computed by the formula

$$\theta = \arccos \sqrt{\lambda}$$

where λ is the eigenvalue of the principal eigenvector.

This methodology is now used widely in structural geological studies, both to specify the statistical preferred orientation of tectonic data and to act as a numerical aid in the description of the general shapes of the normal distributions.

For the Cherokee example, eigenvector analysis was made of the dips in the shales immediately above the sandstone and those within the sandstone. The essential results are

Shale
Principal axis orientation: Dip 6.4°, azimuth 275.9°
Scattering angle: 5.1°.

Sandstone
Principal axis orientation: Dip 3.3°, azimuth 170.3°.
Scattering angle: 2.4°.

The relative sizes of the eigenvalues suggest a slightly more "girdle"-shape orientation to the sandstone dips, when contrasted with the more unimodal distribution of the shale dips, as might be expected from inspection of their Schmidt net plots.

FABRIC ORIENTATION ANALYSIS FROM THE DIPMETER

The traditional application of the dipmeter is focused on the correlation of distinctive microresistivity features across the borehole. By this means, local planes are established as surfaces with dip and strike. The procedure is an operation in topology, in which the data are restricted to increments of relative vertical shift, the orientation of the arms with respect to geographic north, and the deviation of the borehole from the vertical.

A secondary source of information is contained in the resistivity variation of the correlative features. If an individual feature shows the same resistivity value on all dipmeter arms, the rock fabric is suggested to be isotropic. Conversely, if there is variation in resistivity between the arms, the fabric is anisotropic, and the degree of anisotropy is reflected in the range of its directional resistivity. The critical role of shape distribution and shape orientation in the modification of resistivity was discussed extensively in the previous chapter and provides the key to useful geological interpretations of this phenomenon.

Older methods of grain orientation analysis have relied on samples of individual particulate measurement from thin sections. These methods are direct, but exceedingly tedious in practice, since they require relatively large samples of observations. By contrast, aggregate methods of fabric measure generally utilize variations in physical properties such as light intensity, imbibition of water, acoustic velocity, dielectric constant, and thermal and electrical conductivity. These approaches are indirect, but have generally shown good correlations with observed grain orientations. They also have the advantage that they measure samples of grains whose number usually run into the millions.

Shelton et al. (1974) studied orientations of parting lineations and small-scale and medium-scale cross beds in sands of the Cimarron River in Oklahoma. In addition, they made measurements of conductivity anisotropy using a field probe as an indirect assessment of grain orientation. Azimuth frequency measurements of the observations are collectively shown in the diagram of Figure 23. The statistical average of the conductivity variation shows good agreement with both the field observations and the valley trend.

The grain orientation of a sediment is obviously a function of the hydraulics of transport and deposition. The shape of most sand grains can be closely approximated by a triaxial ellipsoid. When transported in suspension, their long axes tend to

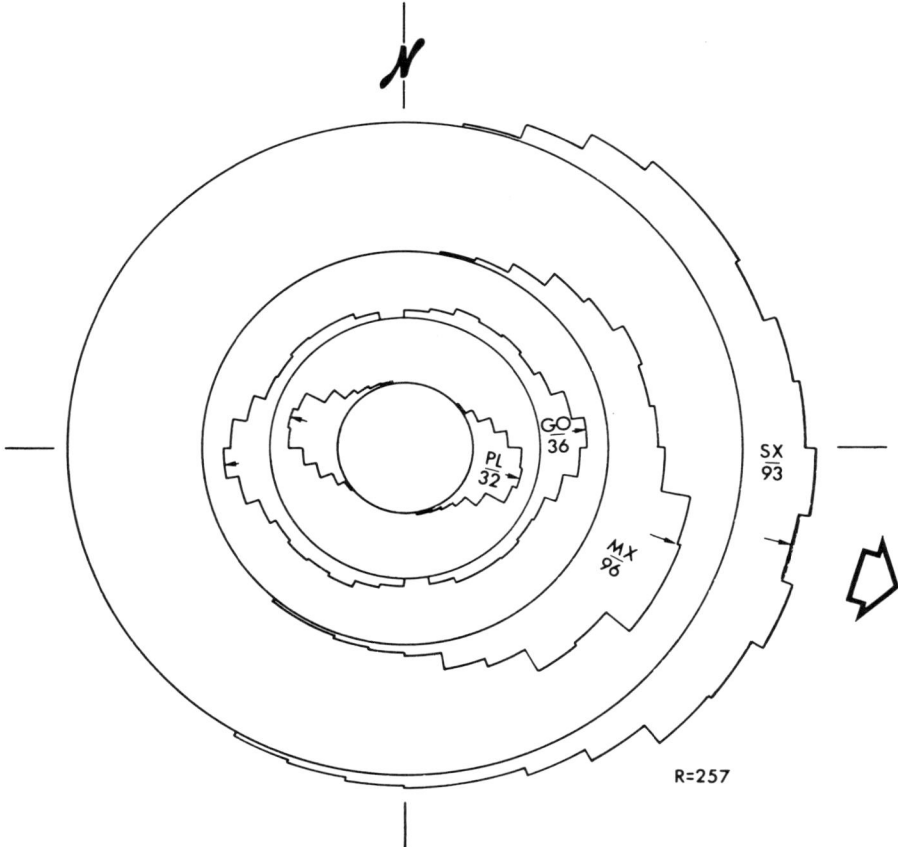

FIGURE 23. Directional measurements of features in Cimarron River deposits, Oklahoma. PL = parting lineation; GO = grain orientation; MX = medium-scale cross-bedding; SX = small-scale cross-bedding. From Shelton et al. (1979).

parallel the current direction. In traction, the grains roll about their long axes with orientation at right angles to transport. The final orientation of the grains following deposition is the product of a number of factors. Many grains falling from suspension will be rotated by traction in the bed load; some traction-rolled grains will be reoriented at their resting sites to parallelism with the current, in order to provide the minimum surface area of resistance.

Shelton and Mack (1970) and other authors summarize the expectations of gross grain orientations associated with different depositional environments. These are based both on theoretical considerations and generalized support from field studies. Traction movement in braided streams is suggested to result in a tendency for grain orientation perpendicular to stream axes. Several published studies in channels of meandering streams report a gross alignment parallel with the current trend. Grains on modern beaches are generally aligned with the direction of wave action, and so perpendicular to the shoreline. Nearshore marine sands deposited by longshore currents show orientations parallel with the shoreline. Modern barrier islands show offshore sand orientations overlain by beach sand orientations. Sands within turbidites are commonly aligned parallel to the basin trend. In all cases, the grains show imbrication upcurrent and this is the necessary key to specify current direction rather than merely axis of movement.

Rodriquez and Pirson (1968) studied the practical application of the dipmeter to the determination of fabric anisotropy in the subsurface. On a three-arm dipmeter, a single feature will be recorded as three different resistivity readings whose orientations are separated by 120° in the depositional plane. These three resistivity vectors define the trace of a resistivity ellipse as a realization of a model of resistivity variation around the entire azimuth cycle (Fig. 24). The equation of the ellipse in polar coordinates is

$$R^2 = \frac{a^2 b^2}{a^2 \sin^2 \theta + b^2 \cos^2 \theta}$$

where R is the resistivity, a is the length of the semimajor axis, b is the length of the semiminor axis, and θ is the angle that the reference electrode (R_1) makes with respect to the ellipse major axis. The orientation of the ellipse is then given by the formula

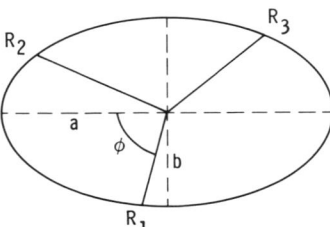

FIGURE 24. Trace of resistivity anisotropy ellipse on a bedding plane prescribed by resistivity measurements from a three-arm dipmeter.

$$\tan 2\theta = \frac{\sqrt{3}(C_3^2 - C_2^2)}{2C_1^2 - C_2^2 - C_3^2}$$

where C_1 is the conductivity reading (the reciprocal of resistivity at the ith arm). The relative elongation of the ellipse is given by the degree of anisotropy:

$$D = \frac{a}{b} - 1$$

This is solved from the equation

$$\frac{a}{b} = \sqrt{\left(\frac{B - A}{A + B}\right)}$$

where

$$A = \frac{2\sqrt{3}(C_3^2 - C_2^2)}{3 \sin 2\theta}$$

and

$$B = \tfrac{2}{3}(C_1^2 + C_2^2 + C_3^2)$$

Using these relationships, Rodriguez and Pirson (1968) computed resistivity ellipse orientations and elongation for correlateable features in a Texas offshore bar sandstone and a Venezuelan marine sheet sand. The ellipse characteristics were integrated with dips and bearings of their host features in composite analyses whose interpretation matched conventional sedimentology models. The authors also recognized that fabric orientation studies from dipmeters were not confined to the examination of sandstone units. Resistivity variations within shales are legitimate measures of their fabric anisotropy. The compression of shales results in plastic yielding with a rearrangement of components which ultimately leads to cleavage and schistosity. They concluded that the major axis of a shale resistivity ellipse would be oriented perpendicular to the local direction of compressional stress. Dipmeter studies of shale fabric patterns therefore offer useful indications of neighboring faults and folds, whose influence could be mapped from dipmeters run in several wells.

REFERENCES

Bigelow, E. L., 1973, High-resolution dipmeter uses in Michigan's Niagaran reefs, *Oil and Gas J.*, Vol. 71, No. 36, pp. 78–88.

Campbell, R. L., Jr., 1968, Stratigraphic application of dipmeter data in mid-continent, *Am. Assoc. Petrol. Geolog. Bull.*, Vol. 52, No. 9, pp. 1700–1719.

Cox, J. W., 1968, Interpretation of Dipmeter Data in the Devonian Carbonates and Evaporites of the Rainbow and Zama Areas, Petroleum Soc. Canad. Inst. Min. Paper No. 6820, 11 pp.

Gilreath, J. A., and Maricelli, J. J., 1964, Detailed stratigraphic control through dip computations, *Am. Assoc. Petrol. Geolog. Bull.*, Vol. 48, No. 12, pp. 1902–1910.

Gilreath, J. A., and Stephens, R. W., 1971, Distributary front deposits interpreted from dipmeter patterns, *Trans. Gulf Coast Assoc. Geol. Soc.*, Vol. XXI, pp. 223–243.

Glennie, K. W., 1972, Permian Rotliegendes of northwest Europe interpreted in the light of modern desert sedimentation studies, *Am. Assoc. Petrol. Geolog. Bull.*, Vol. 56, No. 6, pp. 1048–1071.

Goetz, J. F., Prins, W. J., and Logar, J. F., 1977, Reservoir delineation by wireline techniques, *The Log Analyst*, Vol. XVIII, No. 5, pp. 12–40.

Hamblin, W. K., 1965, Origin of "reverse drag" on the downthrown side of normal faults, *Geol. Soc. Amer. Bull.*, Vol. 76, No. 10, pp. 1145–1164.

Hepp, V., and Dumestre, A. C., 1975, "Cluster," A Method for Selecting the Most Probable Dip Results from Dipmeter Surveys, *Soc. Petr. Eng. 50th Ann. Fall Mtg.*, Dallas, Paper SPE 5543.

Kerzner, M. G., 1982, Analytical Approach to Detailed Dip Determination Using Frequency Analysis, *Trans. Soc. Prof. Well Log Analysts 23rd Ann. Logging Symp.*, Paper J, 17 pp.

McKee, E. D., 1966, Structures of dunes at White Sands National Monument, New Mexico (and a comparison with structures of dunes from other selected areas), *Sedimentology*, Vol. F. No. 1, Special Issue, 69 pp.

Potter, P. E., and Pettijohn, F. J., 1963, *Paleocurrents and Basin Analysis*, Springer-Verlag, Berlin, 296 pp.

Powell, J. W., 1875, *Exploration of the Colorado River of the West*, Govt. Printing Office, Washington, D.C., 291 pp.

Reading, H. G., 1978, *Sedimentary Environments and Facies*, Elsevier, New York, 557 pp.

Robertson, J. M., 1972, Deficiencies of Computer Correlated Dip Logs, *Trans. Soc. Prof. Well Log Analysts, 13th Ann. Logging Symp.*, Paper Y, 15 pp.

Rodriquez, A. R., and Pirson, S. J., 1968, The Continuous Dipmeter as a Tool for Studies in Directional Sedimentation and Tectonics, *Trans. Soc. Prof. Well Log Analysts 9th Ann. Logging Symp.*, Paper G, 25 pp.

Scheidegger, A. E., 1965, On the Statistics of the Orientation of Bedding Planes, Grain Axes, and Similar Sedimentological Data, *U.S. Geol. Survey Prof. Paper* 525-C, pp. 164–167.

Schoonover, L. G., 1974, Computer Recognition of Diplog Patterns—A Tool for Stratigraphic Analysis, *Trans. Soc. Prof. Well Log Analysts 15th Ann. Logging Symp.*, Paper X, 12 pp.

Selley, R. C., 1978, *Ancient Sedimentary Environments* (2nd ed.), Cornell Univ. Press., Ithaca, New York, 287 pp.

Shelton, J. W., Burman, H. R., and Noble, R. L., 1974, Directional features in braided-meandering-stream deposits, Cimarron River, North-Central Oklahoma, *J. Sed. Pet.*, Vol. 44, No. 4, pp. 1114–1117.

Shelton, J. W., and Mack, D. E., 1970, Grain orientation in determination of paleocurrents and sandstone trends, *Am. Assoc. Petrol. Geolog. Bull.*, Vol. 54, No. 7, pp. 1108–1119.

Van Dyke, R. J., 1975, Geology and Depositional Environments of the Reservoir Sandstone, Kincaid Oil-Field, Anderson County, Kansas, unpublished M.S. Thesis, University of Kansas, Lawrence, Kansas, 97 pp.

Van Veen, F. R., 1975, Geology of the Leman gas-field, in *Petroleum and the Continental Shelf of Northwest Europe* (A. W. Woodland, Ed.), John Wiley & Sons, Inc., New York, pp. 223–231.

Vincent, Ph., Gartner, J. E., and Attali, G., 1977, GEODIP—An Approach to Detailed Dip Determination Using Correlation by Pattern Recognition, *Soc. Petr. Eng. 52nd Ann. Fall Mtg.*, Denver, Paper SPE 6833, 18 pp.

Wade, J. A., 1978, The Mesozoic-Cenozoic History of the Northeastern Margin of North America, *10th Ann. Offshore Tech. Conf.*, Houston, Paper OTC 3266, pp. 1849–1858.

Woodcock, N. H., 1977, Specification of fabric shapes using an eigenvalue method, *Geol. Soc. Amer. Bull.*, Vol. 88, No. 9, pp. 1231–1236.

CHAPTER THREE

THE SPONTANEOUS POTENTIAL AND GAMMA-RAY LOGS

The spontaneous potential and gamma-ray logs are subsurface measurements of radically different physical properties. The SP (spontaneous potential) sonde records the strength of intrinsic electromotive forces within the borehole; the gamma-ray device counts the level of natural gamma radiation emanating from formations in the borehole wall. However, the general form of the traces recorded on each log in the same sedimentary sequence is often substantially similar. This broad equivalence marks a common sensitivity in the discrimination of shale beds from relatively shale-free, porous lithologies of sandstone, limestone, and dolomite. For this reason, they are usually recorded (together or individually) in the left-hand log track as a companion trace to either a resistivity or a porosity log, situated to the right.

At the most basic level, both logs are particularly valuable in the recognition of marker horizons and reservoir units, stratigraphic subdivision, and correlation between wells. More detailed examination of trace fluctuations within subsurface formations gives useful indications of profiles in vertical variation in shale content. These features have implications regarding interpretations of sedimentology and the typing of depositional environment. The common objectives of the primary use of these logs in subsurface geology studies is the reason for their collective consideration within the same chapter of this book. However, there are important distinctions in their response to different lithologies which are valuable in both qualitative and quantitative analysis. These various aspects will be emphasized in the following sections.

THE SPONTANEOUS POTENTIAL LOG

In the early days of resistivity logging, natural potentials were observed to occur in boreholes which were measurable in millivolts and found to distinguish permeable formations from intervening shales. The natural "battery" was initiated by the drilling of the borehole itself, in that the use of muds with a salinity contrast with formation waters, juxtaposed solutions with different ion concentrations. The cumulative electromotive force that is generated is comprised of a shale (or membrane) potential, a liquid junction potential, and an electrokinetic potential. The electrokinetic is very minor when compared with the components of shale potential and liquid junction potential, which result from the permeability of shale to sodium ions but not chloride ions, and the greater mobility of chloride ions relative to sodium ions, respectively. These two types of potential are classed as the electrochemical component of the SP log and result in an electrical circuit at the boundaries of shales with permeable beds. The flow of the current is from permeable bed to shale to borehole mud and back to permeable bed (Fig. 1).

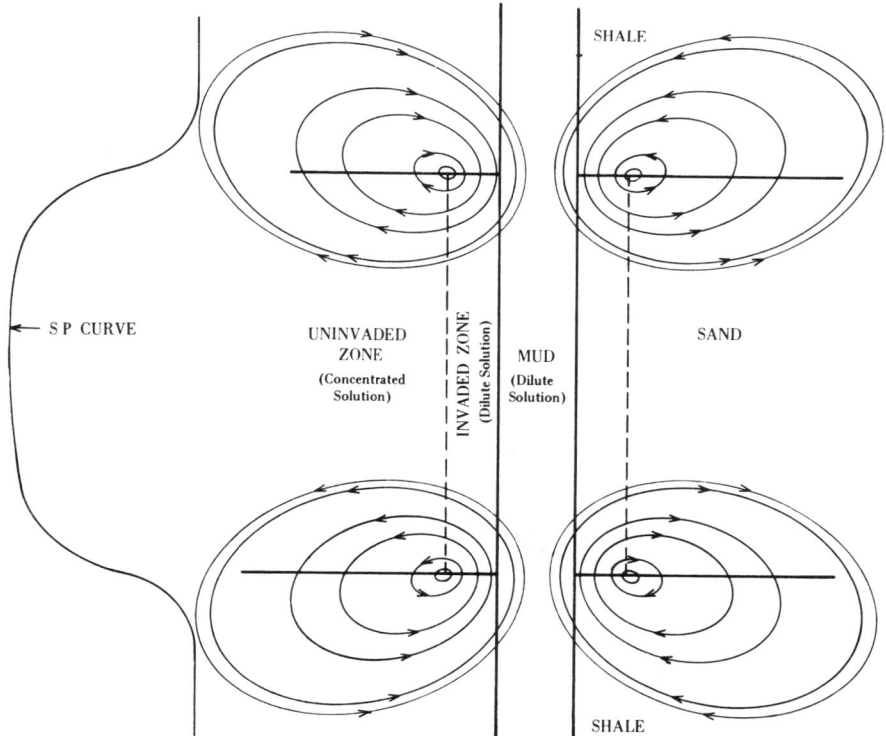

FIGURE 1. Electric current movement at the boundaries of permeable and impermeable beds in the vicinity of a borehole. From Dressser-Atlas publications.

The magnitude of the electrochemical potential is a function of the chemical activities of the formation water (a_w) and the mud filtrate (a_{mf})

$$E = -K \log \frac{a_w}{a_{mf}}$$

The coefficient K is determined by the equation

$$K = 2303 \frac{RT}{F}\left(1 + \frac{u-v}{u+v}\right)$$

where R is the perfect gas constant, T is the absolute temperature, F is the Faraday (96,450 coulombs), and u and v are the mobilities of the anions and cations. It is clear that variation in K is controlled by temperature changes for any given solution.

For the idealized case of a water containing a single salt, the chemical activity of the solution is proportional to the salt concentration. If mud filtrate and formation water can be considered to be closely approximated as sodium chloride solutions then

$$E = -K \log \frac{R_{mf}}{R_w}$$

where R_{mf} and R_w are the resistivities of the mud filtrate and formation water, respectively. This simple equation dictates the general features shown by an SP log:

1. When the drilling mud is fresh water (i.e., $R_{mf} > R_w$), the SP log shows a shift to the negative when moving from a shale to a permeable formation.
2. Conversely, when the drilling fluid is a salt mud (i.e., $R_{mf} < R_w$), the SP displaces in a positive sense in moving from a shale to a permeable formation.
3. The magnitude of the shift is a function of the resistivity contrast between the mud filtrate and the formation water.
4. This gives another method to estimate the formation water resistivity, since the formation temperature can be estimated (giving a value for K), and the mud filtrate resistivity is measured at the well site.

In its modern usage, the SP log is run in conjunction with the deep induction log and short normal as an induction electrical survey (IES). The SP defines the location of permeable formations and intervening shales and, for fresh mud boreholes, shales read as relatively positive readings and permeable formations as relatively negative readings (Fig. 2). Since electrical potential expresses a difference in potential rather than an absolute magnitude, the readings are not scaled with reference to an absolute zero but merely in incremental units of millivolt difference. The measurements are actually made with reference to an electrode at ground level. However, the

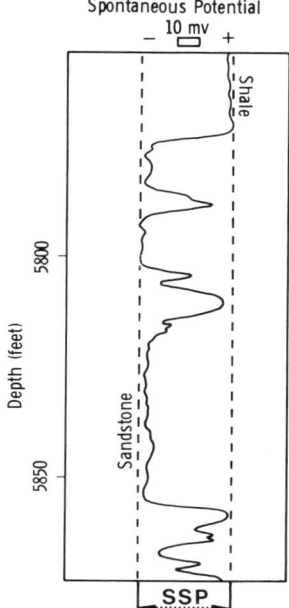

FIGURE 2. Spontaneous potential log of Tertiary sandstones and shales in a Louisiana well.

measurement of interest is the difference between the shale potential and that of a clean (shale-free) permeable formation, since this is an expression of the resistivity contrast of the formation water and mud filtrate (both measured at formation temperature). This quantity is known as the "static self-potential" or SSP and corresponds to the potential value, E, of earlier equations.

Formation water resistivity may be estimated from the SP log in a relatively straightforward procedure which utilizes the formula

$$\text{SSP} = -K \log \frac{R_{mf}}{R_w}$$

Traditionally, the method has employed a rather cumbersome series of charts that also take into account the differences in ionic composition of typical formation waters from an idealized pure sodium chloride solution. These differences appear to be most pronounced in the case of relatively fresh formation waters where calcium and magnesium ions are often a significant component of the total ionic composition.

In shaly, permeable formations, the displacement of the potential from the shale reading is reduced to a pseudostatic potential (PSP) (see Fig. 2). The relationship between PSP and SSP is approximately a function of shale content which is generally quantified as

$$\alpha = \frac{\text{PSP}}{\text{SSP}}$$

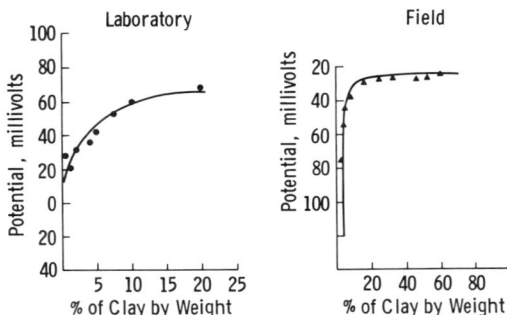

FIGURE 3. Relationships between potential and clay content. After Griffiths (1952).

Used in this way, the SP trace is often used as one of several "shale indicators," where $(1 - \alpha)$ is an estimate of the proportional shale content. In practice, the relationship appears to provide gross overestimates of shale content. Laboratory and field measurements of potential magnitude versus clay content by Griffiths (1952) appeared to conform to a hyperbolic function as shown in Figure 3. Bacon (1948) observed a similar pattern of behavior from laboratory measurements of diffusion potential. He also studied the systematic separate effects of different clay minerals on potential magnitude. The largest potential is caused by montmorillonite and the smallest by kaolinite. The implications of this work should be borne in mind when attempting to visualize an SP log as an expression of vertical variation in shale content. Other indicators of shale content are provided by the gamma-ray log and a combination of the neutron and density logs. They generally give more reasonable estimates and their use for this purpose is described in later sections of this book.

In estimating either formation water resistivity or shale contents of shaly permeable units, the same general procedure is used. If a fresh mud is used in the borehole, shales will register values to the right (relatively positive potential) and will tend to be coincident with a constant value which is drawn on the log as a vertical "shale baseline." Similarly, in clean, permeable units with the same formation water resistivity, the SP will deflect to the left (relatively negative values) to a constant bounding level which is marked on the log as the "clean line" or "sand line." The difference between the shale and clean lines is the SSP measured in millivolts, which may be used in computations of formation resistivity or as the reference interval for shale proportion computations in shaly, permeable formations.

When the borehole mud is saltier than the formation water, the above relations apply, but as a mirror image, in the sense that shales have a more positive potential related to permeable formations and occur to the left on the SP log, rather than to the right.

Factors Influencing the SP Log

The SP log is primarily a "permeability log" which is a response to permeability to ion flow, rather than permeability to fluid movement. Like all batteries, the circuit is either complete or broken, so that it is inappropriate to relate the SP response to the degree of permeability.

Although the procedure for defining shale and clean lines, outlined above, is widely applicable, it is predicated on a constant formation water resistivity over the interval of interest. In a section of wildly fluctuating formation water resistivities, there is a corresponding variation in the SSP separation. This situation occurs in the Tertiary deltaic sands of the Niger Delta and causes problems in SP log interpretation as described by Poupon et al. (1967). The distinction of shaly sandstones from clean sandstones with anomalous formation waters becomes ambiguous, but can be resolved through the use of a neutron-density log combination to differentiate shale content. The recognition of abrupt changes in water salinity between contiguous sandstones is useful diagnostic information since it suggests hydraulic separation and is an aid in the correlation and mapping of sandstone bodies. In "normal" sequences, there will be a gradual drift in the static self-potential with depth, in sympathy with the increasing salinity and decreasing resistivity of deeper subsurface brines.

The ideal succession for the application of the SP log is a sandstone–shale sequence comprised of relatively thick beds. At positions of thin, permeable beds, the amplitude of the SP deflection will be dampened and will be virtually suppressed opposite very thin beds. The ratio of the formation resistivity (R_t) to mud resistivity (R_m) also has a strong influence on the form of the SP trace. For high values of this ratio, the bed boundaries become progressively more rounded, together with some suppression of the potential deflection. As a consequence, the SP log has variable success in the definition of limestone and dolomite beds in carbonate–shale sequences. Deflections will generally register opposite carbonate units with moderate to high porosities. However, at relatively tight zones or thin carbonates, the trace often shows little differentiation from the shale baseline. The gamma-ray log is generally used as a substitute log to define lithological bed boundaries and estimate shale contents in carbonate successions.

The reduction in SP response opposite oil and gas zones in reservoir units has been noted on field logs for many years. A laboratory investigation of the phenomenon was conducted by McCall et al. (1971). They showed that the presence of oil or gas shifted the SP trace towards the shale baseline. The effect was greatest at higher hydrocarbon saturations and was more pronounced in shaly sandstones than in relatively shale-free equivalents.

Use of SP Shapes in the Recognition of Sandstone Depositional Environments

The SP response of a sandstone provides a profile whose shape gives general indications of the nature of the upper and lower contacts, degree of interbedding and gross grain-size variation. Sharp contacts give abrupt changes in potential at shale–sandstone boundaries. The occurrence of alternating beds of shale, shaly sandstone, and sandstone within the main sandstone unit will be marked by a serrated character to the SP trace, provided that those beds are not too thin to be resolved individually by the tool. Since the SP is sensitive to changes in shale content, and

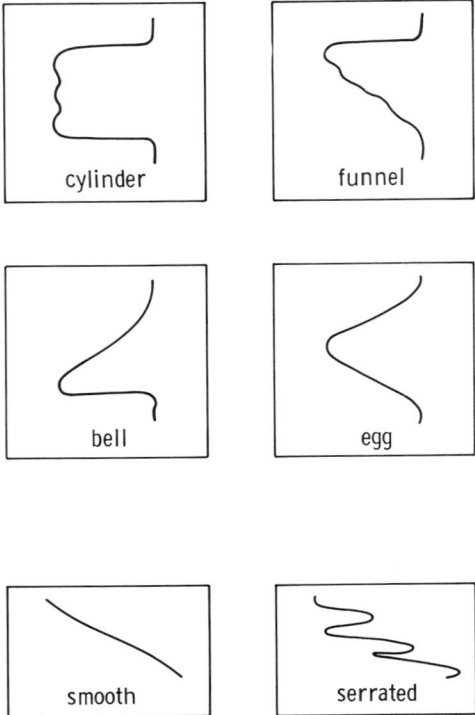

FIGURE 4. Basic descriptive shapes of SP profiles.

this in turn is generally related with differences in grain size, the trace within sandstones may be interpreted in terms of fining-upward or coarsening-upward trends.

The possible variations in basic SP shape for thick sandstone bodies are shown on Figure 4 and are often referred to as "bell," "cylinder," "funnel," and "egg" shapes. An extensive literature exists on the geometry and textural characteristics of sand bodies deposited in different environments which can be translated into hypothetical SP traces. Conversely, the SP profiles of subsurface sandstones may be interpreted in terms of geometry and texture, with subsequent classification of depositional environment (See, for example, Visher, 1961; Ruoff, 1976; Garcia, 1981). Obviously, there are dangers in using the SP log as an unaided device for this purpose, and interpretations of this kind are invariably linked with supporting evidence from nearby outcrop, isopach maps, core and well cuttings, dipmeter analysis, and the regional stratigraphic context.

This is one of the few logging applications that is used widely by both academic and petroleum geologists. For the academics, the technique provides a useful method to extrapolate interpretation of sedimentary facies of units into the subsurface, using profiles which can be related directly to sedimentation models. For petroleum geologists, interpetative studies are made in the same manner, although the ultimate objective is the definition of potential stratigraphic traps.

In the analysis of facies variations, the SP trace is usually coupled with a resistivity profile as a dual curve motif. In general, the resistivity log forms an approximate mirror image of the SP, although it often shows more detailed character, since it is sensitive to porosity changes, unlike the SP. The resistivity log will also pick out coal beds and tight limestones as distinctive resistive spikes, which are missed by the SP as "shale" responses.

Various problems can occur in the application of the SP as a qualitative facies pattern recognition device and these are often overlooked by geologists:

1. If there are significant variations in formation water salinity in a sequence of sandstones, relatively freshwater sands will appear to be shaly. This situation, for example, frequently occurs in sandstone units of the Niger Delta. The effect is usually most pronounced when moving vertically between sandstones which are not linked hydraulically, but may also be registered within correlateable sands which show systematic lateral changes in water salinity.

2. Since the static self-potential is a direct function of the resistivity contrast between formation water and mud filtrate, the use of a wide variety of drilling muds in a set of wells will result in varying accentuations and suppressions of the SP profile. In extreme situations where the mud filtrate and formation water have similar resistivities, the result will be a virtually featureless SP log, suggesting a monotonous shale sequence (see Fig. 5a).

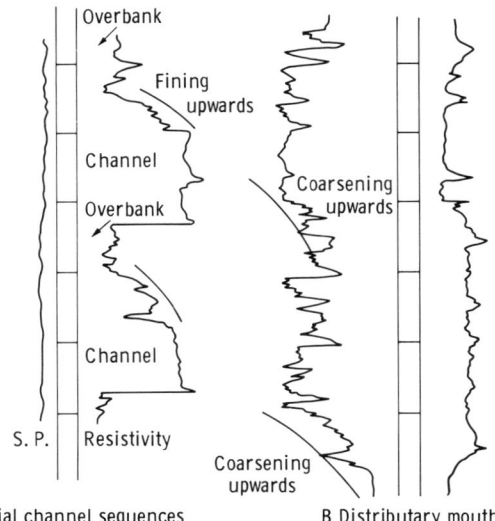

FIGURE 5. SP–resistivity profiles of Tertiary deltaic sequences from the Gulf Coast. After Fisher (1969).

3. In certain cases, the influence of temperature variations (as a function of depth) may be significant. This effect is registered by the quantity K in the electrochemical potential equation.

4. Sandstone units with oil or gas zones will show reductions in SP response toward the shale line.

5. The "mirror image" motif of complementary variation in the spontaneous potential and resistivity curves is the general rule in compacted sandstone–shale sequences. However, the pattern may be less clear-cut and even reversed in relatively uncompacted zones containing highly porous sands with saline formation water (see Fig. 5b).

None of these problems are insurmountable obstacles, but they indicate that detailed work should incorporate the collection of "monitor data" such as zone temperature and mud filtrate resistivity, and (where necessary) independent evaluations of formation water resistivity and hydrocarbon saturations from a resistivity–porosity log combination. In problem areas, "standardized" SP profiles could then be generated through the application of simple log analysis concepts.

A representative collection of SP–resistivity log facies shapes are shown from the Sacramento Basin in Figure 6. The geologic setting has been interpreted as a fore-arc basin whose Cretaceous sediments show fluvial, deltaic, shelf, and slope environments (Garcia, 1981). The contrasts and similarities of these shapes illustrate that, although they tend to mimic grain size variations, this characteristic is not unique to any particular environment and requires additional criteria of regional stratigraphy and core data for intelligent analysis. This interpretation method is most effective in sandstone–shale sequences and is not generally appropriate for carbonate successions, since relatively "tight" carbonates frequently show no response

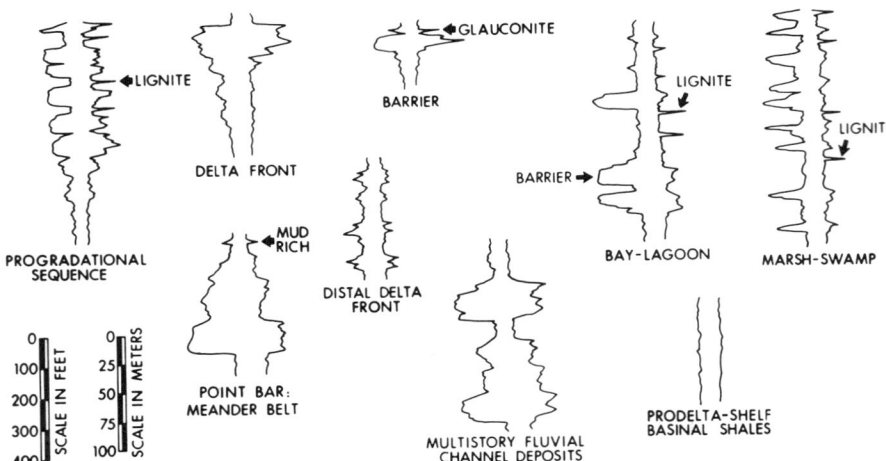

FIGURE 6. Representative SP–resistivity log shapes for facies from the Cretaceous sediments of the Sacramento Basin. From Garcia (1981).

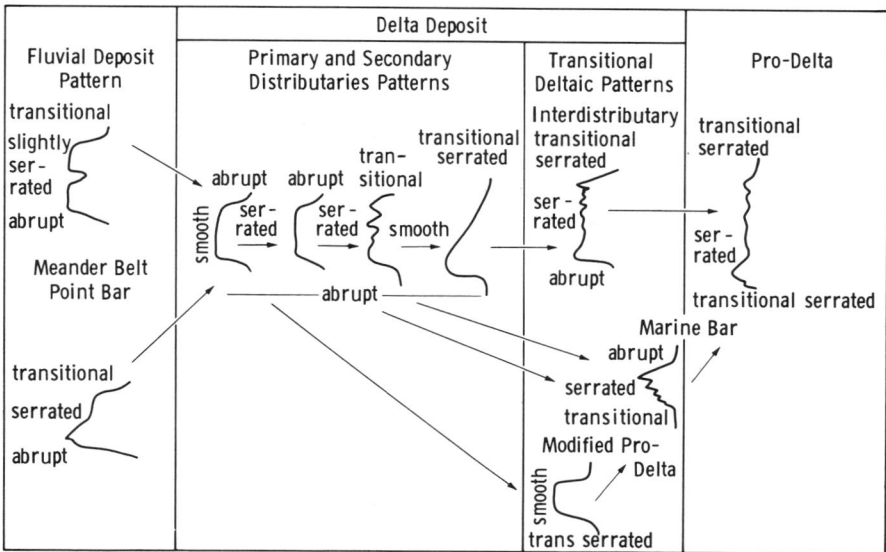

FIGURE 7. Flow diagram of SP shape motif changes across a delta environment. After Saitta and Visher (1968).

on the SP trace. However, analogous methods can be applied using the gamma-ray log to either carbonate or clastic sections, as will be described. Expectations concerning the general shapes of SP profiles associated with various clastic depositonal environments may be deduced from the "formation analysis" profiles of Figures 11 to 18 in Chapter 2.

In areas of dense well control which have been extensively logged by SP–resistivity log combinations, some studies have reached beyond basic environmental recognition to examination of the lateral changes in profile shape within and between environments. Saitta (in Saitta and Visher, 1968) produced a synthesis of "average" curve shapes as a flow diagram of shape change moving from fluvial to prodelta regimes (Fig. 7). His characterization was based on a study of about 4000 logs. Visher and Rennison (1978) extended this conept in an environmental interpretation map of a deltaic sandstone complex in the Coffeyville interval (Pennsylvanian) in the vicinity of Tulsa, Oklahoma (Fig. 8). Their interpretation was based on outcrop studies and isolith maps of the sandstone unit within the interval, aided by analysis of the shapes of SP and resistivity logs.

As a general statement, major distributary channels may be expected to have sharp, erosional basal contacts, fining-upward profiles, and either gradational or sharp upper contacts. The shape of their SP profiles should therefore broadly correspond with "bell" or "cylinder" themes. Enclosed bay sequences should show variations in general shape as a function of the proximity of nearby distributaries, but with an overall displacement toward the shale baseline, puncutated by serrations corresponding to crevasse splays, levee encroachment, and so on. Barrier islands and marine bars are generally contrasted with channels in their vertical development,

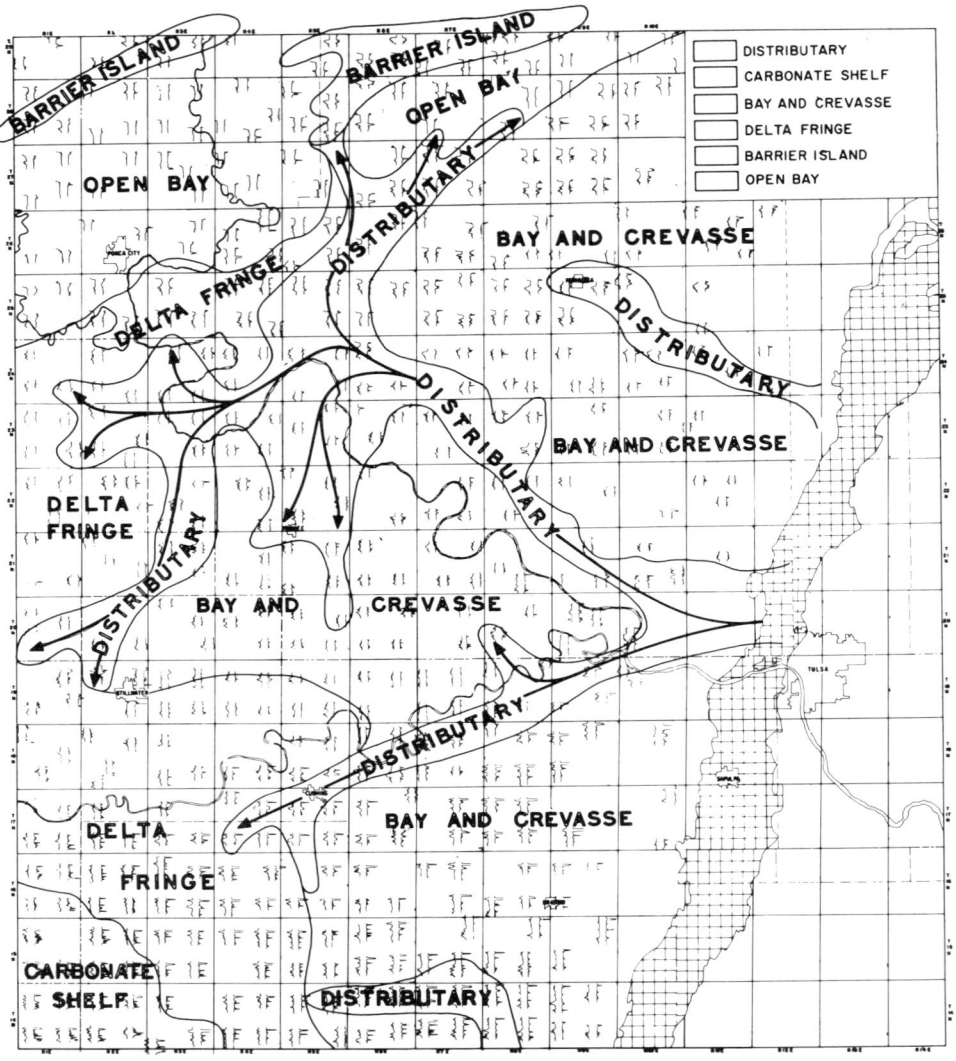

FIGURE 8. Environmental interpretation map of the Coffeyville interval (Pennsylvanian) near Tulsa, Oklahoma, based on isopach data and SP–resistivity profiles. From Visher and Rennison (1978).

with gradational lower contact, upward-coarsening trend and sharp upper contact, which is expressed as a "funnel" theme of SP profile. Pro-delta sections should ideally show an undistinguished profile, closely hugging the shale baseline, but showing some degree of serration introduced by thin regressive sands emanating from the delta front.

When used in this manner, SP logs can be considered to be "mapped" in the most rudimentary sense. The process involved is essentially a posting of SP profiles at their well map coordinates. As such, it is a modest advance on the information contained in a series of cross-sectional profiles of correlated sections. Part of the function of a useful map is to allow trends and regional patterns to be discerned easily. Ruoff (1976) suggested the use of simplified standard symbols to express the major themes of vertical variation on SP logs. His choice of symbols conformed closely with the motifs of "bell," "cylinder," and "serrated" variants described earlier as hypothetical endmembers of a simple classification. When posted on a map (Fig. 9), interpretation of patterns of distribution was generally clarified. As Ruoff (ibid) pointed out, the enhancement by the process was "analogous to viewing a forest of trees from the air rather than looking at one tree at a time while on foot in order to locate the concentration of different species."

It would obviously be highly desirable to characterize each SP profile by a limited set of numerical values which captures its vertical variability in a meaningful manner. These quantities could then be mapped in a similar fashion to methods used for formation thickness and structure. Features of both regional and local significance would then be immediately apparent. Unfortunately, the concise numerical description of shape is by no means a trivial operation. This is a problem shared by geologists who attempt to reduce particle and fossil shapes to a succinct set of measurements. While simple statistics are generally adequate decriptors of gross trends, slight but diagnostic nuances are often lost as "insignificant" deviations.

Magara (1979) proposed a numerical method of log profile classification which used a least-squares procedure of triangle fitting. The triangles were used to distinguish

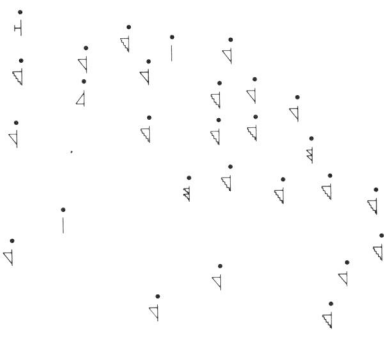

FIGURE 9. Symbolic mapping of SP profile shapes. From Ruoff (1976).

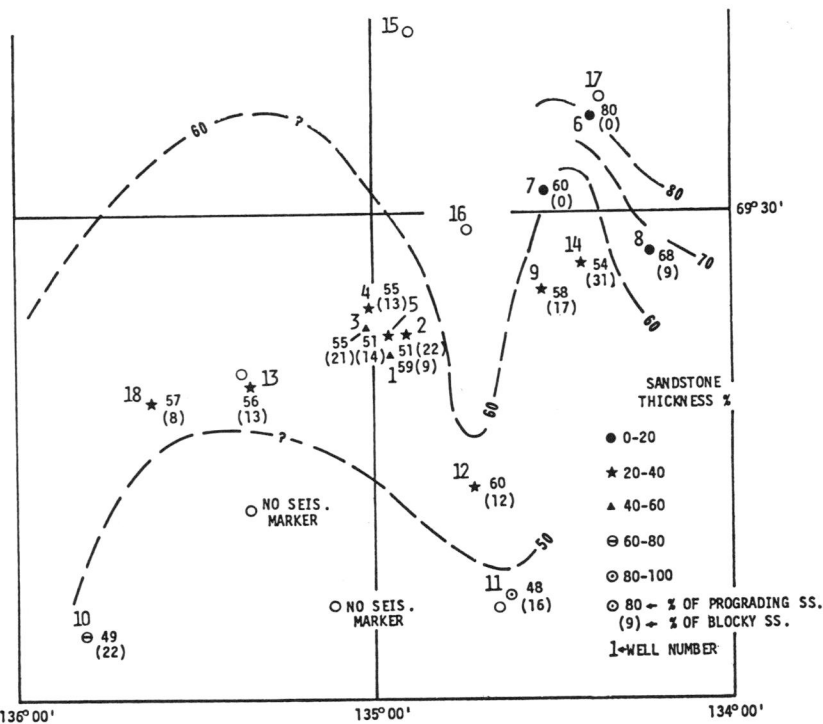

FIGURE 10. Progradational percentage map of a Tertiary interval in the Beaufort Basin. From Magara (1979).

between three possible endmembers of progradational (coarsening-upward), aggradational (coarsening-downward), and "blocky" (uniform grain size) trends. Log profiles of individual sandstones were analyzed numerically and assigned to one of the three possible classes. A sequence within a well could then be summarized in terms of its percentage content of progradational, aggradational, or blocky sandstones. In an application to a Tertiary succession in the Beaufort Basin, Magara (ibid) was able to discern trends in the relative geographic dominance of these classes. When mapped in conjunction with sand isolith maps (Fig. 10), he recognized aggradational and blocky facies which graded seaward to progradational facies and, still further seaward, to aggradational facies interpreted as turbidite sands.

In summary, the typing of sedimentary environments from SP logs and their use in regional mapping of facies is still a highly subjective procedure whose utility draws heavily on supporting evidence from other sources of geological information. On a more positive note, SP logs are far more common than cored sequences and often provide detailed information on the shape of subsurface units expressed either as cross sections or on maps. Furthermore, they give clearer visual impressions of vertical variability than is available from coarsely sampled drill cuttings. Obviously, their intelligent application is as a contributory facet within an integrated geological interpretation. Numerical methods for their analysis have not yet succeeded in making their mark as routine ancillary procedures. However, a new technique that may prove useful in this regard will be reviewed in Chapter 9.

THE GAMMA-RAY LOG

A majority of the elements are found in a variety of isotopic forms. Many of these isotopes are unstable and decay to a more stable form, while emitting radiation of several types. Alpha rays (helium nuclei) and beta rays (electrons) have a relatively short distance of penetration and so are of academic interest for logging purposes. However, gamma rays have significantly larger penetrations and can be detected by simple counter devices. Of the many radioactive isotopes which are known, only three types occur in any appreciable abundance in nature: the uranium series, the thorium series, and the potassium-40 isotope.

The first experimental well-bore instrument for gamma radiation measurement was developed in 1935 and put into commercial operation in 1940. It had the distinction of being the first tool that could measure formation characteristics through casing. The earlier detectors were Geiger counters but these have been replaced by scintillation detectors in most gamma-ray tools. However, while the scintillation detectors are more efficient, they require thermal protection and the added bulk generally precludes their use in slim-hole logging. The gamma-ray reading of a formation is roughly proportional to the weight concentration of the radioactive source. (As a result, the gamma-ray response should be standardized by multiplying by the bulk density when doing detailed analysis, although the correction factor is generally minor for most logging applications and is usually neglected.) It has been estimated that under typical borehole conditions, about 90% of radiation measured comes from the first 6 inches of adjacent formation, which prescribes a radius of investigation for the tool.

Since the emission of gamma rays is a stochastic process, the radiation count is a statistical quantity. The fluctuation in count level is smoothed by the use of an averaging (time constant) circuit. The count rate is also smoothed as a function of logging speed of the tool as it is raised through the borehole. The selection of an appropriate logging speed and time constant is made as a compromise which seeks to minimize the component of statistical noise, while not eliminating the effect of systematic minor variations in radiation as part of the averaging process. On each well, a check is made of the statistical background fluctuation by holding the tool stationary within the borehole for a few minutes. A trace of this statistical check is generally recorded on the log to be used in the evaluation of minor anomalies.

The gamma-ray tool is field calibrated at each well by exposing the tool to a standard calibrator at a fixed distance. While earlier logs were recorded simply in "counts" or equivalent weights of radium per ton, all modern logs are scaled in terms of the API gamma-ray unit. The API test pit at the University of Houston uses standards of reference, which are scaled in a manner such that the average midcontinent shale registers at about 100 API units.

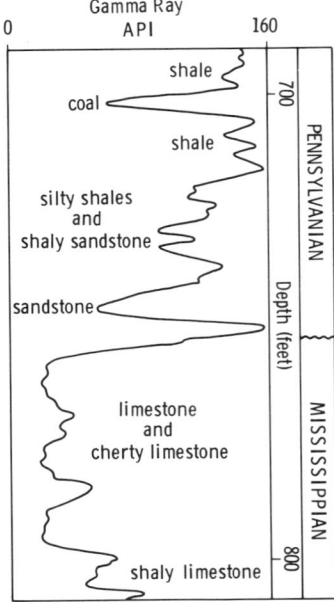

FIGURE 11. Gamma-ray log of a Pennsylvanian/Mississppian section from a well in eastern Kansas.

In the majority of stratigraphic and petroleum geological applications, the gamma-ray log is used as a "shale log," both to differentiate shales and "clean" formations and to evaluate shale proportions in shaly formations. Typical sandstones, limestones, and dolomites have relatively low concentrations of radioactive isotopes as contrasted with shales. As a result, the general form of the gamma-ray log often shows a broad similarity with an SP log run for the same section, with shales to the right and clean formations to the left of the log track (Fig. 11).

Application of the Gamma-Ray Log to Shale Content Evaluation

In evaluating typical sedimentary sequences, the gamma-ray log is used both for the definition of bed boundaries between shales and sandstones or carbonates, and in evaluating the shale content of shaly units. The procedure follows similar lines to that used in evaluating the SP log. A "shale baseline" is drawn which is a best estimate of the reading for "normal shales" (as opposed to uranium-rich "hot shales") in the sequence. Similarly, a "clean formation line" is drawn, which is roughly coincident with the lowest readings that occur in clean units of interest. In each case, these lines would be drawn to satisfy a general plateau of values which appear to correspond to natural boundary levels. If the clean formation and shale lines have API values of C and S, respectively, the "gamma-ray index" (GRI) of any zone, with associated reading, G may be estimated by

$$\text{GRI} = \frac{G - C}{S - C}$$

This equation represents a simple linear interpolation between the two boundary values. The operation is illustrated diagrammatically on Figure 12 in which the shale ratio of a zone is estimated in a Kansas Mississippian Osage section.

Since the flux of gamma-ray emission is directly proportional to the weight concentration of the radioactive source, the GRI is an appropriate measure of shale volume on an arithmetic scale. Implicit in the procedure is the assumption that the radioactive sources in the units evaluated correspond to shale contents whose clay mineralogy and radioactive isotope compositon is not markedly dissimilar from the bounding shales. In practice, the clay mineralogy, volume, and nature of silt content will generally vary between thick shales (used as the calibration standard) and thin shales within sandstone and carbonate beds. This variation reflects the changes of clay mineral associations that characterize different depositional facies and the results of diagnetic changes. An example of typical lateral and vertical patterns of clay mineralogy is given in Figure 13 as a summary of marine to nonmarine depositional environments in the Pennsylvanian of northern Missouri (Brown et al., 1977).

In the final analysis, the GRI is an empirical measure of shale volume content. It generally provides an acceptable estimator which is useful in stratigraphic studies and for mapping purposes as discussed in Chapter 9. Use of the index certainly compares very favorably with methods of shale content estimation based on cuttings, since these suffer from the problems of shale caving infiltration and selective comminution by the drill bit. If felspars, micas, heavy minerals, or other sources of radioactivity occur in appreciable quantities, the GRI becomes suspect. However, independent evaluations of other logs run in the same section often highlight the presence of additional disturbing components. The SP log reviewed earlier in this chapter and a combination of the neutron and density logs (see Chapter 5) are particularly useful for this purpose.

The primary interest of the industrial log analyst in the estimation of shale content focuses on the need to compensate for shale effects in the reservoir analysis of porosity and hydrocarbon saturations. Apparent porosity readings from sonic, density, and neutron logs must be corrected to estimations of effective porosity through the elimination of distortions introduced by shale. Resistivity log readings must be modified to accommodate the contribution of cation exchange conductivity effects of clay minerals. For this purpose, the shale ratio is either used directly as an estimate of clay content or is corrected to a lesser value, using a chart devised by Larionov. Some type of correction is necessary when estimating the proportion of *clay minerals for resistivity calculations*, since the shale ratio is, at best, an estimate of the *shale content*. There appears to be a great deal of confusion regarding the distinction between "shale" and "clay" and the two terms are often used interchangeably by log analysts (see Heslop, 1974). The confusion is seen on the Larionov chart itself (Fig. 14) which is based on laboratory measurements of gamma-ray levels and clay mineral volumes measured by X-ray diffraction techniques. The GRI on the vertical axis *shale* volume is translated to an estimate of clay content volume on the horizontal axis. The curves on the Larionov chart are often approximated by the formula

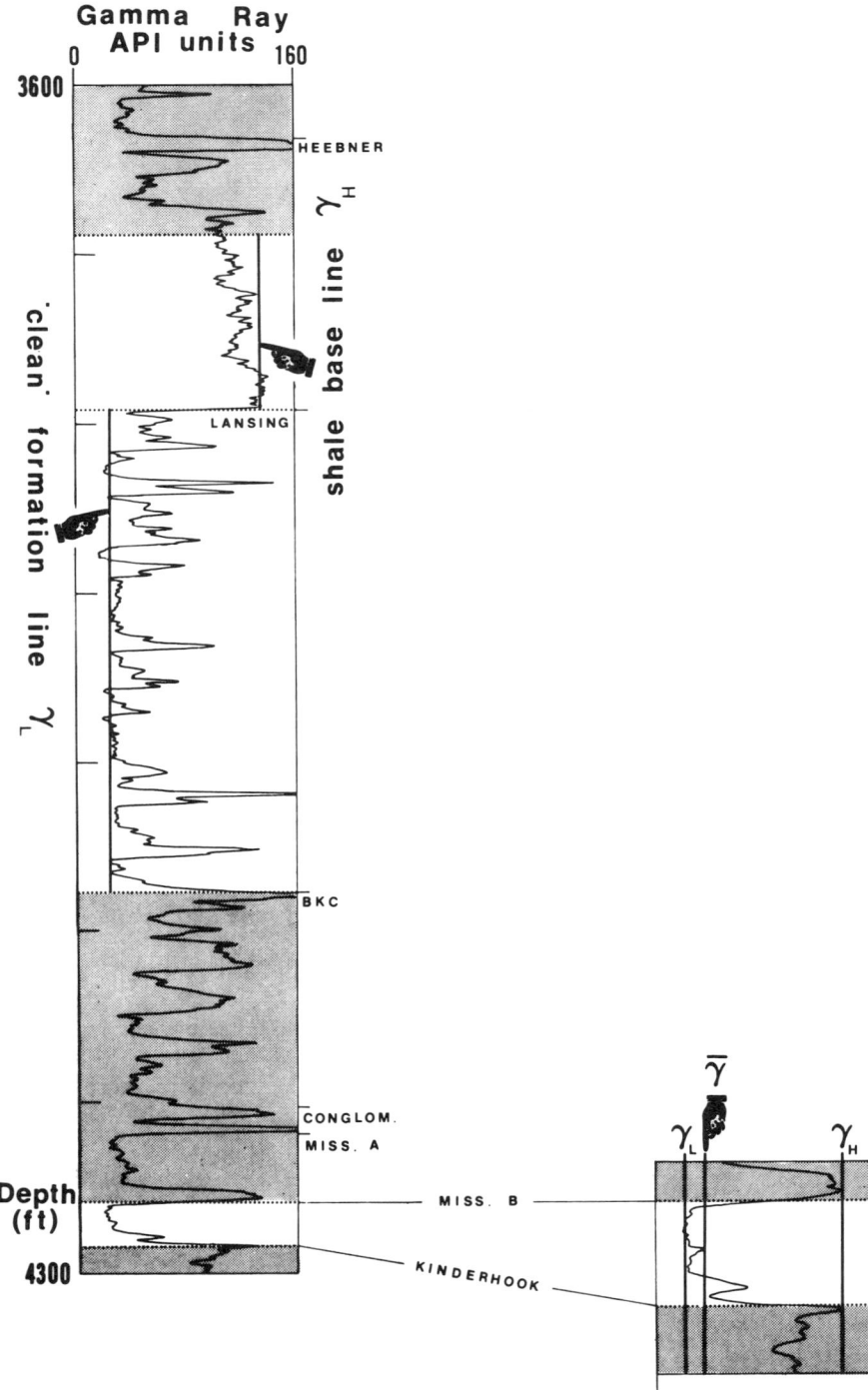

FIGURE 12. Computation of the shale ratio of a zone by linear interpolation on a gamma-ray log.

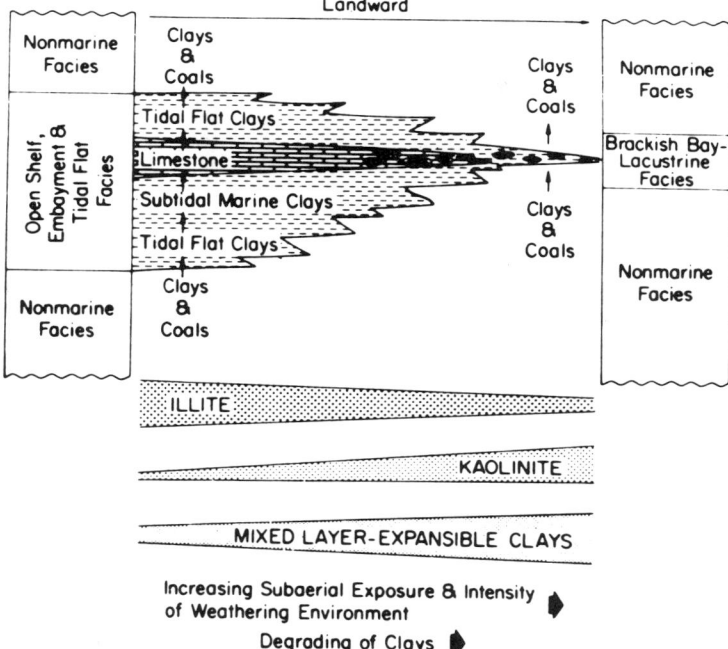

FIGURE 13. Lateral variation of clay minerals in a marine/nonmarine pinchout, Desmoinesian Series (Pennsylvanian), northern Missouri and southern Iowa. From Brown et al. (1977).

$$V_{CL} = \frac{0.5 \times GRI}{1.5 - GRI}$$

where V_{CL} is the volume of clay, and GRI is the gamma ray index estimate of the volume of shale. If the formula is rewritten in terms of proportion of clay within the shale fraction, the equation is

$$P_{CL} = \frac{0.5}{1.5 - GRI}$$

The integration of this equation between the GRI limits of zero and 1 gives an average estimate of the clay mineral content of an "average shale." The value of this quantity is 55%—remarkably close to the average clay content composition of 10,000 shales reported by Yaalon (1962) as 59%.

In summary, most of the current log analysis techniques can only provide estimates of shale content rather than specific clay mineral content. The reason is that they use the log properties of a "typical shale" in the analytical section to function as the extreme endmember of variation. Since shales composed entirely of clay minerals are rare, any calculations of volume must contain an implicit silt content of accessory minerals. While not impossible, the use of individual clay minerals as limiting

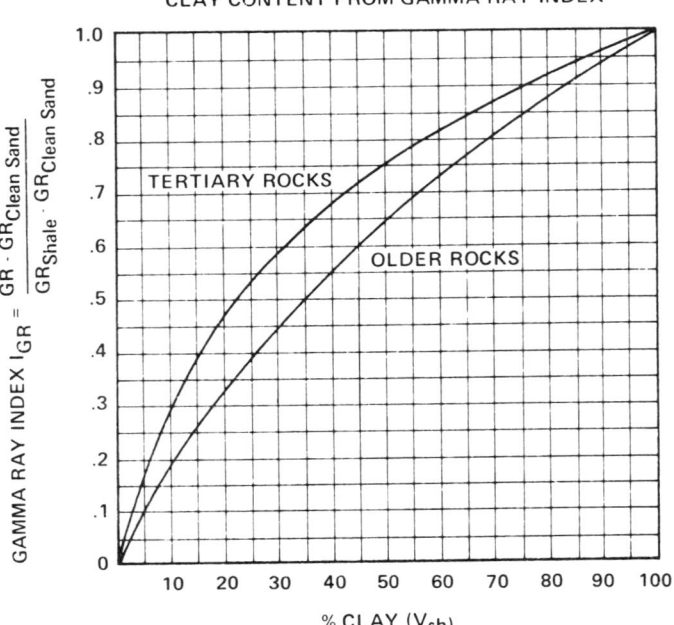

FIGURE 14. The Larionov chart used for estimation of corrected shale volumes based on the gamma-ray index. From Dresser-Atlas publications.

endmembers of variation is complicated by their variations in composition and physical properties. The controls of subsurface temperature and overburden and hydrostatic pressures are additional perturbing factors.

Simple Sedimentary Environmental Interpretation from Gamma-Ray Logs

Gamma-ray log profiles may be applied in an analogous manner to the SP log for the interpretation of depositional environments of both sandstone–shale and carbonate–shale sequences. In common with the SP log, the gamma-ray trace records the nature of upper and lower contacts as sharp or gradational, and gives indications of shale content variation. Unlike the SP, the gamma ray can be used in the distinction of uranium-rich black shales and possibly in the recognition of arkosic sandstones and zones of high heavy mineral concentration. Selley (1976) advocated a "cowboy geology" approach to the interpretation of North Sea sections, using the gamma-ray trace in conjunction with levels of glauconite and carbonaceous matter observed in the associated well cuttings. An illustrative example is given from his paper in Figure 15. Again, the gamma-ray log should be used in conjunction with more orthodox geological data in order for it to be a realistic aid in environmental interpretation.

General Applications of the Gamma-Ray Log

Chemical analyses and work with the gamma-ray spectral log have been useful aids in tagging the sources of radiation within various lithologies to specific isotopes. The clay minerals of shales contain small amounts of potassium, of which 0.02% is the radioactive isotope K-40 and may account for anything up to 20% of the total radiation emitted. More important, however, are concentrations of thorium which appear to be fixed to the clay minerals by processes of absorption and cation exchange. Small amounts of uranium also occur in almost all shales and provide the second major source of radiation. In certain shales uranium is concentrated in

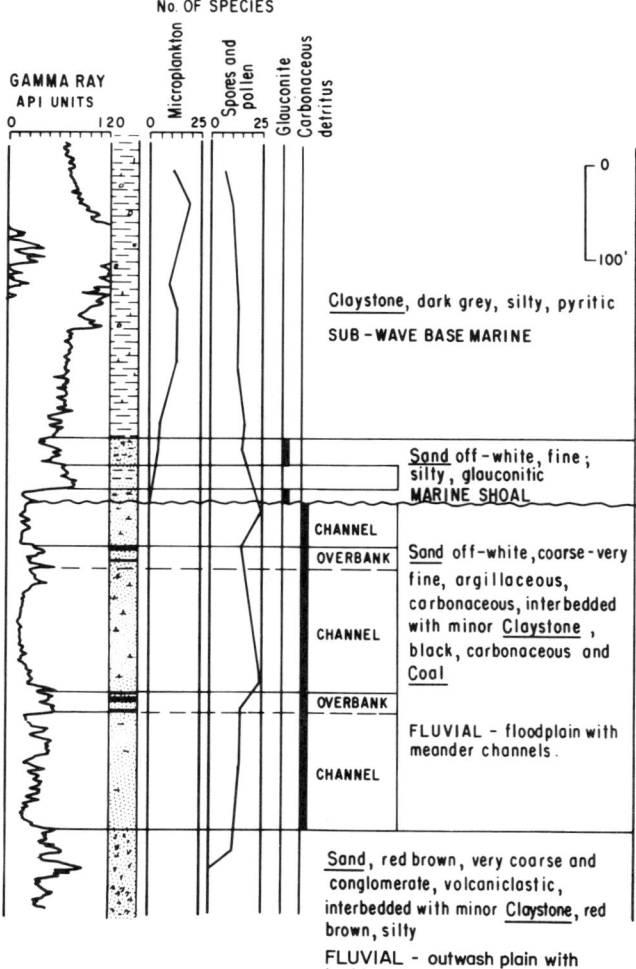

FIGURE 15. Environmental analysis of a North Sea gamma-ray log. From Selley (1976).

high proportions (e.g., Chattanooga Shale, Tennessee: 50–60 ppm), which is generally a reflection of depositional conditions of a highly reducing environment, where the uranyl ion is readily fixed by organic and mineral matter. "High uranium" shales are easily recognized on the gamma-ray log by their excessively high readings and are particularly useful as stratigraphic markers for correlation purposes.

Most carbonates show very low levels of radioactivity unless they contain disseminated shale or have been mineralized by uranium-bearing solutions. Simple orthoquartzites show similarly low values, although relatively high readings may be introduced by significant amounts of shale, felspar, mica, or heavy minerals such as zircon.

The distinction of clay minerals from micas has only recently been given the recognition it deserves, largely as a result of experience in Jurassic sections in the North Sea. In the absence of other disturbing components, appreciable variations of radioactivity in sandstones have traditionally been attributed to shale content, whose magnitude is computed in order to make necessary corrections to porosity and resistivity readings. As mentioned earlier, the cation exchange of contaminating clay minerals tends to reduce the resistivity reading and must be corrected for in water saturation calculations. However, since micas are basically insulators, their quantification by gamma-ray log measurements as clay minerals results in major errors in water saturation estimations. Gamma-ray spectral logging runs have indicated that the major source of radioactivity in typical sedimentary micas is a moderate amount of potassium, with only very minor contributions of uranium and thorium.

In addition to petroleum applications, the gamma-ray tool is widely applied in uranium exploration, where it is generally run with a density log. The sensitivity of the tool to the potassium-40 isotope has also made it extremely useful in the search for minerals with high potassium content, such as sylvite and other salts in evaporite sequences. When appropriate calibrations of the tool are made, direct translation of the API count scale to equivalent concentration of potassium oxide is possible.

Gamma-ray logs are commonly used for stratigraphic correlation in the subsurface, both because of the simplicity of their measurement and their discrimination of many beds of interest to the geologist. Ettensohn et al. (1979) describe an interesting extension of this subsurface procedure to outcrop studies. The correlation and stratigraphic subdivision of outcrops of "monotonous shales" is especially difficult with the absence of faunal zonation or fortuitous marker beds. Devonian and Mississippian black shales in Kentucky can be correlated fairly easily in the subsurface because distinctive changes in uranium concentrations associated with organic content variations give rise to characteristic gamma-ray logs. Ettensohn et al. (1979) measured radioactivity profiles of shale outcrops in eastern Kentucky using a scintillometer. As a result, they were able to subdivide zones which could be correlated both between outcrops and with units recognized by Provo (1977) in the subsurface on the basis of gamma-ray logs. A comparison of an outcrop radioactivity profile with subsurface gamma-ray logs from their study is shown in Figure 16. In their textbook, *Sedimentology of Shale*, Potter et al. (1980) advocate that gamma-ray scans of outcrop "should be a standard part of every description of a shale."

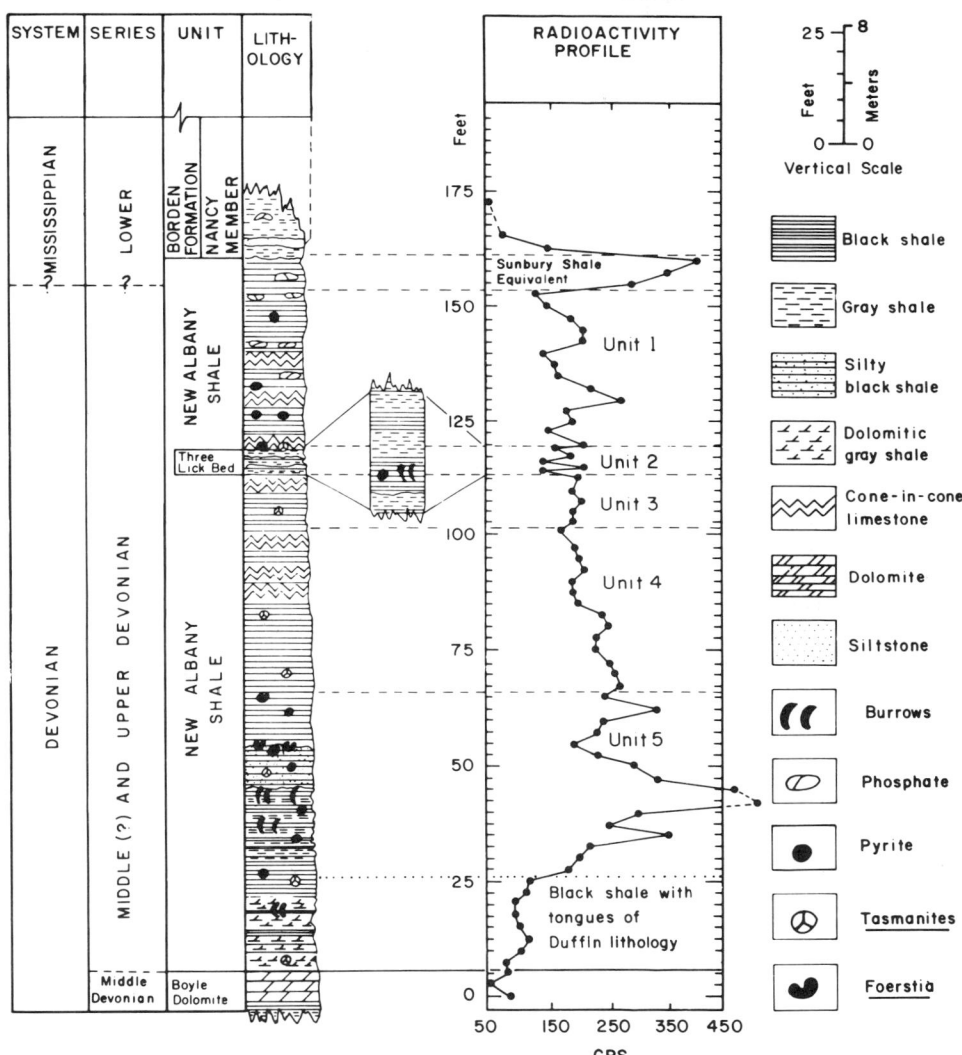

FIGURE 16. Radioactivity profile of Devonian-Mississippian outcrop near Clay City, Kentucky. From Ettensohn et al. (1979).

Geologic Radioactive Sources of the Gamma-Ray Log

The introduction of the spectral gamma-ray log in the early seventies has promoted renewed interest in the geological application of subsurface radioactivity measurements. The essential limitation to detailed analysis of the conventional gamma-ray log is a consequence of the indiscriminate pooling of potassium-40, uranium, and thorium series element sources in a composite reading.

The spectral device makes use of the fact that gamma rays from these three sources are characterized by distinctly different energy levels (Fig. 17). Scattering effects that intervene between the gamma-ray creation and its eventual detection by the tool "smears" the crisp theoretical spectra into a diffuse train of peaks. However, by subdividing the total energy range into restricted intervals or "windows," estimates can be made of the separate contribution of potassium-40, uranium, and thorium sources. As is the case with all gamma-ray measurements, the process of radioactive decay is stochastic. Consequently, statistical estimation theory must be applied to produce coherent values which reflect real variation rather than statistical counting error. The results are recorded as a spectral gamma-ray log which shows three separate traces referred to a radioactive source as a function of depth (Fig. 18).

The relationships between radioactive elements and geological hosts have been studied, based on both subsurface and laboratory measurements. Interpretations are sometimes complex or even ambiguous, since the emplacement of radioactive materials

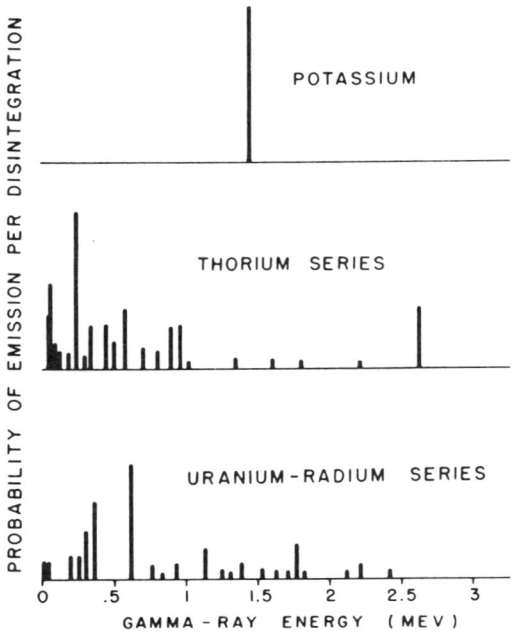

FIGURE 17. Gamma-ray emission spectra of radioactive sources. From Serra et al. (1980).

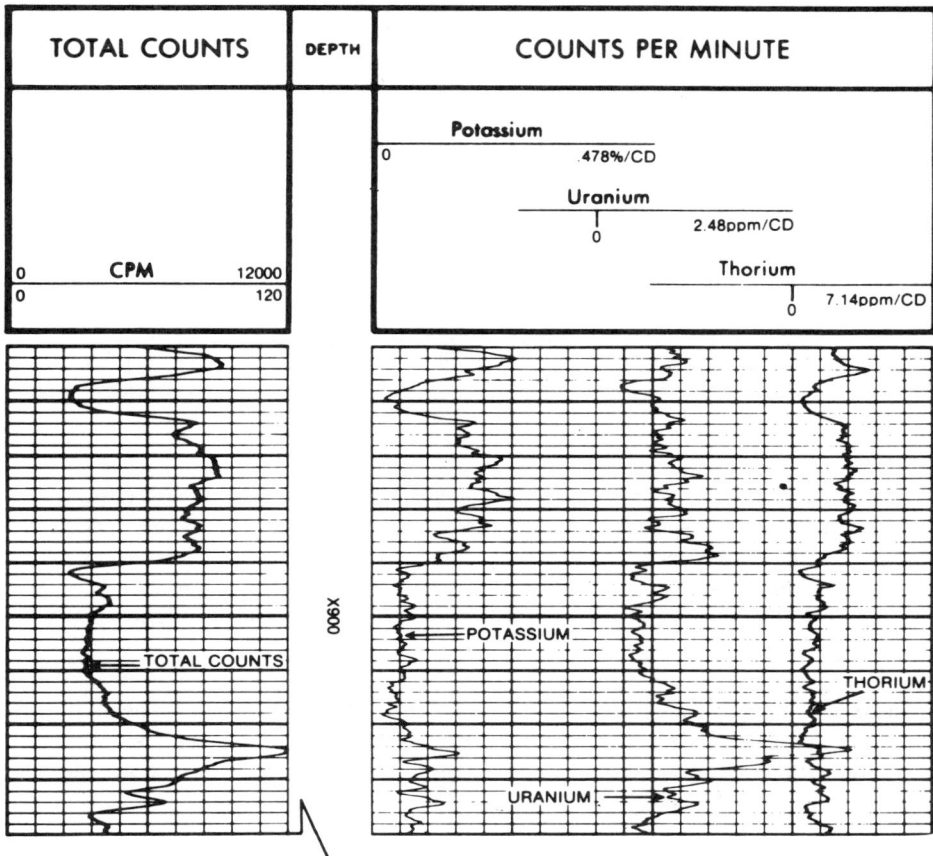

FIGURE 18. Spectral gamma-ray log of Permian carbonate section in Kansas. From Fertl (1979).

is the result of a variety of mechanisms, acting singly or in concert. These include depositional environment controls and diagenetic effects, which involve an interaction of geochemical processes operating on their relative solubilities, migration, precipitation, and fixation properties.

Hassan et al. (1976) made comprehensive geochemical analyses of over 500 samples drawn from limestone–marl, sandstone–shale, and organic shale sequences. They focused their statistical analysis on the interrelationships between levels of potassium, uranium, and thorium, and the concentration of organic carbon and major elemental oxides. A correlation coefficient matrix of their findings is shown in Table 1. The matrix shows a distinctive pattern of associations, but is hampered by its consideration of two variables at a time.

The authors extended their study by the application of factor analysis to the correlation matrix. Essentially, the method considers the chemical variables as a bundle of vectors in a conceptual multidimensional space. The eigenvectors of a

TABLE 1
Coefficients of Correlation

	Th	U	K_2O	Al_2O_3	TiO_2	Fe_2O_3	CaO	SiO_2	MgO	C_{org}
Th	1.00									
U	0.51	1.00								
K_2O	0.72	0.29	1.00							
Al_2O_3	0.91	0.35	0.66	1.00						
TiO_2	0.93	0.40	0.66	0.95	1.00					
Fe_2O_3	0.64	0.21	0.43	0.74	0.73	1.00				
CaO	−0.79	−0.21	−0.74	−0.87	−0.19	−0.64	1.00			
SiO_2	0.57	0.12	0.61	0.64	0.59	0.39	−0.92	1.00		
MgO	−0.19	−0.28	0.12	−0.17	−0.13	—	—	−0.12	1.00	
C_{org}	—	0.86	—	—	—	0.63	−0.25	−0.22	−0.25	1.00

correlation matrix are "principal components" matched with a new system of orthogonal axes which successively absorb the variation expressed by the dispersion of chemical vectors. In a "rotated factor analysis" solution, the eigenvector axes are rotated in a search for a position which tends to either minimize or maximize associations with the chemical variables. A more extensive discussion of this approach will be deferred until Chapter 7.

Listed in order of importance, the first four rotated factors of the correlation matrix are shown in Table 2 and account for 95% of the total variation. The numbers express the relative "loadings" of the raw variables of each factor and allows a geological interpretation of each factor. The first (and major) factor shows strong positive loadings on thorium, potassium, aluminum, and titanium oxides. The second and third factors show solitary loadings of silica and potassium oxide, respectively, while the fourth factor is matched by a strong uranium–organic carbon association. The following conclusions seem reasonable:

TABLE 2
Rotated Factor Matrix (R Mode)

Variable	Factor 1	Factor 2	Factor 3	Factor 4
Th	0.940	0.127	−0.088	0.152
U	0.032	−0.092	−0.092	0.910
K_2O	0.713	0.287	0.482	0.101
Al_2O_3	0.939	0.127	−0.032	−0.045
TiO_2	0.942	0.153	−0.018	0.150
Fe_2O_3	0.583	−0.068	−0.070	0.350
CaO	−0.675	−0.666	−0.067	−0.208
SiO_2	0.298	0.950	0.050	−0.078
MgO	−0.074	−0.233	0.420	−0.004
C_{org}	0.019	0.149	0.358	0.826

1. Clay minerals are linked most strongly with thorium contents, and secondarily with potassium, whose residual variation is picked up by an independent factor (factor 3).
2. The concentration of silica (most commonly as sand) has no distinctive associations with the other variables.
3. Uranium shows a strong correlation with organic carbon alone and is not associated with clay mineral concentration.
4. Calcium generally has a negative correlation with the radioactive elements, probably reflecting the low radioactivity associated with most limestones.

The interpretation of the rotated factor analysis is readily compatible with modern ideas concerning the genesis and geological habitat of radioactive elements. The original source of potassium is chiefly in acid igneous rocks within potassium felspars, micas, and some minor accessory minerals. During weathering, feldspars and micas are transformed to a variety of clay minerals. However, a large proportion of the potassium content is dissolved by water and transported ultimately to the oceans. Compaction of clay minerals during burial results in the progressive destruction of montmorillonite and replacement by illite which takes up some potassium from dissolved ions in subsurface pore waters.

Thorium and uranium are also derived from acid igneous rocks. Thorium compounds are insoluble, have a limited mobility, and are mostly transported in suspension. As a result, they tend to concentrate in residual minerals such as the clay minerals, where the thorium ion is probably fixed between platelets by absorption. Uranium is readily oxidized by the action of bacteria to the uranyl ion, which is extremely soluble and mobile. In a reducing environment, the ion forms complexes with organic matter which fix it in the host sediment. Consequently, most organic shales tend to be anomalously radioactive. Postdepositional migration of uranium is common and controlled by the chemical environment. Leaching and dissolution will mobilize uranium, with later fixation by organic matter or precipitation in rock fracture systems.

Applications of the Spectral Gamma-Ray Log

Gamma-ray spectrometry of outcrop samples has been recognized recently as a useful method to evaluate shales in terms of organic carbon content, paleoenvironmental interpretation, and stratigraphic correlation. Zelt (1984) described gamma-ray spectral studies of outcropping marine shale and chalk in the Cenomanian–Turonian Greenhorn cyclothem of Colorado, New Mexico, and Utah. He concluded that variations in the ratios of gamma-ray sources could be correlated for hundreds of kilometers across lithofacies boundaries. Results could be used to deduce sediment transport directions, suggest proximity of paleoshorelines, and aid in the interpretation of lateral and temporal changes in paleosalinity. He further emphasized that direct comparison could be made between outcrop profiles with gamma-ray spectral logs of equivalent units in the subsurface. The combination of outcrop spectral data and

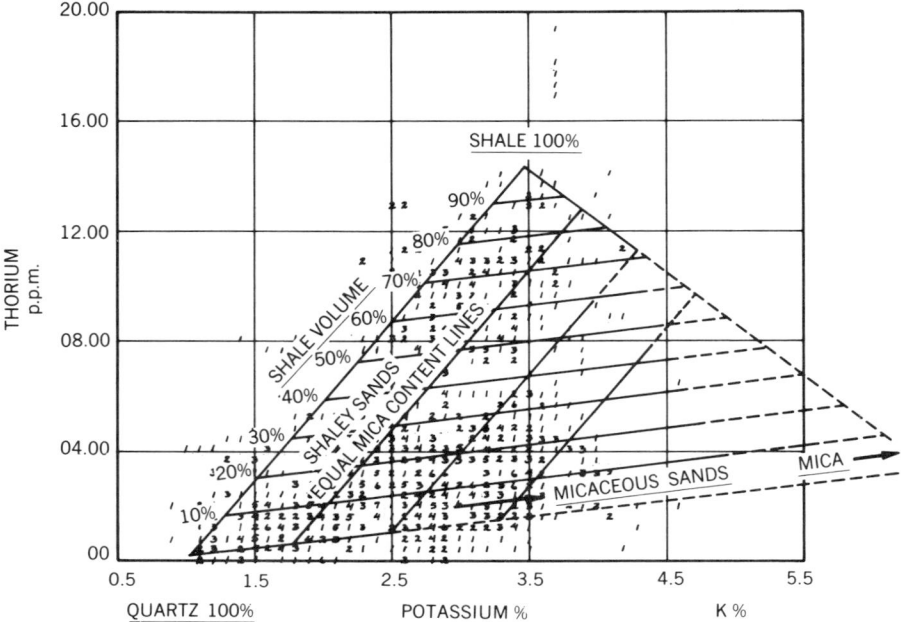

FIGURE 19. Crossplot of zones from a North Sea Jurassic Sandstone sequence in terms of contents of potassium and thorium measured by a spectral gamma-ray log. From Hodson et al. (1976).

spectral logs shows great promise for detailed geological interpretation of shale units in basin studies.

As indicated earlier, the spectral gamma-ray log has found extensive use in the interpretation of Jurassic sandstones of the North Sea (Hodson et al., 1976). The primary purpose is the distinction of mica fractions, characterized by relatively high potassium levels, from clay minerals, which are discriminated by their thorium content (Fig. 19). The spectral log also allows the specific recognition of felspathic zones and heavy mineral concentrates within sandstones (Fertl, 1979). The log has proved to be a distinct asset in logging carbonate sequences. On a conventional (pooled radiation) gamma-ray log, radioactive peaks within limestones or dolomites are often interpreted as thin shales. However, the spectral log can be used to distinguish fractures containing mineral precipitates (high uranium–low potassium) from shales (high thorium–moderate potassium). The interpretation concept has been applied successfully to the reservoir engineering of Austin chalk wells in Texas (Fig. 20), with multifold increase in oil production (Fertl, 1980). With the recent interest in the development of geothermal reservoirs, spectral gamma-ray measurements have assumed a special importance in the logging of crystalline basement rocks. Based on tests in the deep holes of the Los Alamos Dry Hot Rock Geothermal Project, West and Laughlin (1976) concluded that the log was valuable in the determination of rock types, detection of fracture zones, and studies of the mobility of the radioactive elements (Fig. 21).

The relative novelty of this tool suggests that its full potential as a major source of subsurface geological information has not yet been realized. A particularly intriguing goal in sedimentary sequence studies is the quantitative analysis of clay mineral species in thin shaly sections. Early attempts by Hassan et al. (1976) to compute clay mineral profiles were widely considered to be ambitious, largely because of the count statistic vagaries which are compounded with the spectra measurements.

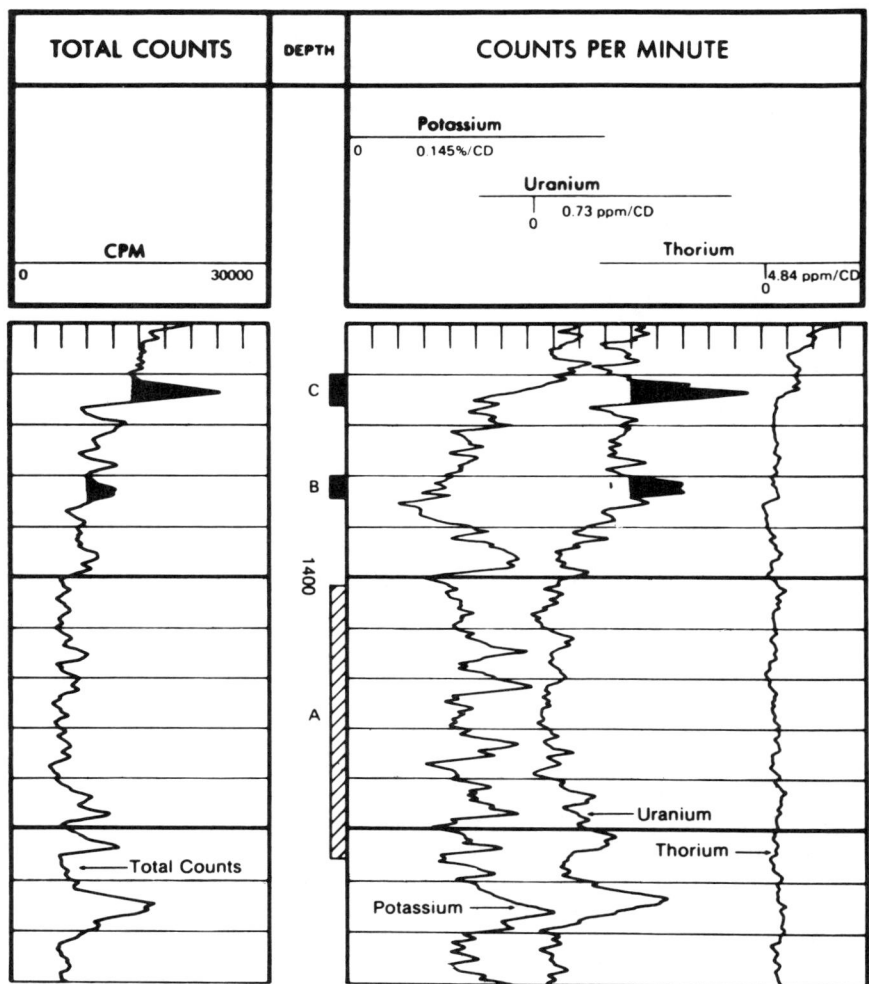

FIGURE 20. Distinction of fractured zones from shale in the Austin Chalk through use of the spectral gamma-ray log. From Fertl (1979).

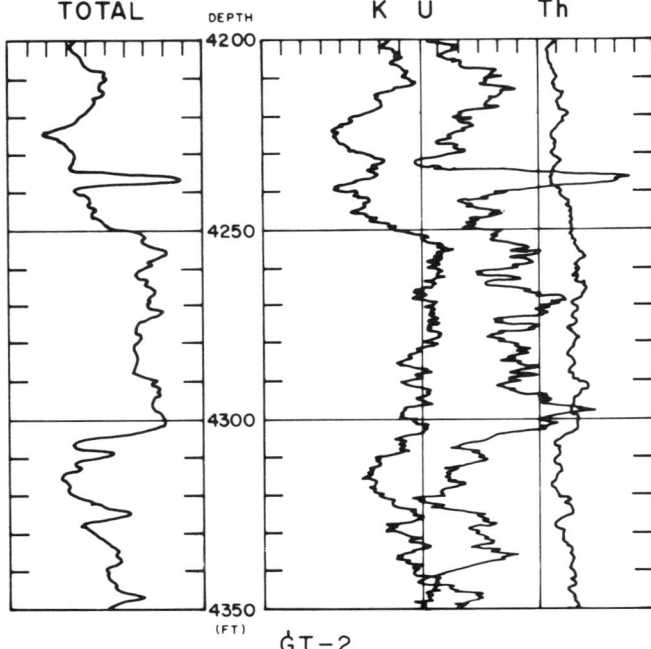

FIGURE 21. Spectral gamma-ray log of a leucocratic monzogranite dike in a crystalline basement rock section at Los Alamos, New Mexico. After West and Laughlin (1976).

However, sophisticated data processing described by Serra et al. (1982) has improved measurement precision significantly. The coordination and calibration of spectral log data with other sources of information, such as X-ray diffraction analyses, may mark a major advance in the detailed log interpretation of subsurface sequences. Recent work in this area is described in connection with the mineral estimation methods discussed in Chapter 6.

REFERENCES

Bacon, L. O., 1948, Formation clay minerals and electric logging, *Pennsylvania State College Min. Indus. Exper. Sta. Bull.*, Vol. 52, pp. 53–75.

Brown, L. F. Jr., Bailey, S. W., Cline, L. M., and Lister, J. S., 1977, Clay mineralogy in relation to deltaic sedimentation patterns of Desmoinsian cyclotherms in Iowa-Missouri, *Clays Clay Minerals*, Vol. 25, pp. 171–186.

Ettensohn, F. R., Fulton, L. P., and Kepferle, R. C., 1979, Use of scintillometer and gamma-ray logs for correlation and stratigraphy in homogeneous black shales, *Summary: Geol. Soc. America Bull.*, Part 1, Vol. 90, No. 5, pp. 421–423.

Fertl, W. H., 1979, Gamma ray spectral data assists in complex formation evaluation, *The Log Analyst*, Vol. XX, No. 5, pp. 3–37.

Fertl, W. H., Stapp, W. L., Vaello, D. B., and Vercellino, W. C., 1980, Spectral gamma-ray logging in the Texas Austin Chalk Trend, *J. Petrol. Tech.*, Vol. 32, pp. 481–488.

Garcia, R., 1981, Depositional systems and their relation to gas accumulation in Sacramento Valley, California, *Am. Assoc. Petrol. Geolog. Bull.*, Vol. 65, No. 4, pp. 653–673.

Griffiths, J. C., 1952, Grain-size distribution and reservoir-rock characteristics, *Am. Assoc. Petrol. Geol. Bull.*, Vol. 36, No. 2, pp. 205–229.

Hassan, M., Hossin, A., and Combaz, A., 1976, Fundamentals of the differential gamma ray log, *Trans. Soc. Prof. Well Log Analysts 17th Ann. Logging Symp.*, Paper H, 18 pp.

Heslop, A., 1974, Gamma-ray log response of shaly sandstones, *Trans. Soc. Prof. Well Log Analysts 15th Ann. Logging Symp.*, Paper M, 11 pp.

Hodson, G., Fertl, W. H., and Hammack, G. W., 1976, Formation evaluation in Jurassic sandstones in the northern North Sea area, *The Log Analyst*, Vol. XVII, No. 1, pp. 22–32.

Magara, K., 1979, Identification of sandstone body types by computer method, *Mathematical Geology*, Vol. 11, No. 3, pp. 269–283.

McCall, C., Von Gonten, W. D., and Osoba, J. S., 1971, The effect of hydrocarbons on the SP opposite sands, *Trans. Soc. Prof. Well Log Analysts 12th Ann. Logging Symp.*, Paper C, 20 pp.

Potter, P. E., Maynard, J. B., and Pryor, W. A., 1980, *Sedimentology of Shale*, Springer Verlag, New York, 303 pp.

Poupon, A., Strecker, J., and Gartner, J., 1967, A review of log interpretation methods used in the Niger delta, *Trans. Soc. Prof. Well Log Analysts 8th Ann. Logging Symp.*, Paper Z, 53 pp.

Provo, L. J., 1977, Stratigraphy and Sedimentology of Radioactive Devonian-Mississippian Shales of the Central Appalachian Basin, Grand Junction, CO., *U.S. Dept. of Energy* Report No. GJBX-37077, 102 pp.

Ruoff, W. A., 1976, A technique for interpreting depositonal environments of sandstones from the SP Log utilizing the computer, *The Log Analyst*, Vol. XVII, No. 4, pp. 3–10.

Saitta, S., and Visher, G. S., 1968, Subsurface study of the southern portion of the Bluejacket delta, *Oklahoma Geol. Soc. Guidebook*, 33 pp.

Selley, R. C., 1976, Subsurface environmental analysis of North Sea sediments, *Am. Assoc. Petrol. Geolog. Bull.*, Vol. 60, No. 2, pp. 184–195.

Serra, O., Baldwin, J., and Quirein, J., 1980, Theory, interpretation and practical applications of natural gamma ray spectroscopy, *Trans. Soc. Prof. Well Log Analysts Ann. Logging Symp.*, Paper Q, 30 pp.

Visher, G. S., 1961, How to distinguish barrier bar and channel sands, *World Oil*, May issue, pp. 106–113.

Visher, G. S., and Rennison, J., 1978, The Coffeyville Format (Pennsylvanian) of Northern Oklahoma: A model for an epeiric sea delta, *South-Central Section Geol. Soc. America Guidebook*, 24 pp.

West, F. G., and Laughlin, W. A., 1976, Spectral gamma logging in crystalline basement rocks, *Geology*, Vol. 4, pp. 617–618.

Yaalon, D. H., 1962, Mineral composition of average shale, *Clay minerals Bull.*, Vol. 5, No. 27, pp. 31–36.

Zelt, F. B., 1984, Gamma-ray spectrometry of marine shales in outcrop—a tool for petroleum exploration and basin analysis (abs.), *Am. Assoc. Petrol. Geolog. Bull.*, Vol. 68, No. 4, p. 542.

CHAPTER FOUR

POROSITY LOGS: SONIC, DENSITY, NEUTRON

"Porosity log" is an informal generic term for log traces of subsurface sections recorded by either the sonic, density, or neutron tools. The designation reflects the main function of these logs as the quantitative evaluation of porosity. Each tool measures a radically different physical property: compressional wave velocity, electron density, and relative hydrogen concentration. However, in each case there is a pronounced distinction between the tool response to fluid in the pore space and those for minerals of the rock framework. Consequently, the basic measurements may be transformed to estimations of proportional fluid content and these correspond to pore volume.

There are smaller, but systematic, differences between the sonic, density, and neutron properties of the common reservoir rock minerals. The different values for the zero porosity endpoint determine separate porosity calibration scales for each type of reservoir rock. As a result, some prior knowledge concerning the lithology type (generally either sandstone, limestone or dolomite) is required for accurate porosity estimations. In the absence of such information, log overlays and crossplots of several porosity logs in combination can be used to deduce lithology, as described in Chapter 5.

In this chapter, the theory of each tool is described individually, together with the computational methods for porosity estimation. The response of these logs to a variety of mineral assemblages is also reviewed, as well as their application to less traditional log interpretations of subsurface geology. In particular, the variation in log properties of shale beds with depth is considered in the analysis of compactional trends and their information concerning rates of subsidence and tectonic history. Finally, the form of porosity trends and patterns are discussed as data diagnostic of depositional fabric and diagenetic modification.

THE SONIC LOG

The sonic or acoustic log was introduced in the 1950s to provide exploration geophysicists with borehole velocity profiles. The sonic tool provides a continuous velocity survey of formations within a borehole which can be processed as a reference match for comparison with nearby seismic field records. It is still heavily used for this purpose as a means to create "synthetic seismograms." The stratigraphic subdivision of a borehole sequence is a routine procedure for the geologist who draws on information from both well cuttings and logs. The geologist's picks of formation tops often coincide with distinctive velocity changes, particularly when these match boundaries between thick shales and relatively rigid sandstones, limestones, or dolomites. An example is shown in Figure 1 where velocity is graphed as a function of cumulative travel time, and can be thought of as "acoustic stratigraphy" as perceived by a sound wave. Reflections of seismic energy are most pronounced at surfaces where there is a marked change in velocity. The velocity profile is readily transformed into a sequence of reflection coefficients, which are then convolved with a wavelet whose frequency character matches that of the energy source used for field seismic work. The end result is a synthetic seismogram in which reflection events can be tagged with specific formations and used as a key to identify corresponding events on seismic records (Fig. 2).

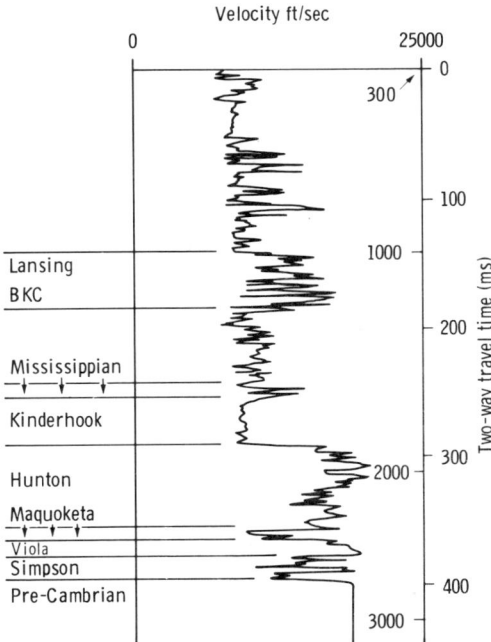

FIGURE 1. Sonic velocity profile of a borehole stratigraphic sequence from northern Kansas.

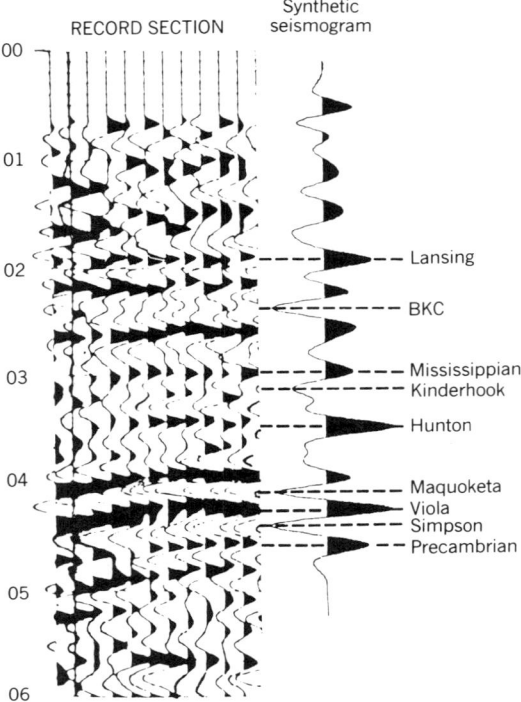

FIGURE 2. Synthetic seismogram generated from the velocity profile of Figure 1, and matched with a nearby field seismic record section.

The potential of the sonic log for porosity evaluations was recognized soon after its introduction. This was made possible by the strong contrast in sound velocity between the mineral framework of a reservoir rock and the fluids contained within its pore space.

The schematic design of the earliest sonic tools is shown in Figure 3a. A transmitter emits sonic pulses at a rate of 10 per second. The pulse frequency is centered around 20,000 cps and is in the low ultrasonic range which is beyond human hearing but within the frequency transmission of Ivory Coast bats (Fenton and Fullard, 1981). Acoustic energy emanates from the transmitter as a compressional wave which crosses the relatively slow velocity medium of the borehole fluid. It is refracted at the borehole wall and travels through the formation at a greater speed caused by the increase in density and rigidity. Compressional waves are about twice as fast as other wave forms and are the first to reach the receiver. The path of the first arrival at the receiver is sketched on (Fig. 3a). The receiver records the time between the emission of the signal and its reception.

The conversion of the data into estimates of velocity within the formation is complicated by the elimination of the slow transit time through the borehole fluid and the variable angle of refraction. Consequently, the sonic tool was modified to new designs which culminated in the modern "borehole compensated tool" (Fig.

3b). The use of two transducers which alternate signal transmission and two receivers has several advantages. The design automatically compensates for varying degrees of tilt angle of the tool within the hole and is less sensitive to effects of washed-out horizons. The differences between transit times registered at the receivers effectively results in a measurement of formation velocity through cancellation of common paths of waves through the borehole fluid. The thickness of formation sampled corresponds with the "span" (distance) between the two receivers (Fig. 3b).

When the sonic tool is drawn upward through the borehole, its measurements are converted to a velocity profile of the sequence as a function of depth. However, sonic logs are conventionally recorded in terms of "transit time." This is the time in microseconds taken for the compressional wave to travel through 1 foot of formation and is the reciprocal of its velocity (Fig. 4).

The velocity of compressional acoustic waves through minerals is not a difficult laboratory measurement and tables of representative values are listed in handbooks of physical constants. Figures generally used for interpretation of sonic logs in common reservoir lithologies are:

> Quartz   55.5
> Calcite   47.5
> Dolomite 43.5

The units are in microseconds per foot and are averages for ranges of variation. By contrast, the corresponding transit times for pore fluids are markedly higher:

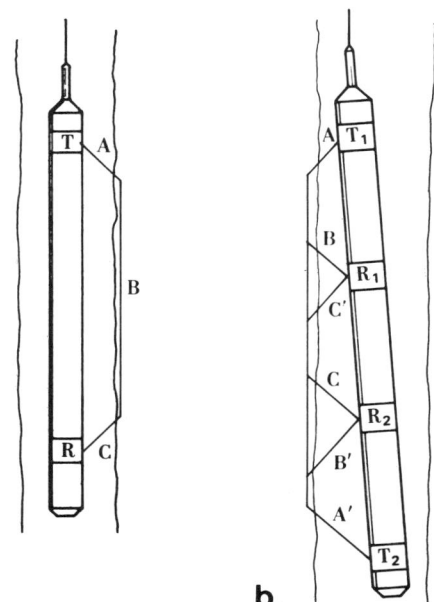

FIGURE 3. Design of the earliest sonic tool (*a*) and the modern borehole compensated sonic tool (*b*). From Dresser-Atlas publications.

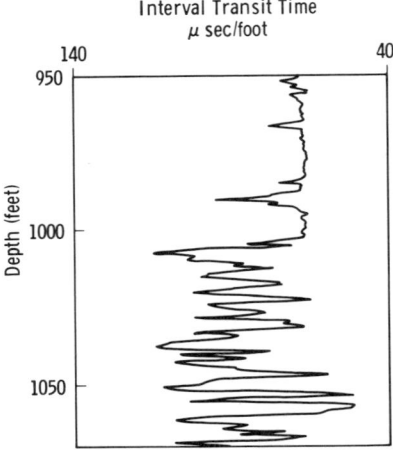

FIGURE 4. Sonic log of a Permian section from central Kansas.

Brine (20% NaCl)	189
Oil	238
Gas (methane)	626

The seeming precision of these figures is, of course, illusory, since the actual times are strongly influenced by compositional variation, temperature, and confining pressure. However, the judicious use of mineral and fluid transit times as limiting endmembers of a porosity range allows the calculation of porosity from sonic logs.

As a result of laboratory measurements of transit times of lithologies with variable porosities, Wyllie et al. (1956) demonstrated that there was an approximate linear relationship between transit time and porosity. Porosities could be estimated from a linear interpolation between the transit time of the matrix mineral (zero porosity) Δt_{ma} and the transit time of the porosity fluid, (100% porosity) Δt_f. In other words, for any zone of shale-free reservoir lithology,

$$\Delta t = \phi \Delta t_f + (1 - \phi) \Delta t_{ma}$$

where Δt = transit time of the zone

ϕ = porosity of the zone

Expressed as the "Wyllie time average equation," this becomes

$$\phi = \frac{\Delta t - \Delta t_{ma}}{\Delta t_f - \Delta t_{ma}}$$

Selection of an appropriate value of transit time for porosity fluid is keyed to the nature of the fluid within the flushed zone, rather than the virgin formation. This follows, because the radius of investigation of the sonic tool is extremely shallow since it is dictated by the path of the fastest wave to be transmitted from transducer to receiver. For fresh muds, this transit time will generally correspond to that of the mud filtrate (approximately 189 μsec/foot). Significant residual oil saturations will modify this figure slightly and any residual gas saturations, such as might be observed in shallowly invaded Gulf Coast gas sands, can result in a pronounced "hydrocarbon effect."

The time average equation generally provides reasonable porosity estimates in both clean sandstones and carbonates, although the sonic log has several limitations. The first arrival wave to reach a receiver will generally bypass larger porosity features such as vugs and fractures. As a result, the porosity estimate of the transit time is the sum of intergranular and intercrystalline porosities, but excludes vugs and fractures. This problem is generally very minor in sandstones, but is often important in carbonate sequences. However, this drawback can be turned to an advantage if the sonic log is run simultaneously with either a neutron or density tool, both of which are sensitive to all porosity types. The sonic estimate then corresponds to a "primary porosity" (sensu petrophysica) and the other tool estimate to a "total porosity." The difference between the two porosities is the "secondary porosity" (sensu p.), which is an estimate of the proportion of vugs and/or fractures.

The presence of vugs and fractures is also often registered by the generation of "cycle skips." These are obvious on a sonic log as sudden thin events with excessively anomalous transit time values whose thickness is less than the span of the receivers. They are caused by the loss in acoustic energy and consequent attenuation of the compressional wave. The suppressed wave slips below the bias gate set at the receiver to guard against triggering by extraneous borehole noise. The result is a late trigger of the receiver and recording of an anomalous transit time.

In unconsolidated or semiconsolidated sands, an additional correction must be applied to porosities estimated by the time average equation, since its application presupposes a rigid matrix framework. The transit time through a sand grain pack with 20% porosity will be longer than the time through a cemented sandstone with the same porosity, due to the extra loss in mechanical energy introduced by the loose grains. The apparent porosities of the time average equation are converted to true porosities through multiplication by a constant. This constant is generally derived from numerical comparisons of apparent porosities with porosities from either neutron, density, or resistivity logs, or from available core porosities.

The matrix transit times of lithologies such as sandstones, limestone, and dolomites are defined by narrow ranges, since these correspond to values for quartz, calcite, and dolomite, which may be measured in the laboratory. However, "shale" is a general term for a variety of clay mineral mixtures with accessory constituents, whose physical character can range from a mushy gumbo to a highly compacted mudstone. As a result, the transit times of shales are drawn from the sonic log itself, in shale intervals indicated by the SP or gamma-ray logs, rather than from standard tables.

THE DENSITY LOG

The first commercial density logging tool was introduced in the 1950s as a device to estimate the bulk density of formations within a borehole. It was originally intended to be used to gather data useful in the interpretation of surface gravity surveys. However, its applicability to porosity estimation was soon realized and this is its primary function on most exploration well logging runs.

The density tool typically consists of a radioactive source of gamma rays (e.g., cesium-137) which is mounted on a pad that is pressed against the formation by the tool. These constant energy gamma rays radiate into the formation and interact with the electron clouds of the atoms they encounter by Compton scattering, photoelectric effect, and pair production processes. As a consequence, there is a reduction in the gamma-ray flux which is measured by a short-spaced and a long-spaced detector (Fig. 5). This reduction is proportional to the average electron density of the formation. The electron density measurements may be converted to apparent bulk density through considerations of simple atomic theory.

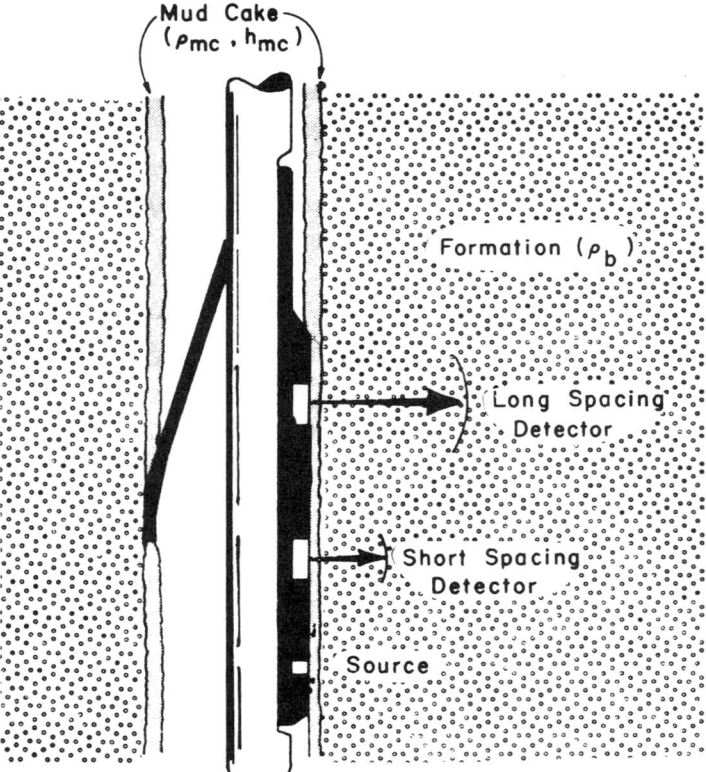

FIGURE 5. Schematic design of the dual-spacing density tool. From Wahl et al. (1964), JPT, © SPE-AIME.

Each element can be characterized by an atomic structure of a nucleus comprised of protons and neutrons surrounded by concentric shells of electrons. The atomic number Z is the number of protons in the nucleus and is characteristic of the element. This number is equivalent to the number of electrons surrounding the nucleus. The mass of the atom is contained in the nucleus and is effectively the sum of the number of protons and neutrons. This number is characterized by the atomic mass number, A, which will vary slightly for different isotopes of the same element, in accordance with the varying number of neutrons in the nucleus. As a general rule, the number of protons closely matches the number of neutrons, with the result that the Z/A ratio of most elements is close to 0.5. This is especially true for elements which are in the lower part of the periodic table, and it is these elements which occur in any appreciable abundance in nature. Now, since the number of electrons matches the number of protons, electron densities measured by the density tool may be translated to an apparent bulk density in grams per cubic centimter, by using a Z/A ratio of 0.5. In logging typical sedimentary lithologies, the difference between apparent and real densities is trivial, since the real Z/A ratios for quartz, calcite, and dolomite are 0.4993, 0.4996, and 0.4994, respectively. However, there is a measurable difference for some commonly encountered minerals, for which halite provides the best example. The density log apparent density for halite is 2.03 while its real density is 2.16, owing to its actual Z/A ratio of 0.4799. These differences are of little concern in log analysis, as calculations are made consistently in terms of apparent density. However, errors may be introduced by the use of real densities, unless they have been modified to their apparent equivalents through correction by a Z/A factor appropriate to their atomic composition.

The density log is conventionally presented as a trace of apparent density scaled between limits of 2 and 3 g/cm^3. This range contains the densities of most lithologies encountered within a borehole, although a secondary scale of 1 to 2 g/cm^3 allows the recording of anomalous zones such as coal beds or excessive washout zones. An example of a modern density log is shown on Figure 6. This log records densities of a Permian evaporite section from central Kansas. In this instance, the strong density contrasts between the component lithologies (halite, 2.03; shale 2.5; anhydrite 2.95) makes the density log a particularly effective lithology discriminator. The vertical resolution of the density tool is about 2 feet, with the result that features thinner than this dimension tend to be averaged on the record.

Since the density tool is a "contact device" (pressed against the borehole wall), measurements can be adversely affected if the borehole is markedly rugose or punctuated by deep washout features. The log is run in conjunction with a caliper. The caliper measures the diameter of the borehole and its readings are used to make any necessary corrections to the density record. Problem zones are usually obvious from their excessively low densities and high diameter readings on the caliper. The anomalies mark points where the pad fails to make reasonable contact with the formation and so becomes heavily influenced by the density of the borehole mud.

The estimation of porosity in reservoir units employs a simple mass balance relationship, in which the bulk density of a zone is the sum of the densities of the zone multiplied by their proportions. For a clean zone, the appropriate equation is

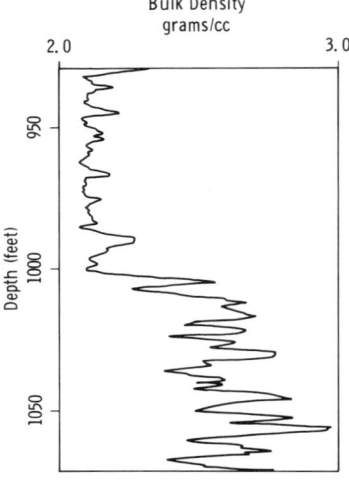

FIGURE 6. Density log of a halite unit above interbedded anhydrite and shale beds in the Permian of central Kansas.

$$\rho_b = \phi\rho_f + (1 - \phi)\rho_{ma}$$

where ρ_b = the bulk density of the zone

ρ_f = the density of the pore fluid

ρ_{ma} = the density of the matrix mineral

Since the density tool has a limited radius of investigation, the pore fluid relates to the flushed zone rather than the virgin formation. For fresh mud filtrate, this figure will be close to one and will not be modified dramatically by minor residual oil saturations. However, significant residual gas saturations will result in a pronounced "hydrocarbon effect."

Typical figures used for apparent grain densities of reservoir lithologies are:

Quartz	2.65
Calcite	2.71
Dolomite	2.87

When evaluating shaly reservoir units, allowance must be made for the disturbing influence of the contaminating shale in an expansion of the basic equation to

$$\rho_b = \phi\rho_f + (1 - \phi - V)\rho_{ma} + V\rho_{sh}$$

where ρ_{sh} = density of the shale

V = proportion of the shale

Aside from its application to the estimation of porosity, the density log has a variety of uses ranging from the estimation of oil yields from oil shales to its application with sonic logs in the computation of impedance values for use in generating synthetic seismograms. It is a particularly valuable aid in the recognition of certain lithologies and minerals which have appreciable density contrasts with typical sandstones, carbonates, and shales. It is used widely in the evaluation of coal-bearing sequences both in the identification of lithologies and determinations of coal character, related to rank and ash content.

In recent years, the older type of density tool is beginning to be phased out and replaced by an improved model which includes a measurement of photoelectric absorption cross-section index. The nuclear reactions of gamma rays with borehole section materials is controlled to a large degree by energy level. At moderately high energies, interactions are dominated by Compton scattering, whose effect is determined by the number of electrons per unit volume and leads to conventional density measurements. However, at low energies, photoelectric absorption becomes an important phenomenon and this is regulated by the number of electrons per atom, which is the atomic number, Z. Detector count rates at a low-energy window can be used in the computation of a photoelectric absorption cross-section index, Pe, and recorded as an ancillary log in conjunction with the conventional bulk density trace.

The index shows marked differences in value between minerals with the result that its log trace is a good lithology discriminator. Sandstones, limestones, and dolomites are generally readily distinguished, while the index is sensitive to the occurrence of minor amounts of heavy minerals. Conversion of the index to a photoelectric macrosection is particularly useful when crossplotted with density for volumetric evaluation of mixed mineral lithologies as discussed in Chapter 5.

THE NEUTRON LOG

The neutron tool contains a source of high-energy neutrons which radiate spherically into the adjacent formation. These neutrons experience a progressive reduction in energy through collisions with nuclei of the atoms they encounter. The major losses in energy occur when a neutron encounters a nucleus of approximately similar mass. As a result, the primary reduction in energy occurs in collisions with hydrogen atoms within the formation. Ultimately, each neutron is reduced in energy to a thermal state, at which point it is captured by a nucleus and capture gamma rays are emitted.

One or two detectors are placed at a short distance from the source and either measure the quantity of slow neutrons or the capture gamma rays (Fig. 7). The reduction in the neutron flux is mainly a function of the hydrogen concentration within the formation which, for typical reservoir lithologies, can be equated with the pore fluid fraction. A high porosity results in a low count rate at the detector since most of the neutrons will be reduced to a thermal state and captured close to the source. Conversely, a low porosity leads to a high count rate, as a higher

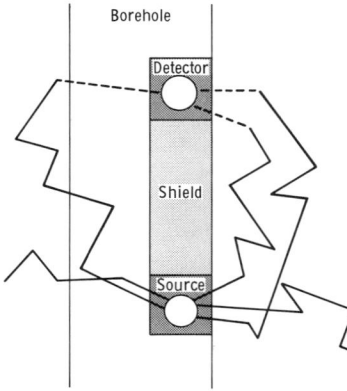

FIGURE 7. Schematic diagram of the operation of a single-detector neutron tool. Solid lines indicate hypothetical paths of individual neutrons deflected at locations of collisions; dashed lines represent gamma rays emitted at points of neutron capture.

proportion of neutrons survive the distance to the vicinity of the detector. This consideration dictates that the radius of investigation of the tool can range from several inches to several feet as determined by the amount of porosity.

Modern tools are scaled in either API neutron units or directly in porosity units related to an assumed limestone matrix with a pore fluid of water. The older tools had both source and detector mounted in the tubular housing of the tool itself, while the neutron measurement was recorded as counts. Computations of porosity are then a function of the specific hardware characteristics of the tool, the diameter of the borehole, and the nature of the formation lithology. There is an exponential relationship linking the porosity with the neutron reading of the general form

$$e^{-K\phi} = CN$$

where ϕ = porosity

N = neutron reading of the zone in counts

K, C = constants related to tool, borehole size, and lithology

This relationship has been approximated for many years by a logarithmic equation of the type

$$\log \phi = C - KN$$

This equation is used in the calibration of the neutron scale in terms of porosity units, and requires at least two points for its definition. A systematic calibration may be made by matching neutron counts of zones with their estimated porosities from other sources such as logs or core measurements. As a pragmatic alternative, the "40:1 method" and other hoary variants have been widely used over the years. This method consists of finding a shale zone, using the gamma-ray or SP log, and ascribing a neutron porosity value of 35–40% to its neutron reading (not an unreasonable value for midcontinent shales). Then a relatively tight zone within a carbonate (recognized by its very high neutron count reading) is equated with a porosity of either 1 or 2%. The remainder of the scale is divided between these two boundary values on a logarithmic system of porosity divisions.

Happily, the modern tools are directly and arithmetically scaled in equivalent porosity units, related to either a limestone, sandstone, or dolomite matrix (Fig. 8). This has been achieved by new tool design features which mount the source and detectors in a rubber pad which is pressed directly against the formation, thereby reducing the influence of varying borehole diameters. At the same time, the recording device is linked directly with a small computer which continuously converts neutron readings into equivalent porosity units by monitoring neutron counts, hardware characteristics, and varying borehole conditions.

The three main neutron tools in the Schlumberger stable are the GNT (an older instrument which has been largely superseded), the SNP (sidewall neutron porosity—a pad device with one detector), and the CNL (compensated neutron log—a pad device with two detectors). Their equivalents, with slightly different names, are run by the other major companies. If a single porosity scale is used, it is generally referred to a limestone matrix as an arbitrary reference point. When logging other lithologies a correction must be applied to the porosity reading to a true porosity reading for the appropriate matrix mineral. This is most easily done by use of a conversion chart for the appropriate tool (GNT, SNP, or CNL) which is listed in standard chartbooks. The reasons for this are that the correction factor is in part specific to the tool and is nonlinear in form. The vertical resolution of modern neutron logs is of the order of about 2 feet.

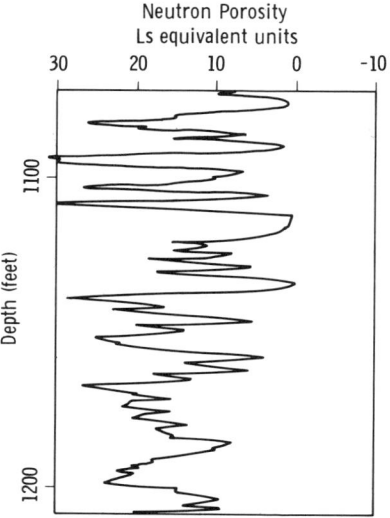

FIGURE 8. Sidewall neutron porosity log of a Permian section from a test hole in central Kansas.

The collisions of neutrons with hydrogen nuclei make no distinctions as to the molecular form that contains them. As a result, the neutron log should be regarded primarily as a measure of hydrogen concentration, and only secondarily as a porosity tool. So, for example, the neutron log is sensitive to hydrogen bound in water of crystallization and is useful in evaporite sequences for discriminating minerals such as gypsum. The neutron log also shows a characteristically high-porosity response in coals, which has been ascribed to moisture content, although it may be related to the high neutron moderating characteristics of concentrated carbon. When encountering porosity with high gas saturations, the neutron log records an extremely low porosity, owing to the relatively low hydrogen density of the pore space. The low porosity reading is produced because the calibration is predicated on a given matrix mineral whose porosity is identified with hydrogen-rich water. This feature allows the neutron log to be used as a "gas detector" and its value is particularly enhanced if used simultaneously with a density log which will tend to record anomalously *high* apparent porosities in gas zones.

In general, a hydrogen index may be computed for any material from simple molecular theory, when using water as a reference value of 1. The atomic weights of hydrogen, oxygen, and carbon are approximately 1, 16, and 12, respectively. The "hydrogen density" of water may be calculated as

$$\frac{\rho_w \times 2H}{2H + O} = \frac{1}{9}$$

where ρ_w is the density of water in grams per cubic centimeter. If a "hydrogen index" for water is equated with 1, then, for any component

$$H_h = 9 \times \text{hydrogen density/g}$$

The molecular composition of an oil may be approximated by the general formula $n(CH_2)$, where the value of n reflects the degree of molecular complexity of the average hydrocarbon within the oil. The hydrogen index of the oil is then

$$H_h = \rho_o \times 9 \times 2/14$$

where ρ_o is the density of the oil in grams per cubic centimeter. For an oil density of 0.85, the hydrogen index is therefore 1.09.

Since the more recent neutron tools are calibrated in terms of a freshwater pore fluid, a neutron porosity reading of 100% corresponds to the reading for a component with a hydrogen index of 1. A sample of an oil with hydrogen index 1.09 would have a corresponding reading of 100%. This consideration shows that zones containing significant residual oil saturations will result in neutron porosity readings which deviate slightly from the true porosity readings.

The "hydrocarbon effect" is generally very minor for most oils, but becomes important in cases where there are appreciable gas saturations. If a zone is encountered which contains methane (CH_4) with a density of 0.1 g/cm^3 at the zone temperature and pressure, its hydrogen index can be estimated as 0.225. As a result the neutron porosity readings will be reduced to values markedly below the true porosity values. As noted before, this property makes the neutron log a valuable aid in the detection of potential gas zones.

By using the hydrogen index formula, apparent neutron porosities may be calculated for various hydrated minerals from their molecular compositions and atomic weights. These figures are useful when applying the neutron tool to the search for certain salts within evaporite sequences. The hydrogen index of materials that do not contain hydrogen will be close to zero, but will vary from this figure by amounts which are dictated by the neutron capture cross sections of the component atoms.

As with the transit time and density, the neutron response of shales will be variable as dictated by such considerations as the quantity of hydrogen associated with the lattices of the range of possible clay minerals, proportion of associated bound water, and other factors. As a result, neutron porosity values for shales are derived from inspection of the log in the interval of interest. Typical values for midcontinent shales run at 30–40% neutron porosity, and generally exceed the levels recorded in intervening lithologies.

ANALYSIS OF SHALE COMPACTIONAL TRENDS

Even in the absence of gamma-ray, SP, or resistivity logs, shale units are recognized easily on the sonic, density and neutron logs. Shale zones generally register high "porosity" values which are a response to water content rather than the effective porosity of potential reservoir lithologies. Water within the shales is the major control on variation in the physical properties of density, bulk modulus of elasticity, and hydrogen concentration. The properties show a broadly monotonic change with depth which reflects compactional effects as the result of mechanical, chemical, and thermal mechanisms.

The burial of muddy sediment and its progressive compaction in subsidence by loading of later sediments is a well-known phenomenon (Hedberg, 1926). As a general observation, the degree of compaction is matched by decreasing transit times, increasing densities, and decreasing neutron porosity readings. The basic trend confirms to a nonlinear function of depths, with rapid changes at shallow depths which curve downward toward an asymptotic value (Fig. 9).

The essential process is one of sediment dewatering, which was originally thought to be purely a mechanical phenomenon driven by gravity. More recent work by Powers (1967) and Burst (1969) has incorporated diagenetic changes in clay mineralogy as a crucial element of more realistic compaction models. Geothermal controls are considered to be more influential than overburden pressure. Alterations in clay mineralogy, primarily smectite to illite, are a substantial component of dewatering and occur at several discrete stages of temperature and pressure (Burst, 1976). If dewatering fails to keep pace with hydrostatic pressure, overpressured zones can develop. In cases of marked differential loading, there is upward diapiric movement of shales, and even penetration of the surface as "mud lumps," described by Morgan

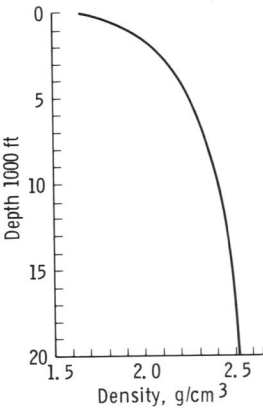

FIGURE 9. Variation of average Gulf Coast shale bulk densities with depth. After Eaton (1969), JPT, © SPE-AIME.

et al. (1968). The complex interplay of physical controls in compaction has been modeled mathematically by Sharp (1976) as a system of momentum and energy balance equations.

The link between many shale properties and their degree of compaction has been recognized for many years as a useful relationship in the reconstruction of the structural history of a sedimentary basin. Athy (1930) noted that if the degree of compaction could be related to the depth of burial, it would be possible to estimate the maximum depth to which a shale had been buried. In this way, a shale could be considered as a geobarometer which recorded its maximum overburden load. Athy (ibid) measured densities of thousands of shale samples in northeastern Oklahoma and established a relation between compaction (density or porosity) and depth. From these data he concluded that overburden sediments had been eroded to a depth of between 4000 and 5400 feet.

In more recent studies, laboratory measurements of shale sample densities have been replaced by logging data which provide continuous records of thousands of feet of section. Dallmuss (1958) observed a systematic increase in shale densities moving from the margin toward the center of a sedimentary basin in South America. He also noted a similar increase related to increasing geologic age, and an increase within a syncline contrasted with an anticline. Fertl (1977) reported a decrease of shale densities on flanks of structures relative to values at structurally high positions, presumably as a result of differential compaction. Compositional variations of shales are obviously an important secondary factor in density variation. High degrees of organic content or shale gas can cause a marked decrease in shale density, while concentrations of calcareous material or heavy minerals (such as pyrite or siderite) increase densities (Fertl, 1977).

Sonic logs have also been used in shale compaction studies. Jankowsky (1962) applied shale transit times to the reconstruction of the structural history of the northwestern German Basin. He mapped areas which showed evidence of "residual lifting" reflecting uplifted zones whose compaction had not been modified by subsequent subsidence. Magara (1976) used transit times to determine the thickness of sedimentary rocks removed by erosion in western Canada.

Analysis of Shale Compactional Trends

Empirical studies of shale compactional effects registered by sonic logs generally use plots of logarithmic transit time versus linear depth (Lang, 1978). In basin wells that penetrate a virtually continuous section of sediment record, the plotted points tend to follow a linear trend and can be equated with a type compaction gradient. In practice, the type compaction gradient is often reconstructed piecemeal from the integration of the transit time records of several wells (Fig. 10). Careful examination

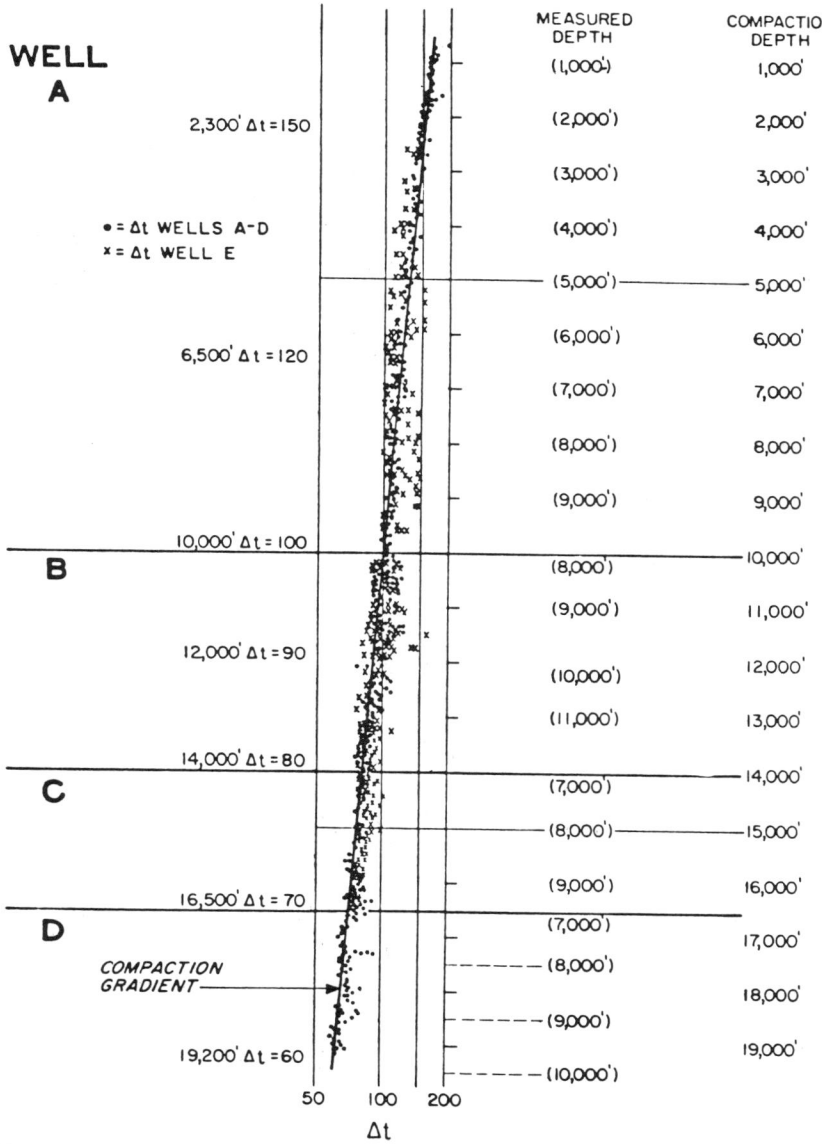

FIGURE 10. Typical compaction gradient of transit times in a California Tertiary basin. From Lang (1978).

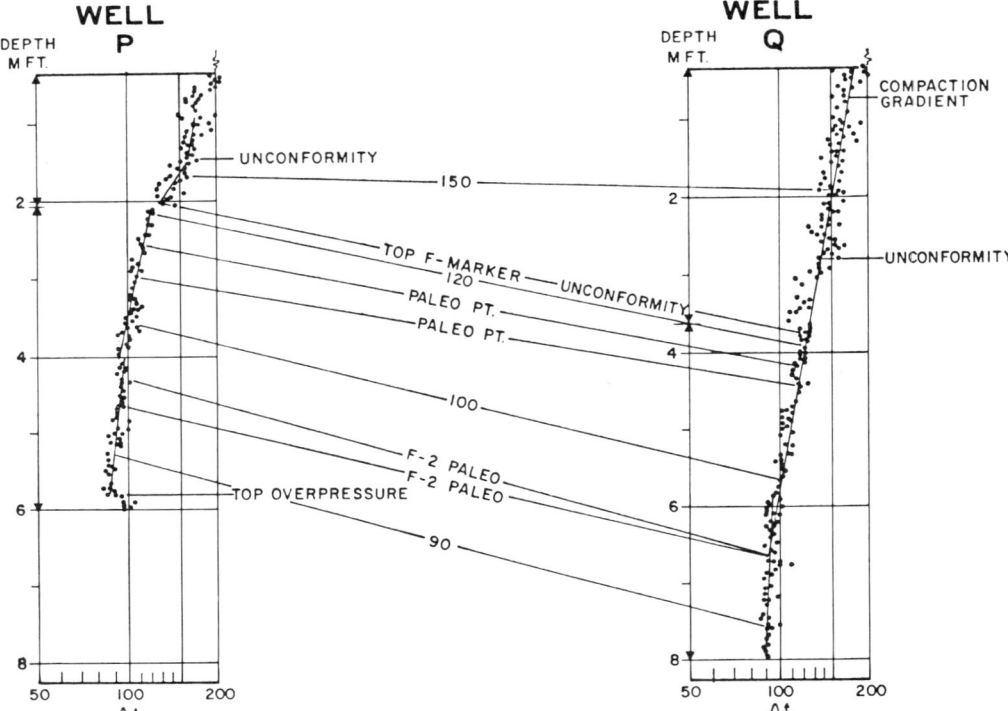

FIGURE 11. Comparison of shale transit time gradients in two wells from a Tertiary sequence in California. Gradient breaks mark the position of unconformities. From Lang (1978).

of individual well gradients and comparison between wells may reveal distinctive breaks and correlative gradient legs which reflect uplift, erosion, and nondeposition. The shale transit time–depth record from a Tertiary sequence in California (Fig. 11) shows several systematic breaks in gradient. These breaks match the position of unconformities deduced from paleontological stratigraphic control (Lang, 1978). The abrupt reversal to higher transit times at the base of the well record marks the top of an overpressured zone.

Shale compactional trends revealed by sonic, density, and resistivity logs have been of interest to the drilling industry for a number of years (Hottman and Johnson, 1965). Deviations from normal compactional trends in thick, relatively young sediments are often diagnostic of overpressured zones whose early detection is crucial in the avoidance of blowout hazards. The weights used for mud programs in well drilling generally presuppose a normal formation pressure which approximately equals hydrostatic pressure. Hydrostatic pressure corresponds to the product of the density and vertical height of a water column with salinities typical of the formation penetrated. A widely used figure for the hydrostatic pressure gradient is 0.465 psi/foot and matches a salinity of 80,000 ppm sodium chloride and temperature of 77° F (Fertl, 1977). This pressure is much less than the overburden pressure, which is caused by the weight of the rock overburden and is of the order of 1 psi/foot depth. If sediment dewatering maintains an equilibrium with compactional gradient, the formation pressure of fluids in subsurface reservoir units approximately equals hydrostatic pressure. These fluids do not experience the overburden pressure, just as people are not crushed by the weight of the buildings they work in but carry the burden of the atmosphere on their frail shoulders.

Overpressured zones often occur at depth in the thick Tertiary sediments of the Gulf Coast. In the shallow part of the section, the high proportion of coalesced sandstones provide effective hydraulic conduits that transmit the fluid released in the dewatering of shales. At greater depths, the sedimentary facies consists of deep-water marine shales with thin sandstone bodies. Subsidence and compaction may proceed at a rate that exceeds the capacity of the sandstones to remove fluid, particularly if the sandstones are isolated hydraulically. Consequently, excessive pressures may be generated within the sandstones while the adjacent shales are characterized by high fluid pressures. Shales in the overpressured zone can be recognized as "undercompacted" relative to their expected degree of compaction on a "normal" compaction-depth trend typical of sequences in which shale dewatering has kept pace with overburden stress. The sonic log has been widely used in the

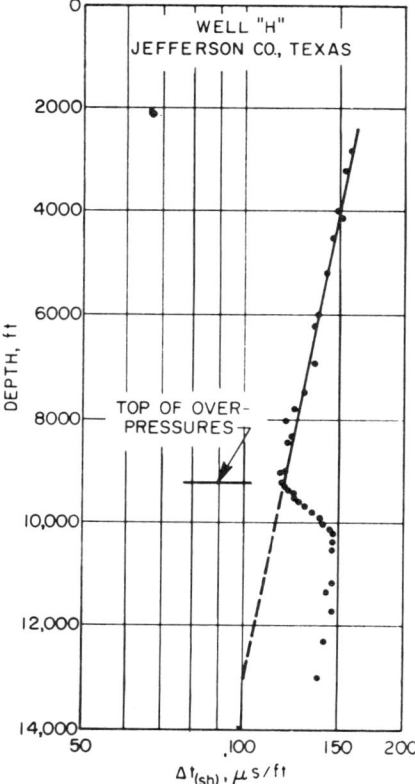

FIGURE 12. Transit time–depth plot of shales in a Gulf Coast Tertiary sequence which shows a break to higher transit times in an overpressured zone. From Hottman and Johnson (1965), JPT, © SPE-AIME.

detection of overpressured zones (Hottman and Johnson, 1965). An example of a transit time–depth plot in a well with an overpressured interval is shown in Figure 12. The distinctive break of shale transit times in a reversal to higher values is obvious on the plot, while the normal compactional trend compares well with a type trend for Gulf Coast Tertiary shales.

The detection of overpressured zones by logging techniques has immediate importance in the engineering of exploration wells in thick sequences of relatively young sediments. At the same time, it gives important insights into the nature of the compaction mechanism as a current dynamic process. The integration of geochemical, geothermal, and hydrodynamic theories with earlier empirical methods is likely to refine conclusions from studies of compactional trends in ancient basins.

GEOLOGIC IMPLICATIONS OF POROSITY TRENDS AND PATTERNS

The study of porosity in rocks was at one time virtually the preserve of the petroleum geologist and hydrologist. More recently, the extraordinary costs of off-shore exploration and the exacting engineering requirements of stimulation projects in depleted reservoirs have encouraged renewed efforts in research. Simultaneously, the widespread use of the scanning electron microscope has literally provided a new insight into aspects of the pore phase which are often obscured in conventional thin section. The pore network is more clearly perceived in terms of depositional fabric geometry, function as a conduit to modifying solutions, and host to diagenetic minerals. The Medium is indeed the Message. Recent studies have attempted to unravel petrography as a multiple record of deposition, tectonism, hydrodynamic, and geochemical regimes, which are basic historical elements of basin development and stratigraphy. Based on work with porosity rhythms in sedimentary sequences, Vistelius (1967) proposed that, since porosity provided the link between the fluid and solid phases, sedimentary formations should be viewed in terms of phase differentiation. This unconventional thesis may be a useful unifying concept for the processes of mechanical compaction and diagenesis.

Early work on porosity variations in sandstones tended to emphasize the role of the fabric as an inert pack of quartz grains subjected to compactional rearrangement, with a degree of pressure solution and redistribution of dissolved silica. In experiments with unconsolidated sands, Beard and Weyl (1973) showed positive relationships between porosity and textural parameters of sorting, low sphericity (shape), and angularity (roundness). Selley (1976) related increasing textural maturity with porosity in Jurassic sandstones of the North Sea in a diagnostic model, ranging from distal turbidite to shoal. More recently, studies by Hayes (1979) and Schmidt and McDonald (1979) have demonstrated that diagenesis is not restricted to the destruction of primary porosity, but ultimately enhances pore volumes in the creation of significant secondary porosity.

From measurements on Holocene carbonate sediments, Enos (1979) showed a systematic inverse relationship between grain size and porosity (in common with carbonate rocks), but a negative correlation between porosity and permeability (in

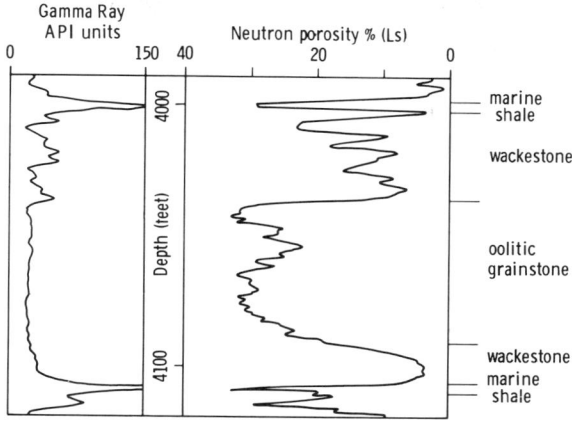

FIGURE 13. Gamma-ray–neutron logs of the J zone, Kansas City Group (Pennsylvanian) from a well in central Kansas. Note the high-porosity oolitic grainstone which overlies the low-porosity wackestone.

contrast with carbonate rocks). Porosity developments in ancient limestones are the cumulative result of a diagenetic sequence which may involve multiple histories of cementation, dissolution, and mineralogy changes. In cases where the original sediment framework showed marked differentiation between contiguous depositional facies the characteristic may be presented as distinctive porosity zonations on wireline logs. An example is shown on Figure 13 which is a gamma-ray–neutron log combination in the J-zone of the Kansas City Group (Pennsylvanian) from a well in central Kansas. The gamma-ray log shows a basal marine shale overlain by a limestone. Although this limestone is characterized by fairly uniform low gamma radiation throughout its development, the neutron log shows a dramatic change in porosity. Well samples identify the low-porosity (approximately 4%) Lower Limestone as wackestone, and the high-porosity (approximately 28%) Upper Limestone as oolitic grainstone. In a regional study of this unit, Watney (1985) concluded that the Lower Limestone was deposited as open-marine micrite sediments on a broad open shelf. The Upper Limestone represents a later stage in regression as the product of oolite shoals. The recognition of the porosity motif on wireline logs enabled Watney (ibid) to map the thickness of the porous regressive carbonate (Fig. 14) which outlines the distribution of oolite grainstone facies. The map shows northward-pointing fingers in a broad oolite facies which resemble recent spillover lobes of oolite in the Bahamas. Watney (ibid) suggested that the cause of the J-zone oolite distribution was a break in the Pennsylvanian shelf paleoslope striking approximately west to northwest, which acted as a focus for waves and currents during shoaling conditions operating late in the deposition of the limestone.

Processes of compaction and chemical diagenesis cause a wide variation in pore volumes and porosity types within carbonate rocks. The transformation of calcite to dolomite involves a contraction on the order of 12%, with associated increase in porosity (Chillingar and Terry, 1954). Although porosity changes are often observed, more recent work suggests that molecule-for-molecule replacement is

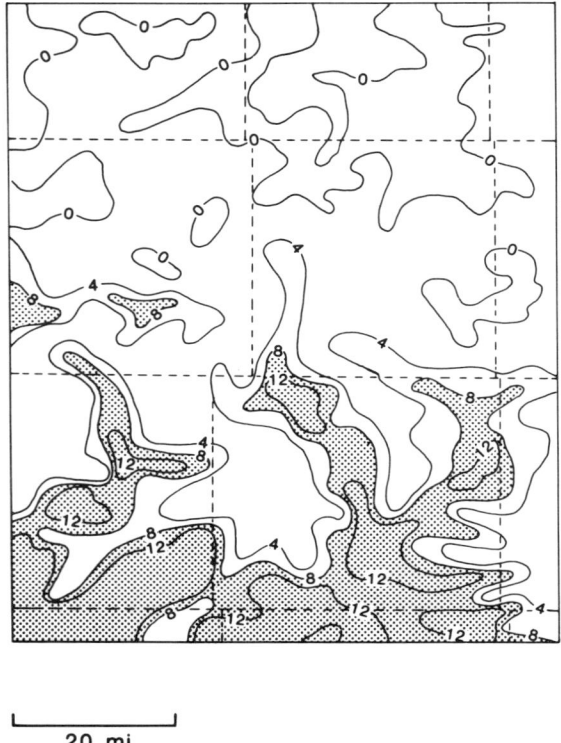

FIGURE 14. Isopach map of the high-porosity oolitic grainstone phase of the Kansas City Group *J* zone in central Kansas. Note the thick lobate features in the south which probably represent oolite shoals. From Watney (1985).

only one aspect of the complex processes of dolomitization. While the precise mechanisms of diagenesis are still in dispute, field studies such as by Murray and Lucia (1967) indicate that dolomitization and pore development often appear to be strongly fabric selective, and result in a concordant diagenetic overprint of contrasts in depositional facies. Lucia (1962) studied a suite of Devonian rocks from Texas which showed differing degrees of dolomitization. He found a direct relationship between the extent of dolomitization and the amount of micrite associated with crinoid fragments within the original sediment. Crinoidal limestones with rim cement surrounding the crinoids showed partial occlusion of pore space. These lithologies were generally not dolomitized and had a porosity of 1.6%. When micrite contents were associated with the crinoidal material, the crinoidal limestones were dolomitized to varying degrees. Highest porosities were developed in rocks where the micrite fraction ranged between 5 and 20%. Dolomitization of micrite and some crinoid fragments combined with leaching of calcite in partially dolomitized rocks resulted in porosities from between 5 and 30%.

Geologic studies of the interrelationships between porosity development, depositional facies, and diagenetic modifications are becoming increasingly common in the literature. Their conclusions are of great importance in the interpretation of porosity trends shown by sonic, neutron, and density logs, as well as resistivity traces in water-saturated sections. As is the case with other logs, conventional sources of geologic information (generally core and cuttings) provide the necessary key to both initiate and examine alternative genetic models. However, "hard data" of this kind is generally restricted both areally and vertically when compared with the extensive coverage supplied by logging suites. The intelligent coordination of both types of data extends the use of porosity logs beyond their traditional roles as stratigraphic correlation traces and evaluators of potential reservoir units.

REFERENCES

Athy, L. F., 1930, Density, porosity and compaction of sedimentary rocks, *Am. Assoc. Petrol. Geolog. Bull.*, Vol. 14, No. 1, pp. 1–24.

Beard, D. C., and Weyl, P. K., 1973, Influence of texture on porosity and permeability of unconsolidated sand, *Am. Assoc. Petrol. Geolog. Bull.*, Vol. 57, No. 2, pp. 349–369.

Burst, J. F., 1969, Diagenesis of Gulf Coast clayey sediments and its possible relation to petroleum migration, *Am. Assoc. Petrol. Geolog. Bull.*, Vol. 53, No. 1, pp. 73–93.

Burst, J. F., 1976, Argillaceous sediment dewatering, *Ann. Rev. Earth Planetary Sci.*, Vol. 4, pp. 293–318.

Chillingar, G. V., and Terry, R. D., 1954, Relationship between porosity and chemical composition of carbonate rocks, *Pet. Eng.*, Vol. 26, pp. 341–342.

Dallmus, K. F., 1958, in *Habitat of Oil* (L. G. Weeks, Ed.), AAPG Memoir 36, pp. 2071–2124.

Eaton, B. A., 1969, Fracture gradient prediction and its application in oilfield operations, *J. Petrol. Tech.*, Vol. 21, pp. 1353–60.

Enos, P. 1979, Pore space in Holocene carbonate sediments (Abs.), AAPG-SEPM Annual Convention, Houston, p. 81.

Fenton, M. B., and Fullard, J. H., 1981, Moth hearing and the feeding strategies of bats, *American Scientist*, Vol. 69, pp. 266–275.

Fertl, W. H., 1977, Shale density studies and their application, in *Developments in Petroleum Geology* (G. D. Hobson, Ed.), Applied Science Publications Ltd., London, pp. 293–327.

Hayes, J. B., 1979, Sandstone diagenesis—the hole truth, in *Aspects of Diagenesis* (P. Scholle and L. Schluger, Eds.), SEPM Spec. Publ. No. 26, pp. 127–139.

Hedberg, H. D., 1926, The effect of gravitational compaction on the structure of sedimentary rocks, *Am. Assoc. Petrol. Geolog. Bull.*, Vol. 10, No. 11, pp. 1035–72.

Hottman, C. E., and Johnson, R. K., 1965, Estimation of formation pressures from log-derived shale properties, *J. Petrol. Tech.*, Vol. 17, No. 6, pp. 717–722.

Jankowsky, W., 1962, Diagenesis and oil accumulation as aids in the analysis of the structural history of the Northwestern German Basin, *Zeitschrift der Deutschen Geologischen Gesellschaft*, Vol. 114, pp. 452–460.

Lang, W. H., 1978, The determination of prior depth of burial (uplift and erosion) using interval transit time, *Trans. Soc. Prof. Well Log Analysts 19th Ann. Logging Symp.*, Paper C, 16 pp.

Lucia, F. J., 1962, Diagenesis of a crinoidal sediment, *J. Sed. Pet.*, Vol. 32, No. 4, pp. 848–865.

Magara, K., 1976, Thickness of removed sedimentary rocks, paleopore pressure, and paleotemperature, southwestern part of Western Canada basin, *Am. Assoc. Petrol. Geolog. Bull.*, Vol. 60, No. 4, pp. 554–565.

Morgan, J. P., Coleman, J. M., and Gagliano, 1968, Mudlumps: diapiric structures in Mississippi Delta sediments, in *Diapirism and Diapirs*, AAPG Memoir 8, pp. 145–161.

Murray, R. C., and Lucia, F. J., 1967, Cause and control of dolomite distribution of rock selectivity, *Geol. Soc. Amer. Bull.*, Vol 78, No. 1, pp. 21–36.

Powers, M. C., 1967, Fluid release mechanisms in compacting marine mudrocks and their importance in oil exploration, *Am. Assoc. Petrol. Geolog. Bull.*, Vol. 51, No. 7, pp. 1240–54.

Schmidt, V., and McDonald, D. A., 1979, The role of secondary porosity in the course of sandstone diagenesis, in *Aspects of Diagenesis* (P. Scholle and L. Schluger, Eds.), *Soc. Econ. Pal. Min. Spec. Publ.* No. 26, pp. 125–207.

Selley, R. C., 1976, The habitat of North Sea oil, *Proc. Geolog. Assoc.*, Vol. 87, Part 4, pp. 359–387.

Sharp, J. M., Jr., 1976, Momentum and energy balance equations for compacting sediments, *Mathematical Geology*, Vol. 8, No. 3, pp. 305–322.

Vistelius, A. B., 1967, *Studies in Mathematical Geology*, Consultants Bureau, New York, 294 pp.

Wahl, J. S., Tittman, J., Johnstone, C. W., and Alger, R. P., 1964, The dual spacing formation density log, *J. Petrol. Tech.*, Vol. 16, No. 12, pp. 1411–16.

Watney, W. L., 1985, Origin of Four Upper Pennsylvanian (Missourian) Cyclothems in the Subsurface of Western Kansas: Application to the Search for Accumulation of Petroleum, unpublished PhD Dissertation, University of Kansas, 506 pp.

Wyllie, M. R. J., Gregory, A. R., and Gardner, G. H. F., 1956, Elastic Wave velocities in heterogeneous and porous media, *Geophysics*, Vol. 21, No. 1, pp. 41–70.

CHAPTER FIVE

GRAPHICAL METHODS OF LITHOLOGY DETERMINATION

The exclusive consideration of one wireline log can only give a single numerical value at any given depth. Clearly, the information content is restrictive and of limited geological value. Spontaneous potential and gamma-ray logs discriminate shale beds from nonshale units. Resistivity logs record conduction carried by pore space brines and cation-exchange mechanisms of clay minerals within shales. Porosity logs are primarily sensitive to the volume of fluids within most lithologies. In practice, the situation is better than these statements would suggest. Interpretations are supplemented by information from core and cuttings, as well as knowledge concerning regional geology and general principles of stratigraphy and sedimentology. At the same time, it is rare to consider a log reading in isolation, and log interpretation is normally made in the context of trace variation over an entire subsurface section.

Considerations of reservoir mineral identification and the recognition of mixed lithologies are of interest to the industrial log analyst. The computation of accurate porosities from either the sonic, density, or neutron log presupposes a knowledge of the reservoir lithology. A single log value will yield different porosity values according to the mineral selected as the zero porosity endpoint. The problem is further complicated if the reservoir unit is comprised of variable mixtures of minerals. However, because each porosity tool responds to separate minerals in slightly different ways, a combination of porosity logs can be used to discriminate several minerals and simultaneously solve for porosity. A variety of graphical techniques have been used successfully for this purpose and are proving to be useful geological tools in their own right.

OVERLAY OF POROSITY LOGS ON A COMMON REFERENCE SCALE

The three types of porosity log which are available (sonic, density, and neutron) are recorded in drastically different units (microseconds per foot, grams per cubic centimeter, and neutron counts or percentage porosity units). However, the logs may be directly compared, if calibrated on a common reference scale. The most widely used scale for this purpose is defined in terms of equivalent units of limestone percentage porosity. Since this is the most common reference of the neutron logs, they generally require no rescaling. The density scale may be transformed by the simple operation of setting the grain density of calcite (approximately 2.71) to zero porosity, the porosity fluid density (approximately 1.00 for freshwater) to 100% porosity, and interpolating intermediate values. By an analogous procedure the transit times of the sonic log can be converted to the limestone porosity equivalents by setting the matrix transit times of calcite and pore fluid as the two porosity extremes for interpolation. Both these operations represent simple linear transformations.

An overlay of any combination of the three porosity logs will give immediate indications of the lithology of logged units, by virtue of the different responses of matrix minerals to the individual porosity logs. (Fertl, 1979). This point may be illustrated by comparing the hypothetical response of a mixed sequence of lithologies to the density and neutron logs as shown on Figure 1.

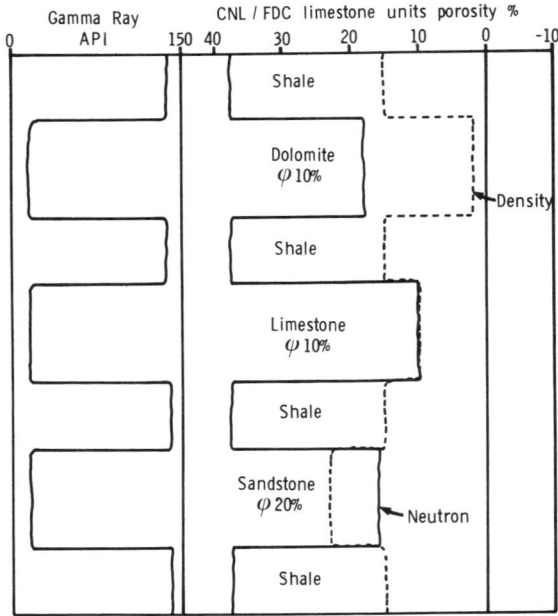

FIGURE 1. Overlay of hypothetical gamma-ray, neutron, and density responses in a simple sequence of lithologies.

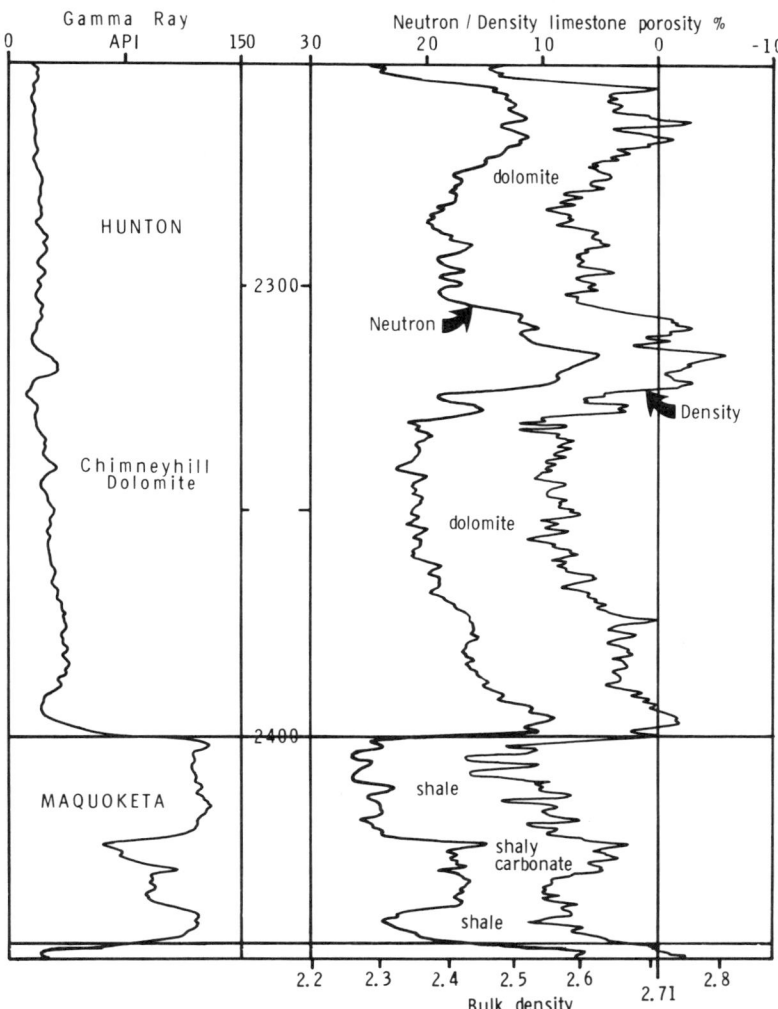

FIGURE 2. Gamma-ray, neutron, and density logs of Ordovician units in a northern Kansas well. (The section overlaps with Fig. 3.)

When viewed together with the gamma-ray log, the relative disposition of the density and neutron logs identifies the common sedimentary lithologies. Shales have a high neutron reading, a low density reading, and a high gamma-ray value. Limestones have a low gamma-ray value, and a coincident neutron and density reading, since the reference scale is calibrated to an assumed limestone. Dolomites have a low gamma-ray value, a relatively low density reading (since the grain density of dolomite is higher than calcite), and a relatively high neutron reading (since the zero porosity value for dolomite is a positive reading in limestone equivalent porosity). Sandstones have a low gamma-ray value, a relatively high density reading (the grain density of quartz is less than that of calcite), and a relatively low neutron reading (the zero porosity value for sandstone is a negative reading in limestone equivalent porosity).

The seasoned log analyst can perform this composite procedure of recalibration and superimposition of the porosity logs, by simply overlaying the field blueprints on a light box. He is aided in this operation by the universal convention of recording porosity logs, which dictates that regardless of the basic units of measurement, porosity increases from the right to the left of the log. By aligning the logs on a common calcite value (zero porosity on the neutron, calcite grain density on the

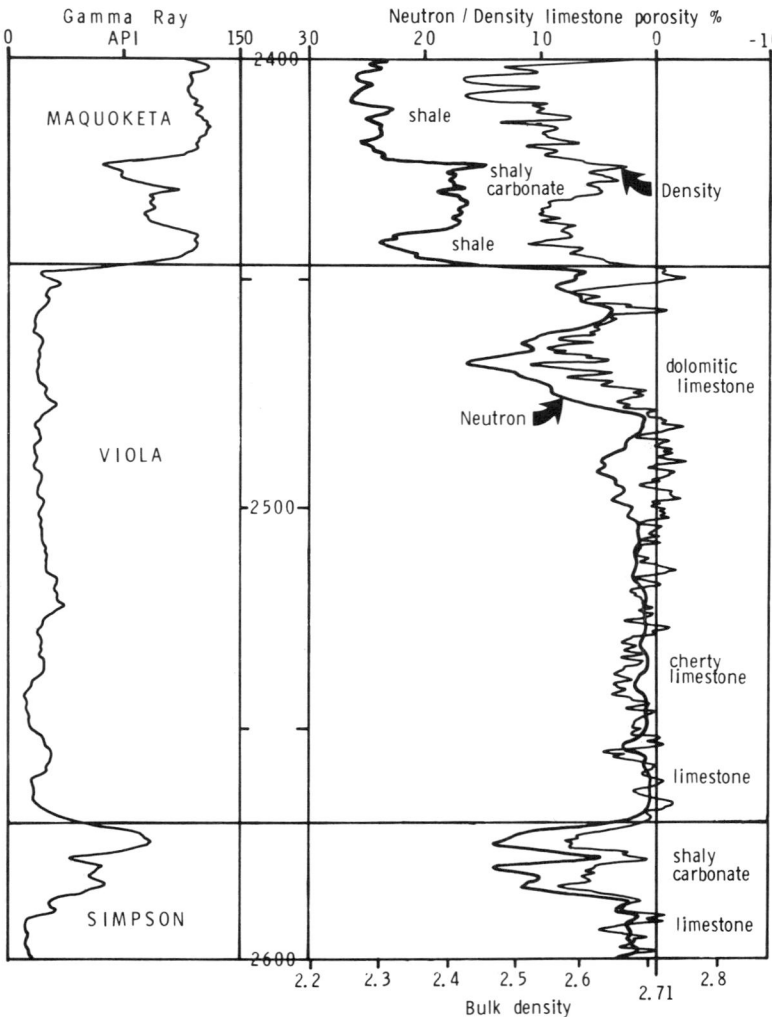

FIGURE 3. Gamma-ray, neutron, and density logs of Ordovician units in a northern Kansas well. (The section overlaps with Figs. 2 and 4.)

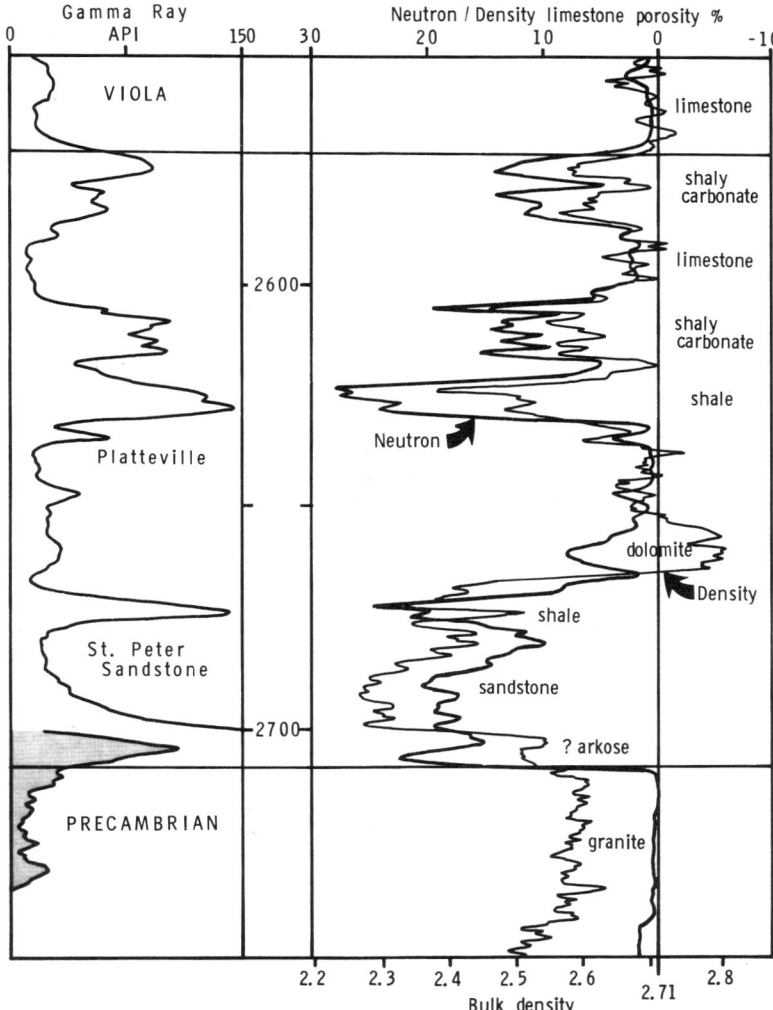

FIGURE 4. Gamma-ray, neutron, and density logs of Ordovician and Precambrian units in a northern Kansas well. (The section overlaps with Fig. 3.)

density, and calcite transit time on the sonic log), lithologic character becomes apparent if appropriate allowance is made for the differing degrees of "stretch" that should be applied to the raw unit scales.

As an example of the practical use of this procedure, gamma-ray, neutron, and density logs are shown on the preceding pages (Figs. 2, 3, 4) for a lower Paleozoic sequence encountered in a well located in Nemaha County, Kansas. The 500-foot section penetrates the lower part of the Hunton Group (Silurian), Maquoketa Shale, Viola Limestone, and Simpson Group (Ordovician), which overlies Pre-Cambrian granite. Following the general principles outlined, lithologies may be assigned within these units through comparison of the three logs.

The procedure is unabashedly a process of simple pattern recognition, and is limited to a basic recognition of gross lithological variation. Composite mixtures such as limy dolomites, dolomitic limestones, and cherty limestones are sometimes subtle to perceive, and are more easily resolved in crossplots of the zone porosity values.

POROSITY LOG FREQUENCY CROSSPLOTS

Crossplots may be made of the porosity log readings of zones recorded in their raw units as simple graphs, whose vertical and horizontal axes correspond to two porosity log scales. Since the theoretical values of pure sandstones, limestones, and dolomites can be computed for any range of porosity, these trends can be drawn on the appropriate crossplot graph to act as boundary endmembers. The location of zone coordinates with respect to these endmembers can be used to identify the zone compositions, both in terms of mineralogy and true porosity, either qualitatively or quantitatively by interpolation.

Examples of neutron-density crossplots for the lower Hunton, Viola, and St. Peter Sandstone (Simpson Group) in the Nemaha well section are shown on the following pages (Figs. 5–7). The plotted zones correspond to readings digitized at 1-foot intervals. Neutron-sonic and sonic-density crossplots are also included for the Viola. (Figs. 8, 9)

Although these crossplots add a degree of definition that is lacking in basic porosity log overlays, there is still a measure of ambiguity concerning the precise mineralogical composition of many zones. As an example, the neutron-density crossplot for the Viola indicates zones of limestone, dolomitic limestone, and cherty

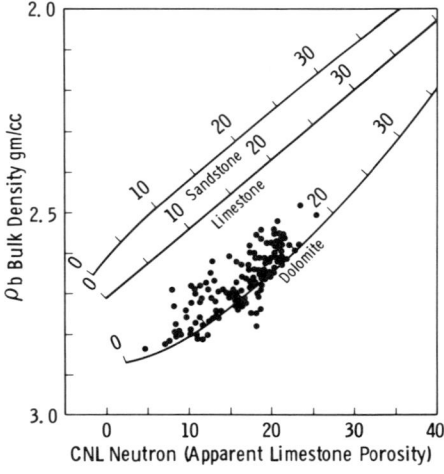

FIGURE 5. Neutron–density crossplot of zones in the Chimneyhill Dolomite (Hunton).

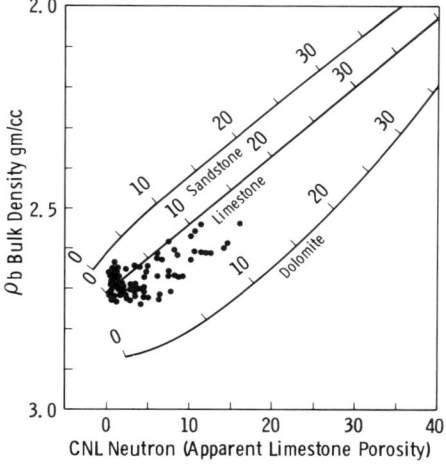

FIGURE 6. Neutron–density crossplot of zones in the Viola Limestone.

limestone. However, there is the possibility that many of the dolomitic limestones may be cherty in part, or even cherty dolomites. Some evidence for this is suggested by the neutron-sonic and sonic-density crossplots for the Viola. The cause of this ambiguity is the relationship between the number of "knowns" which are supplied (the porosity logs) and the number of "unknowns" to be resolved (porosity plus mineral components). Two logs can be analytically solved in terms of three unknowns (since the unknowns collectively constitute a closed system). If pore volume is an unknown, a unique solution in terms of only two minerals can be made. However, three minerals may only be independently evaluated if all three porosity logs are considered simultaneously. The most commonly used methods for this purpose are the M–N crossplot or a matrix algebra solution (both described in later sections).

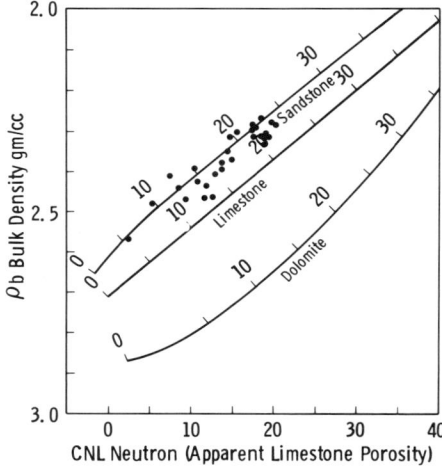

FIGURE 7. Neutron–density crossplot of zones in the St. Peter Sandstone.

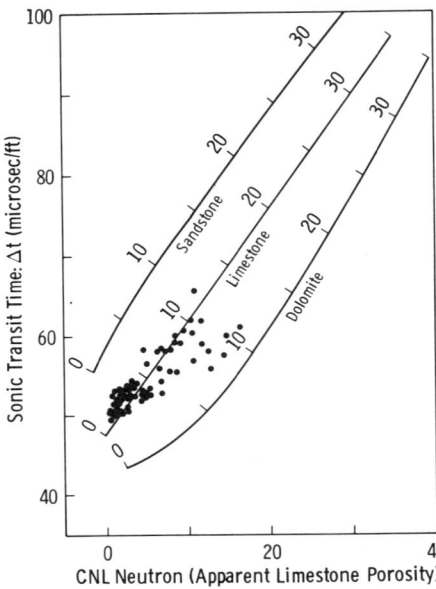

FIGURE 8. Sonic–neutron crossplot of zones in the Viola Limestone.

The location of a large number of zones on porosity crossplots is a tedious operation when done by hand and is greatly expedited if done by simple computer programs. Analyses of this type are made increasingly with the aid of a computer which draws on digitized data, either from field tapes or digitized records of log blueprints. On a line printer graph, many crossplots would show an undistinguished cloud of points, each of which would represent a variable number of similar zones.

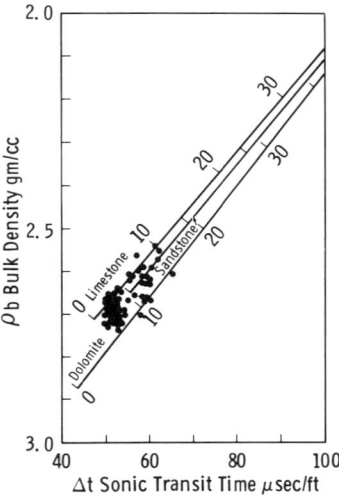

FIGURE 9. Density–sonic crossplot of zones in the Viola Limestone.

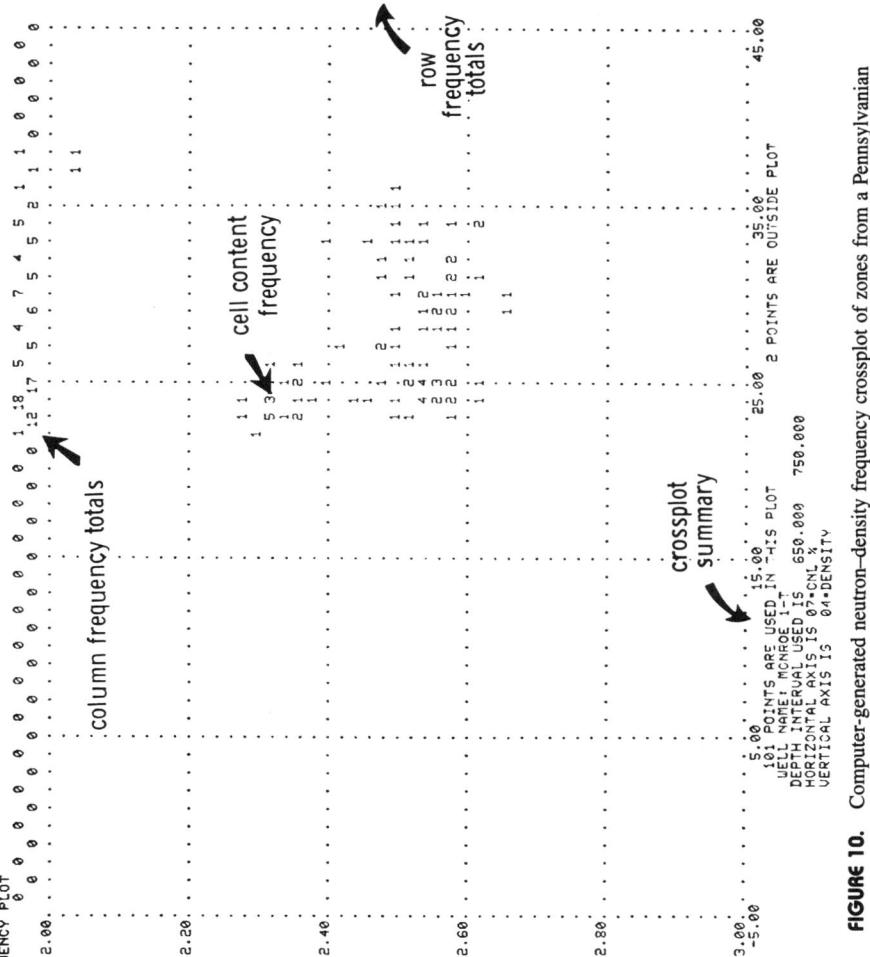

FIGURE 10. Computer-generated neutron–density frequency crossplot of zones from a Pennsylvanian shaly sandstone.

As an aid in interpretation, the number of zones which occur at each point is generally printed rather than an anonymous symbol. This type of crossplot shows the relative densities of zones across the graph field and is known as a "frequency crossplot" (Fig. 10).

A THIRD DIMENSION: Z-PLOTS*

Frequency crossplots are easiest to interpret for shale-free sedimentary lithologies and minerals. If appreciable proportions of shale occur in some or all of the zones, allowance should be made for the introduction of an additional component, namely shale. As noted earlier, there are no distinctive properties for shale that allow it to be located as a unique reference point on the crossplot graph. Instead, a shale point must be defined for the analytical section that best typifies the porosity readings of shale zones. This may be done from observations taken from the raw logs themselves, or by using the variation of the crossplotted zones to define its location.

The Z-plot is a useful variant of the basic crossplot that allows a third variable (Z) to be added to the two variables which correspond to the two axes of the graph. The most common choice for Z is a shale indicator (usually the gamma-ray log). For any zone, the shale content is translated into an integer scale, which ranges from zero for clean zones to 9 for shale zones. Crossplotting is done in the normal manner, but the symbol printed is the appropriate Z value. The resulting crossplot is a two-dimensional mapping of a three-dimensional situation, in which the plotted values represent relative heights above the plane, and the vertical axis is the Z variable.

As an example, a Z-plot is shown for digitized readings in the Maquoketa Shale (Fig. 11), where the Z values correspond to integer-scaled gamma-ray readings.

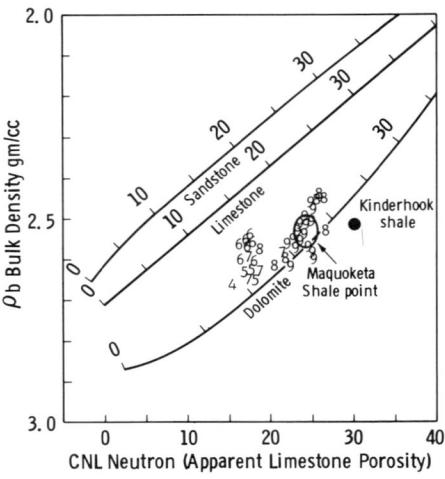

FIGURE 11. Gamma-ray Z–neutron–density crossplot of zones from the Maquoketa Shale.

* Trademark of Schlumberger.

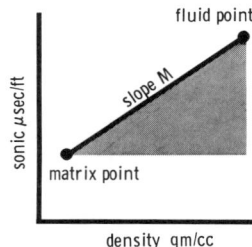
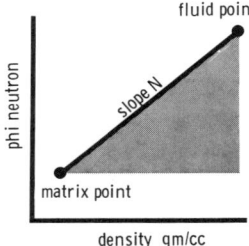

FIGURE 12. Geometry of M and N values as slopes of lines in the neutron–density–sonic triaxial space system.

Without the Z values, the crossplot would suggest the Maquoketa to be a dolomite. However, the gamma-ray values correctly typify the unit as a shale with some zones of shaly carbonate (? limestone). A shale point has been located for the Kinderhook Shale in the same well and indicates that the Maquoketa Shale has an appreciable content of mineralogical impurities, if minor compactional differences between the two shales are neglected. This observation is corroborated by petrographic descriptions of the Maquoketa which generally identify the unit as a dolomitic shale.

CONDENSED PROJECTION PLOTS

When three porosity logs are available, the ideal crossplot would incorporate all three simultaneously. This could be achieved by a Z-plot, using the third porosity log as a vertical axis for zone values. However, the result would be very difficult to interpret requiring, as it does, the simultaneous evaluation of variation in three distinctly different types of measurement unit. The M–N plot* was devised by Burke, Campbell, and Schmidt (1969) as an ingenious solution to this three-dimensional problem that reduces it to two dimensions by the elimination of one of the unknowns. The M–N plot is directed to the task of identifying the mineral unknowns in the zones and estimating their relative proportions. Viewed in this light, the porosity component becomes a redundant unknown. Composite variables, M and N, may be computed, which, by taking pairs of porosity values at a time, eliminate the contribution made by the pore volume component.

The rationale that underlies the calculation of M and N can be easily understood when referring to the diagram on Figure 12. The quantity M incorporates the sonic and density logs

---

* Trademark of Schlumberger.

$$M = \frac{\Delta t_f - \Delta t_{ma}}{\rho_{ma} - \rho_f} \times 0.01$$

where M represents an estimate of the slope of the line linking the matrix and fluid transit times and densities. If M is computed for a sandstone (using appropriate values for quartz and pore fluid, typically freshwater), it will be found that the computation of M for an unknown zone, that is actually a sandstone with any porosity, by the formula

$$M = \frac{\Delta t_f - \Delta t}{\rho_b - \rho_f} \times 0.01$$

where Δt = transit time of the zone

ρ_b = bulk density of the zone

will give substantially the same value of M. The reason for this is that the transit time and density variations with porosity are effectively linear, with the result that the slope value M is constant over the entire range of porosity variation.

Similar arguments can be developed in the computation of the slope value N for a neutron-density combination. Constant values for N for any given matrix mineral will not be perfectly realized, however, due to the nonlinearity of the neutron response, particularly at low-porosity readings. The calculation is given by the equation

$$N = \frac{1 - \phi_n}{\rho_b - \rho_f}$$

The procedure outlined enables the computation of M and N values for a variety of matrix minerals to be computed prior to log analysis and located on a graph of M versus N. (The values of M are conventionally divided by 100 to give units of similar order of magnitude to those of N.) M values result in constants; N values show slight variations in value which are dictated by the nonlinearity of the neutron response of the mineral in question. The technique is not restricted to minerals which constitute the framework of porous lithologies, but can be extended to "non-porous" minerals such as salt and anhydrite. In these cases, the M and N values express the slopes of hypothetical lines linking the mineral with a porosity fluid component. An M–N plot is shown on the following page and illustrates the location for commonly occurring minerals (Fig. 13). In this chart, the neutron log used is the SNP (different N values must be used for the CNL). A critical factor in the computation of these constants is an appropriate selection of fluid characteristics. Since all the porosity tools have a relatively restricted radius of investigation, this fluid will generally correspond to mud filtrate (either fresh or salt, as dictated by the nature of the drilling mud).

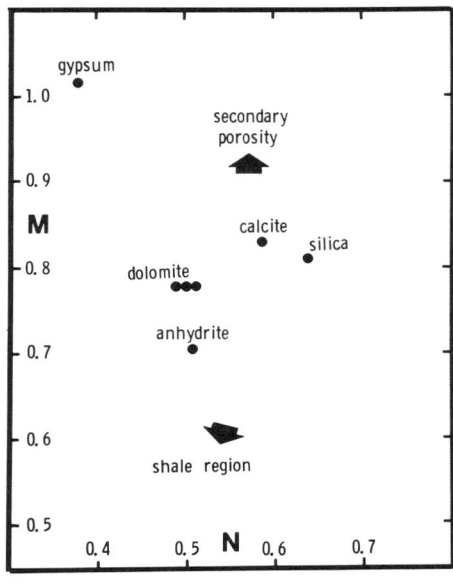

FIGURE 13. Simplified M–N plot, showing locations of common minerals. (The plot is computed for a SNP neutron log and a fresh mud filtrate.)

Having located the theoretical positions of an appropriate suite of minerals on the M–N plot to act as reference points, M and N values may be calculated from each zone porosity reading and located on the plot. The disposition of the zones gives immediate indications of their matrix mineralogies. For any given zone, the composition may be resolved in terms of three minerals as an upper limit to the number of unknowns that may be accommodated uniquely. Again, three porosity logs restrict analysis to four unknowns (including the suppressed unknown, porosity). In zones which have appreciable residual gas saturations, expressively large values of M and N will result, and so will be immediately apparent.

An example of the practical application of the M–N plot is shown for a short sequence within the Viola of the Nemaha well (Fig. 14). The plot succeeds in resolving some of the ambiguities raised by the porosity crossplots of the Viola. The lower part of the sequence is identified as cherty limestone which grades progressively upward into dolomitic, cherty limestone, and cherty dolomite. The relative proportions of these components may be calculated in terms of the composition triangle defined by the calcite, quartz, and dolomite endpoints.

Since the quantities M and N are artificial compound variables of the three porosity logs, it is often difficult to visualize the spatial "meaning" of an M–N plot, which would be useful in the analysis of problematical zones. The M–N plot is a simple example for a "conical projection." If a set of log responses are plotted as points in a three-dimensional space whose orthogonal axes are the three porosity logs, the M–N plot is equivalent to a projection that would be seen by an eye placed at the fluid point. The fluid point eye would perceive porous cherty carbonates within the tetrahedron to be projected as points on the basal quartz–calcite–dolomite triangle, and is a spatial realization of the M–N plot convention.

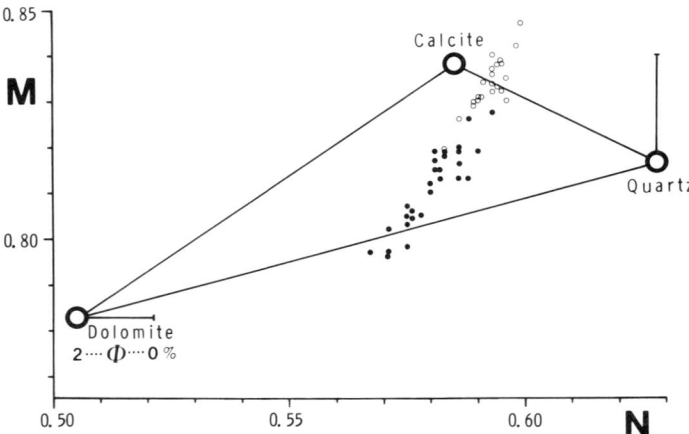

FIGURE 14. Detailed *M–N* plot of zones in the lower part of the Viola Limestone in a northern Kansas well. The zones are drawn from the logged section of Figure 3.

As another example of an application of the *M–N* plot, Bartok et al. (1981) found it useful in distinctions of both lithologies and facies. In a detailed study of the Cogollo Group (Lower Cretaceous) of Venezuela, they were able to recognize depositional environment subdivisions from analysis of core and outcrop. The *M–N* plot was effective on the distinction of glauconitic zones from shaly layers, on the basis of greater *N* values. The shaly beds had lower *N* values, because of their higher neutron porosities caused by kaolinite and chlorite contents. However, the presence of accessory heavy minerals or complex lithological associations were found to make interpretations complex and ambiguous. Some discrimination was observed between depositional facies of grainstone–packstone bars, interbar deposits, and pelecypod biostromes, plotted as fields on an *M–N* plot (Fig. 15).

The composite variables, *M* and *N*, are two of the three possible combinations that can be derived from the sonic, density, and neutron logs, taken two at a time. A third composite variable, designated very reasonably as *P*, is given by the equation

$$P = \frac{100 - \phi_n}{\Delta t_f - \Delta t}$$

This quantity was introduced by Roberts and Campbell (1976) for application to the distinction of micas from clay contents in Jurassic sandstones of the North Sea. The discrimination has been found to be most effective on crossplots of *P* versus gamma ray (Fig. 16).

Although *M*, *N*, and *P* are quantities which are very easy to compute, their interpretation in subsurface studies is not as straightforward as simpler crossplots. The values of *M* and *N* vary slightly as a function of mud filtrate salinity in the flushed zone, since the mud filtrate dictates the position of the fluid point used for the crossplot projection. The location of mineral points on the N axis is also complicated by the nonlinearity of the neutron porosity response. The curvature of the neutron lithology lines for dolomite and (to a minor extent) sandstone results in a range of possible values which represent projections of tangents. Finally, and possibly most important, computed values of M and N are abstract quantities which have no immediate physical interpretation.

In view of these problems Clavier and Rust (1976) proposed an alternative method of projection which they termed the *MID* plot.* The rationale for the *MID* plot is no different from the M–N plot but the methodology attempts to eliminate the drawbacks of the earlier technique. Rather than deal with abstract projection values, the *MID*-plot technique isolates apparent density and apparent sonic travel time as parameters which are independent of porosity. The computation of apparent density

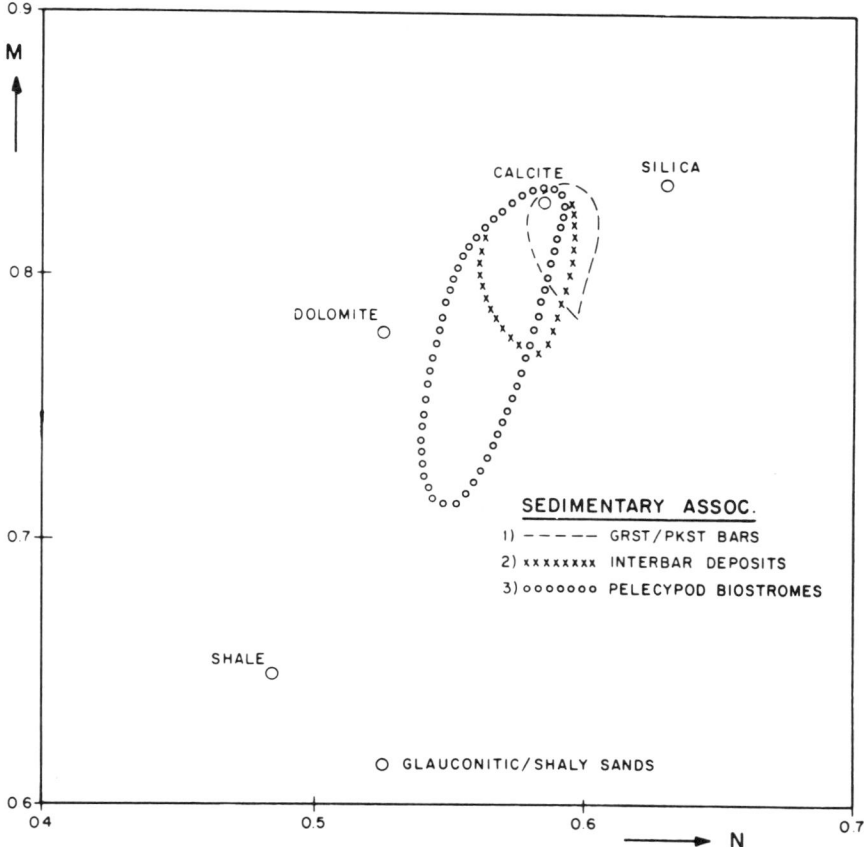

FIGURE 15. Model M–N plot of depositional facies fields from the Cogollo Group (Lower Cretaceous) of Venezuela. From Bartok and others (1981).

* Trademark of Schlumberger.

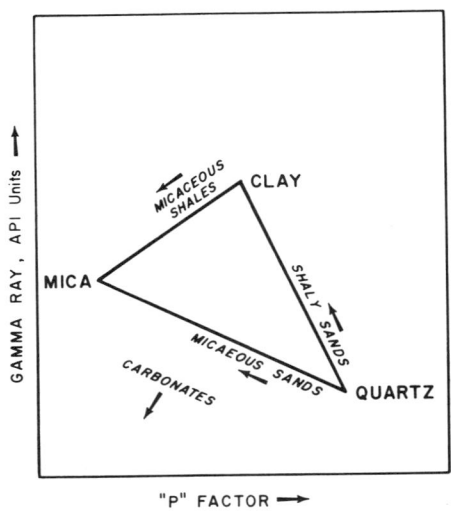

FIGURE 16. Model gamma-ray–*P* plot applicable to Jurassic sandstones in the North Sea. From Hodson and others (1976).

can be understood by reference to Figure 17. On a crossplot of bulk density and neutron porosity, apparent density isoliths are located by interpolation between sandstone, limestone, and dolomite lines and extrapolation beyond their limits. On the three fundamental lithology lines, the apparent densities have real meaning as actual densities of the matrix minerals quartz, calcite, and dolomite. For other minerals, the apparent densities are artificial quantities, but are not grossly dissimilar from actual values. The same methodology is applied in the contouring of apparent sonic travel times on a crossplot of sonic transit time versus neutron porosity.

In practice, the apparent densities and transit times for any logged zone can be estimated from interpolation on *MID*-plot charts published in recent editions of service company chartbooks. They may also be calculated by a computer program, although not with the immediate ease that is the case for *M* and *N* values. Lithology identification of mineral composition estimation is then made by crossplotting zone values on a *MID*-plot grid (Fig. 18). The general disposition of mineral reference points on the grid is not markedly dissimilar to those on the *M–N* plot. Minor differences result in some improvement in gas detection and the resolution of anhydrite mixtures (Clavier and Rust, 1976). Time will tell whether the *MID* plot supercedes the *M–N* plot as the preferred graphical method for lithological analysis from multiple porosity log combinations. Since they both have the same geometrical principle as their basic rationale, the choice between them is a matter of personal preference.

ROMA–UMA CROSSPLOTS

The recent introduction of the lithodensity* tool has added a new measurement, the photoelectric absorption cross-section index, Pe, which shows excellent dis-

---

* Trademark of Schlumberger.

crimination between minerals. The index is determined by the number of electrons per atom, which corresponds to the atomic number. When the index is multiplied by the electron density, the resultant photoelectric macrosection, U, is related volumetrically to mineral composition, with minimal influence by pore fluids.

For any zone in a stratigraphic section an apparent matrix photoelectric macrosection (UMA) may be calculated, together with an apparent grain density ($ROMA$) from bulk density and neutron porosity readings. Location of these coordinates on a $ROMA$–UMA lithology crossplot (Fig. 19) allows identification and volumetric estimation of mineral species, using concepts similar to the M–N and MID plots.

The photoelectric measurement is particularly valuable in that absorption values show a great range between minerals, and are an order of magnitude greater than those for pore fluids of either water or hydrocarbons. Their use in conjunction with apparent matrix density estimates shows improved performance over crossplots which utilize the sonic log, where the broad relationship between matrix transit times and densities can inhibit effective discrimination. The theory and applications of this new crossplot are outlined by McCall and Gardner (1982), while useful examples are given by Suau and Spurlin (1982).

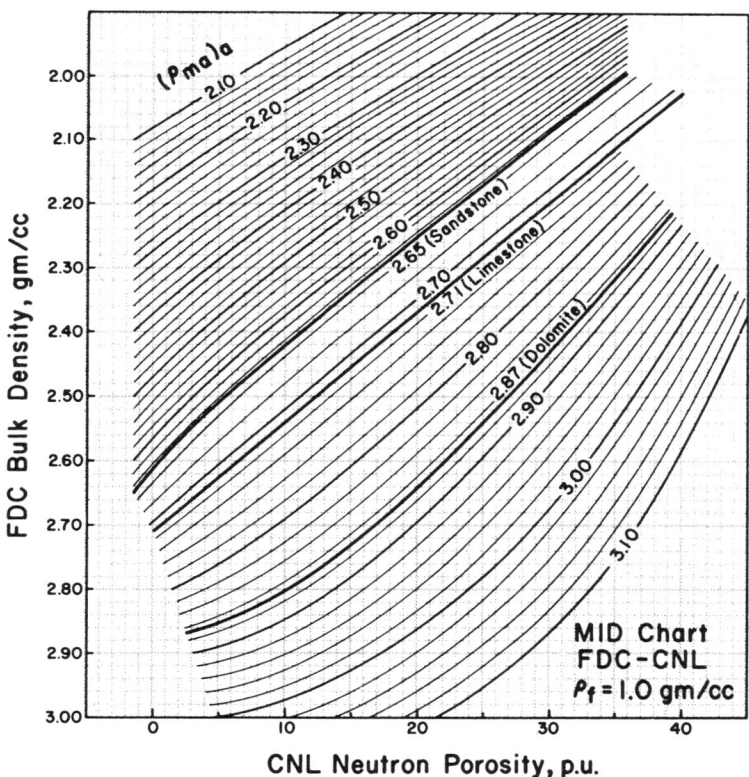

FIGURE 17. Apparent grain density isolines on neutron–density crossplot. From Clavier and Rust (1976).

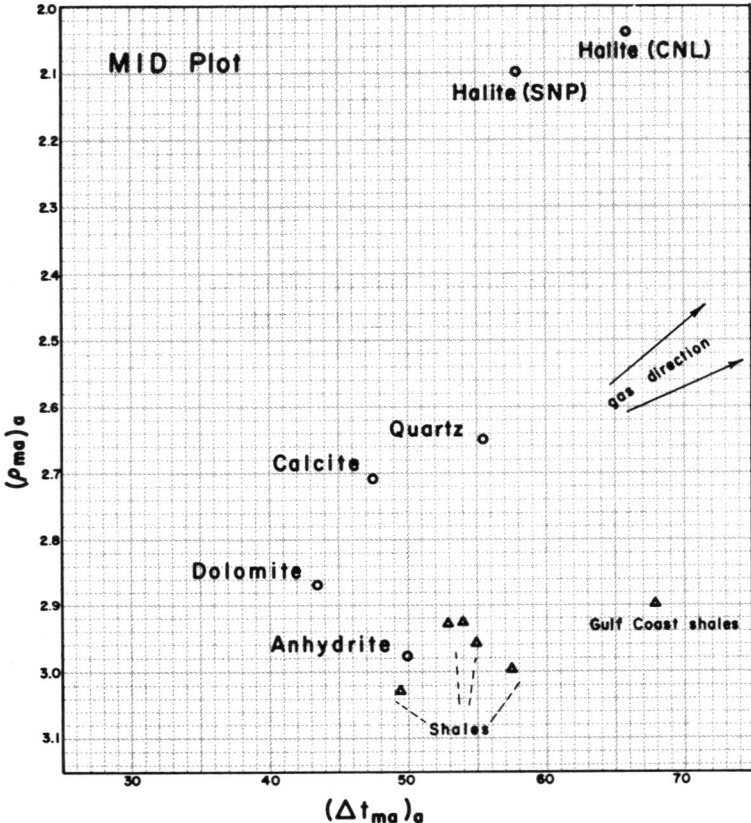

FIGURE 18. *MID* plot grid of apparent grain density and apparent matrix travel time for use in mineral identification. From Clavier and Rust (1976).

GRAPHICAL METHODS FOR SHALY SANDSTONES

The analysis of sequences with significant contents of shale is made complex by the wide range possible for clay mineral composition, nature and volumetric content of the silt-sized fraction, and degree of compaction. This somewhat intransigent character is in stark contrast with simple rock-building minerals, such as quartz or calcite, which have predefined physical properties constrained within relatively narrow ranges of variation. The log analysis of shales and shaly intervals then becomes a two-stage procedure. In the initial phase, shale properties which typify the studied section are identified from the logs themselves. On a second pass through the data, these representative endmember values are used to calculate volumetric estimates of shale content. Crossplots are invaluable in the reconnaissance phase. Their graphic nature is an ideal medium for the pattern recognition skills of the log analyst and geologist. The variability of shales within even short sections requires intelligent decisions concerning what constitutes representative shale properties. The procedure is extremely difficult (and even unnecessary) to simulate in a mathematical decision algorithm. Further the crossplots contain information on trends of variation which are diagnostic of fabric changes and useful in interpretation of sedimentation and diagenesis patterns.

Shaliness affects the various log responses in terms of its volumetric proportion, physical properties, and the manner that it is distributed in the rock. The variety of geometrical forms in which shale is disseminated within a sandstone has been classified by Schlumberger (1972) in terms of three possible endmembers: laminar, structural, and dispersed shale. Laminar shale units consist of thin laminations of shale which separate stringers or beds of sandstone. The occurrence of these laminations is not accompanied by a reduction in the porosities of the sandstone stringers themselves, but there is an overall reduction of the bulk porosity of the total rock. Structural shale describes sandstones in which some of the grains of the framework are shale fragments or clay mineral diagenetic alteration products. A transition trend from a pure orthoquartzite is not necessarily matched by any reduction in porosity. Dispersed shale consists of clay minerals which develop on grain framework surfaces and occlude pore space. Extensive development of a dispersed shale content leads to a progressive reduction in porosity.

These three shale distribution types form natural endmembers of ranges of variation from a theoretical pure orthoquartzite to more shaly counterparts as demonstrated by the diagram of Figure 20. Although the terminology may be unfamiliar to many geologists, it is relatively easy to assign different categories of authigenic and allogenic clays (based on genesis) to the shale distribution types (based on morphology). The reader can verify this for himself by classifying the authigenic and allogenic

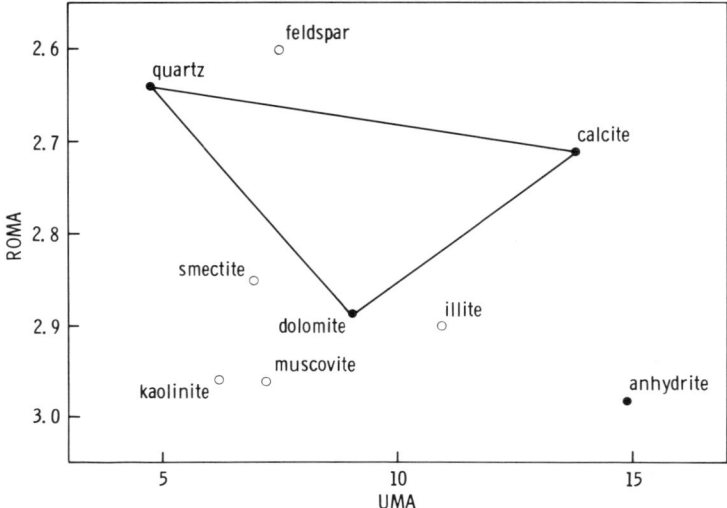

FIGURE 19. *ROMA–UMA* grid of apparent grain density and photoelectric macrosection indexed with common minerals.

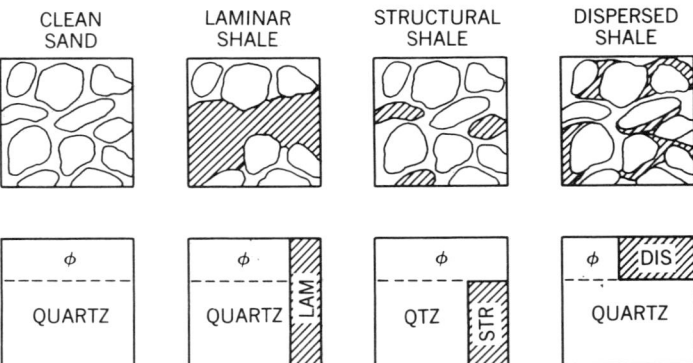

FIGURE 20. Distribution types of shale in shaly sandstones as hypothetical limiting cases of a simple model for log analysis. From Schlumberger publications.

clay types illustrated on Figure 21 in terms of shale types as would be perceived by a logging tool combination.

Shaly sandstones have been a major preoccupation of log analysts for a number of years. By contrast, the widespread distribution of authigenic clays in sandstones has received scant attention in the geological literature until very recently. However, Wilson and Pittman (1977) made an extensive study of clay minerals in 785 siliciclastic sandstones and found that 91.3% contained authigenic clay, while 20% of the remainder contained abundant allogenic clays. A major reason for the recent "discoveries" is the extensive application of the scanning electron microscope. Conventional petrographic methods of thin-section microscopy and examination of sieve fractions tend to overlook clay contents as an important systematic component. Coupled with this problem of observation insensitivity, was the prejudice implicit in older sandstone classifications that regarded all clay minerals as detrital (allogenic) in origin (e.g., Pettijohn, 1957). However, the diagnetic origin of many clays was more fully recognized in the failure to find significant numbers of graywackes with a primary matrix in cores of modern deep sea sediments (Hollister and Heezen, 1964). Graywackes are a major component of many ancient basin sequences, and their clay mineralogy has been interpreted more recently as the product of decomposition of unstable components (Lovell, 1972). These are relatively new ideas concerning the origins and almost ubiquitous occurrence of clay minerals in sandstones. Since most wireline logs are sensitive to clay mineral fractions, there is obviously some potential for novel applications of log combinations in the interpretation of patterns and trends which reflect histories of both sedimentation and diagenesis.

A variety of indicators are used from various logs in the assessment of the proportion of shale in a zone. Most of them follow the same methodology of establishing a value for a clean zone, locating a value that best typifies shales within the interval of interest, and interpolating between these two extremes. The more commonly used shale indicators are the gamma-ray log, SP log, and neutron-density log combination. In sections which have been logged by a variety of tools, separate estimates of the shale content are computed for the shale indicators which are available. It is not uncommon for these estimates to differ from one another to varying degrees, and the lowest estimate is usually taken to be the most valid. This stipulation is made on the grounds that, if anything, a shale indicator will tend to overestimate the shale content. This is not an unreasonable thesis, since accessory components other than shale will generally result in an increased apparent shaliness as recorded by the logging tools.

Although the gamma-ray and SP logs are the most commonly used "shale indicators," a neutron-density log combination is often used as an alternative method to evaluate shale content of shaly sandstones. This approach is particularly useful when disturbing factors hamper the effectiveness of the gamma-ray and SP logs in the discrimination of shale variation. This is the case for many wells in the Tertiary-Recent deltaic sands of the Niger Delta (Poupon et al., 1967). Formation evaluation is complicated by the extremely low-formation water salinities which often vary dramatically between sandstone units. As a result, it is often difficult to establish a consistent static self-potential range for the purposes of shale content assessment.

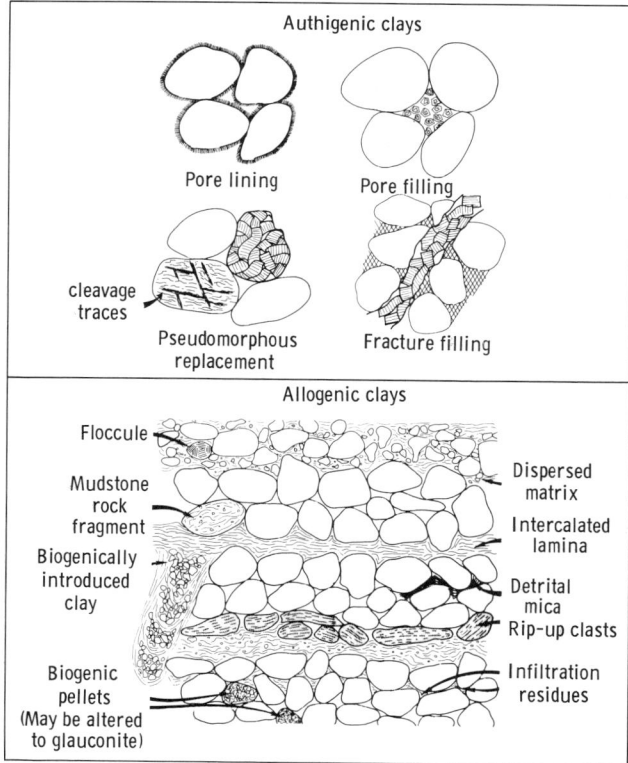

FIGURE 21. Modes of occurrence of authigenic and allogenic clays in sandstones. From Wilson and Pittman (1977).

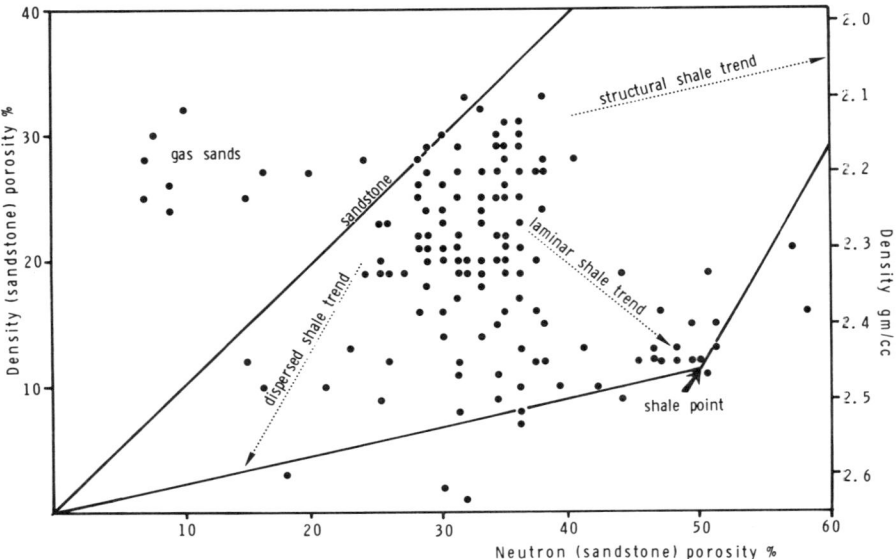

FIGURE 22. Neutron–density crossplot of sandstone and shale zones from the Agboda Formation (Tertiary) in the Niger Delta.

The gamma-ray trace is not always a reliable indicator of shale content due to the occurrence of occasional radioactive concentrations of zircons within some of the sandstones.

However, a crossplot of neutron and density values for zones within the section of interest can be referenced to a composition triangle of quartz, water, and shale. An example of the procedure is shown for a Nigerian well section in the Agboda Formation on Figure 22. A sandstone line may be drawn, which links the log responses of quartz and water (corresponding to an appropriate mud filtrate). A shale point is located at the more extreme values of the plotted zones, in such a manner as to complete the composition triangle by straight line segments which link it to the quartz and water endpoints. Following the precepts of the log analyst model of shale composition, trends may be drawn which correspond to structural shale (isoporosity), dispersed shale (isoquartz), and laminar shale. Referenced to these trends, the disposition of the zones suggests that the shale element of the shaly sandstones is disseminated as a mixture of pore-filling clays and shale laminations. Gas zones are immediately recognized as points which plot to the north–west of the clean sandstone line, due to their high density and low neutron porosities. The crossplot can be used as an initial reconnaissance evaluation that precedes the quantitative solution of porosity and shale content by a matrix algebra solution.

In many sandstones, the use of the gamma-ray log as a shale indicator can be too simplistic if mica content is an appreciable component. The limitations of both the gamma-ray and neutron-density combination as shale indicators have been recognized recently in the evaluation of North Sea sections, where highly micaceous sandstones are common in the Jurassic. New methods have been proposed which coordinate the gamma-ray log with porosity log combinations to differentiate clays and micas as distinctive components.

An example of the general procedure is illustrated with reference to a Cherokee (Pennsylvanian) section in a well located in the Selma field of Anderson County, Kansas. Gamma-ray, neutron, and density logs are shown of the Skinner Sandstone and its overlying silty shales and shales (Fig. 23). The sandstone was cored and its sequence of bedding structures and stratigraphic context suggested its depositional origin as an alluvial channel fill. For purposes of identification on the log profiles and crossplots, the channel sandstone is indicated by squares, the overlying overbank silty shales by triangles, and the open marine shales above a coal bed by circles.

A neutron-density crossplot of digitized zones suggests relatively high shale contents within the Burbank (Fig. 24). However, much of this ''shale'' can be attributed to the high mica content which is immediately apparent in examination of cores from the section. The differentiation of muscovite content from clay minerals is aided by crossplotting the zone values of N (the composite neutron-density variable) and G (the gamma ray standardized by multiplying by bulk density) (Fig. 25). A quartz point may be located from its known log response properties and linked with a ''shale point,'' corresponding to the more extreme shale values. Since zones consisting entirely of mica do not occur in the section, a hypothetical ''mica point'' has been located, using the North Sea values as an analogue. These three points define a composition triangle which serves as a broad reference in typing the clay mineral–mica variation within the Burbank. The pragmatic assumption of equivalence of clay mineral log responses within the shales to those within the

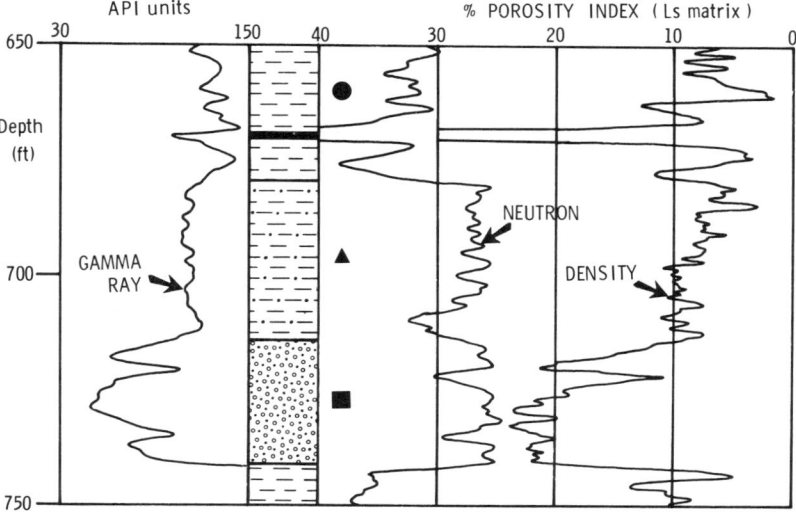

FIGURE 23. Gamma-ray, neutron, and density logs of a Pennsylvanian sandstone–shale sequence in southeast Kansas. Symbol key: squares, channel sandstone; triangles, overbank silty shales; circles, marine shales.

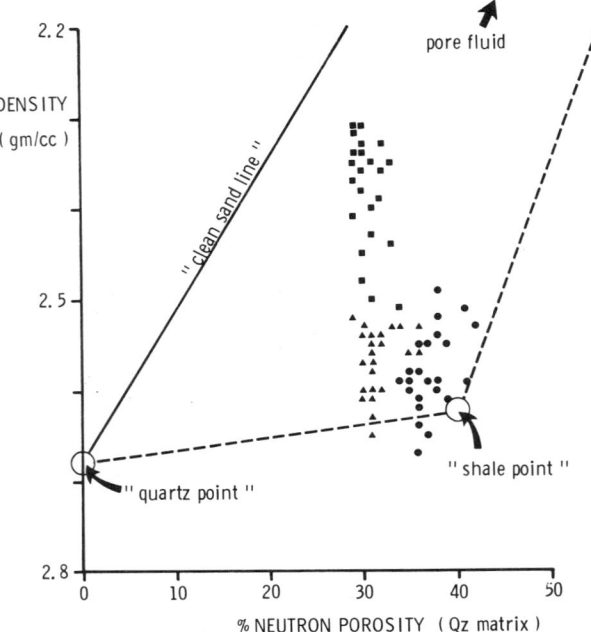

FIGURE 24. Neutron–density crossplot of Pennsylvanian sandstone–shale sequence. Symbols are those used in Figure 23.

sandstone imposes limitations on the method as a qualitative reconnaissance technique, rather than a precise quantitative device for the evaluation of the separate contributions of clay minerals and micas.

The application of log combinations to the identification and estimation of components in mineralogically complex sandstones clearly has its frustrating aspects. In more recent log studies, emphasis has moved to the consideration of results from the natural gamma-ray spectroscopy tool and the lithodensity tool. These offer great promise in the distinction of clay mineral species as well as recognition of accessory minerals such as felspars and micas. Peveraro and Russell (1984) made a detailed study of log responses from these tools in Middle Jurassic sandstones of the North Sea and compared crossplot results with volumetric determinations from X-ray diffraction analysis of core samples. Thorium–potassium crossplots were found to be the most useful, with reasonable discrimination of potassium felspars and micas, and distinction from clay minerals. *ROMA–UMA* plots were used to recognize heavy minerals, such as biotite, siderite, and pyrite. It was found that, with the exception of zones with high concentrations of heavy minerals, scattered occurrences tended to merge into an apparent generalized "shale trend." However, the plot did show a gratifying sensitivity to zones cemented by calcite. The authors concluded that the use of log crossplots showed a "limited but useful capacity for downhole mineral identification."

CROSSPLOTS AS TOOLS OF GEOMETRY

Crossplots can be considered from the narrow viewpoint as simply pictorial presentations of log data. Taking the broader view, the properties of projective geometry allow crossplot techniques to be devised as problem-solving procedures. In any given subsurface study, appropriate logs may be selected and their combinations projected in a manner which best accentuates or discriminates trends and patterns which are of interest to the analyst. Used in this sense, geometry is applied as the analytical medium that found so much favor with the ancient Greeks. In the analysis of log data, crossplots are a natural precursor to numerical analysis of compositional variation by an algebraic methods described in the next chapter. This preliminary phase is necessary to relate the complexity of subsurface variations to the basic framework of a descriptive model with geologically meaningful properties.

The simple frequency crossplots of raw logs represent orthogonal projections of data variation. By contrast, M–N and MID plots are oblique projections in three dimensions. Rather than view the data clouds from a hypothetical infinite viewpoint,

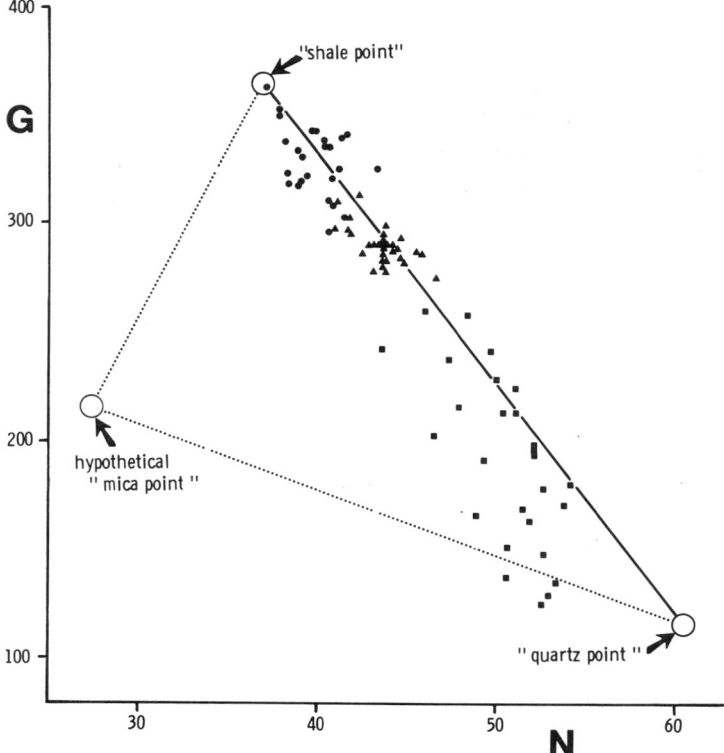

FIGURE 25. Standardized gamma-ray–N crossplot of Pennsylvanian sandstone–shale sequence. Symbols are those used in Figure 23.

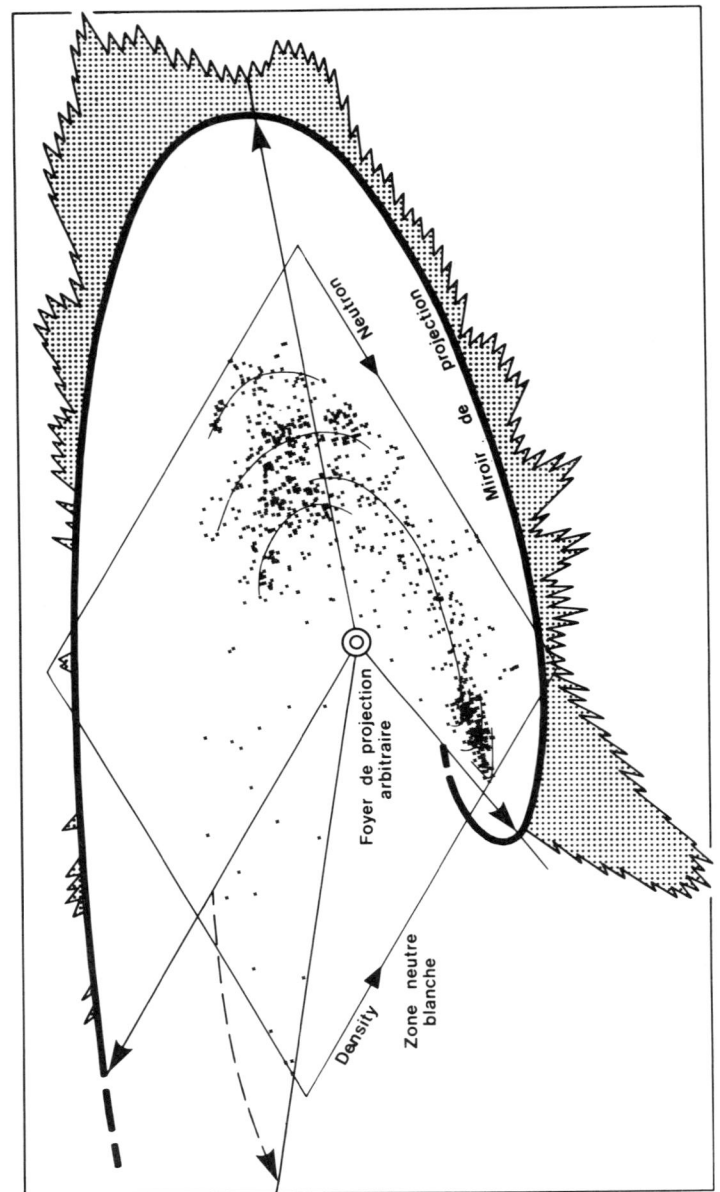

FIGURE 26. Spiral projection of neutron–density paired values from a sandstone–shale sequence. From Maurin and Riguidel (1978).

they are conical projections in which the observer's eye is moved to the fluid point. The fluid point is a useful focus, since the projection causes an effective suppression of porosity variation and so is particularly useful in discriminating mineralogy changes in the common reservoir rocks.

Alternative projections and focal points can be chosen to meet the requirements of different types of studies. For example, Maurin and Riguidel (1978) describe a spectacular instance of the use of a spiral projection centered on an arbitrary focal point of a neutron-density crossplot (Fig. 26). The crossplotted log data were drawn from a sandstone–shale sequence and the projection selected to emphasize variations associated with the repetitive sequence. The input bivariate data are condensed to a single variable expressed as the projection angle onto the encompassing spiral envelope. The authors suggest that the condensation might prove especially useful in the production of "abstract logs" which would aid in the analysis of potential rhythmic sequences.

Application of this concept in a modified form is shown by the following example. In work with the Lansing–Kansas City Groups (Pennsylvanian) of western Kansas, Watney (1980) was able to recognize a systematic cyclic development of lithofacies from core sequences. This was matched by a sequence of distinctive log signatures on both gamma-ray and neutron logs. Each cyclic unit consists of a lower transgressive limestone (absent locally), succeeded by a marine shale and upper regressive limestone, terminated by an upper shale of continental clastic facies. Gamma ray and neutron logs are shown in Figure 27 of the J and K cyclic units. A crossplot of the gamma-

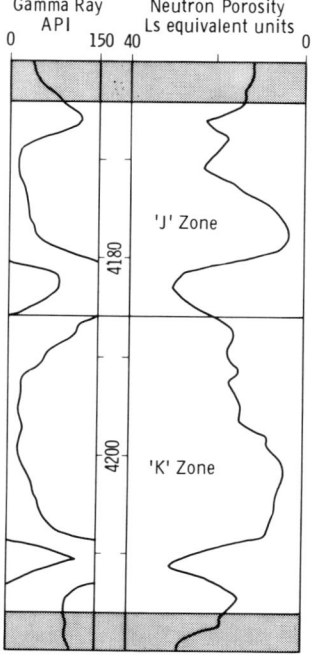

FIGURE 27. Gamma-ray and neutron logs of J and K zones, Kansas City Group (Pennsylvanian) from a central Kansas well.

ray and neutron log readings from the *J* zone is shown in Figure 28, together with their radial projection. The focus of the projection was chosen at a location which discriminated facies subdivisions in a useful manner. The projection reduces the two logs to a single variable which was scaled as azimuth angle. The cyclic form of the projection shows a remarkable concordance with the cyclic model of deposition, but this is a fortuitous circumstance. The azimuths were plotted as a function of depth as a realization of an abstract log for both the *J* zone and the underlying *K* zone (Fig. 29). Although the *J*-zone crossplot was used as the basis for design of the projection, the abstract log clearly shows the cyclic repetitive character in both zones.

The underlying moral of the preceding discussion is that geometrical projection theory is the foundation for useful crossplot conventions whose design can be set by the user. This freedom allows graphic transformations of raw log measurements into a variety of reduced compound variables which can have specialized meanings for different geological contexts. In the examples of this chapter, geometrical methods match either the commonly used conventions or are controlled by a user's choice of focal point and projection type. The operations involved are straightforward to execute and present little difficulty to visualize, when dealing with combinations of either two or three logs.

Graphic displays of projections from larger suites of logs present obvious difficulties, because of the location of log data points in an axis space of many dimensions. However, the numerical methods of principal components and discriminant functions (described in Chapter 7) are useful aids. These techniques locate orthogonal axes in multidimensional space which match the major sources of variation expressed

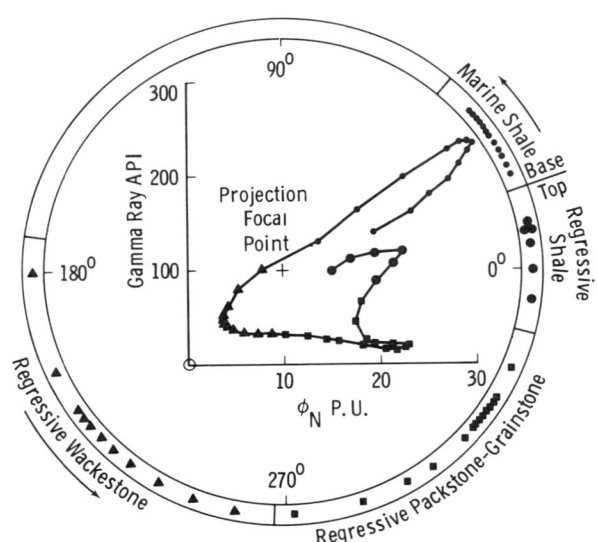

FIGURE 28. Gamma-ray–neutron crossplot of readings from the *J* zone together with their radial projection referenced to an "azimuth" scale.

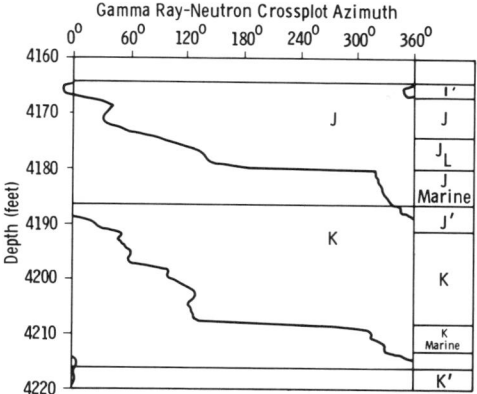

FIGURE 29. Gamma-ray–neutron crossplot azimuth log of *J* and *K* zones.

by the cloud of data points. In this way, they condense the total variation to a smaller number of descriptive composite variables which coincide with a new framework of axes. The projection of the data cloud onto the axes are computed as scores. These can be crossplotted to give views of the logging data in terms of its cumulative natural variability. So, although the transformations are computed from numerical estimates of statistical parameters, their realization is strictly a matter of geometry.

REFERENCES

Bartok, P., Reijers, T. J. A., and Juhasz, I., 1981, Lower Cretaceous Cogollo Group, Maracaibo Basin, Venezuela: Sedimentology, diagenesis, and petrophysics, *Am. Assoc. Petrol. Geolog. Bull.*, Vol. 65, No. 6, pp. 1110–1134.

Burke, J. A., Campbell, R. L., Jr., and Schmidt, A. W., 1969, The lithoporosity crossplot, *Trans. Soc. Prof. Well Log Analysts 10th Ann. Logging Symp.*, Paper Y, 29 pp.

Clavier, C., and Rust, D. H., 1976, MID plot: A new lithology technique, *The Log Analyst*, Vol. XVII, No. 6, pp. 16–24.

Fertl, W. H., 1979, Lithology, other effects on porosity logs, *Oil Gas J.*, March, pp. 68–70.

Hollister, C. D., and Heezen, B. C., 1964, Modern graywacke-type sands, *Science*, Vol. 146, pp. 1523–1574.

Lovell, J. P. B., 1972, Diagenetic origin of graywacke matrix minerals: a discussion, *Sedimentology*, Vol. 19, pp. 141–143.

Maurin, A. F., and Riguidel, M. J., 1978, Éléments de morphologie généralisée, *Total Notes et Memoires*, No. 14, 134 pp.

McCall, D. C., and Gardner, J. S., 1982, Lithodensity log applications in the Michigan and Illinois basins, *Trans. Soc. Prof. Well Log Analysts Logging Symp.*, Paper C, 21 pp.

Pettijohn, F. J., 1957, *Sedimentary Rocks* (2nd ed.) Harper, New York, 718 pp.

Peveraro, R. C. A., and Russell, K. J., 1985, Interpretation of wireline log and core data from a mid-Jurassic sand/shale sequence, *Clay Minerals*, in press.

Poupon, A., Strecker, J., and Gartner, J., 1967, A review of log interpretation methods used in the Niger delta, *Trans. Soc. Prof. Well Log Analysts 8th Ann. Logging Symp.*, Paper Z, 53 pp.

Roberts, H. V., and Campbell, R. L., Jr., 1976, The application of CORIBAND to the micaceous Jurassic sandstones of the northern North Sea Basin: *The Log Analyst*, Vol. XVII, No. 1, pp. 33–40.

Schlumberger, 1972, *Log Interpretation: Volume 1—Principles*, Schlumberger Ltd., New York, 113 pp.

Suau, J., and Spurlin, J., 1982, Interpretation of micaceous sandstones in the North Sea, *Trans. Soc. Prof. Well Log Analysts 23rd Ann. Logging Symp.*, Paper G, 32 pp.

Watney, W. L., 1980, Cyclic sedimentation of the Lansing-Kansas City groups in northwestern Kansas and southwestern Nebraska, *Kansas Geol. Survey Bull.*, No. 220, 72 pp.

Wilson, M. D., and Pittman, E. D., 1977, Authigenic clays in sandstones: recognition and influence on reservoir properties and paleoenvironmental analysis: *J. Sed. Pet.* Vol. 47, No. 1, pp. 3–31.

Watney, W. L., 1980, Cyclic sedimentation of the Lansing-Kansas City groups in northwestern Kansas and southwestern Nebraska, *Kansas Geol. Survey Bull.*, No. 220, 72 pp.

Wilson, M. D., and Pittman, E. D., 1977, Authigenic clays in sandstones: recognition and influence on reservoir properties and paleoenvironmental analysis: *J. Sed. Pet.* Vol. 47, No. 1, pp. 3–31.

CHAPTER SIX

NUMERICAL METHODS FOR LITHOLOGY ESTIMATION FROM WELL LOGS

Graphical methods of overlays and crossplots are extremely useful as means of pattern recognition of lithological variation. However, when all is said and done, a graph is a piece of paper whose two dimensions restrict both the number of logs that can be plotted realistically and the number of components that can be extracted. Numerical techniques can handle any number of logs and provide immediate quantitative solutions which are cumbersome to read from graphs. The mathematics for lithofacies analysis from logs may be drawn from simple matrix algebra. From an operational standpoint, matrices are easily stored and manipulated in a computer and most programming languages are structured in terms of a matrix representation. Consequently, lithofacies analysis programs can be written which are cheap and fast to run, even for extensive logged sections.

Before considering the theory behind matrix algebra solutions, a most important point must be made: graphic techniques and matrix representations of log data are one and the same. A log crossplot indexed with mineral reference points is a picture of a matrix algebra system of equations. A matrix algebra solution of component compositions is equivalent to an eyeball compositional estimation of crossplotted zones. Rather than being made redundant by computer methods, crossplots play an integral role in comprehensive analysis. In order to solve for mineral components, they must be explicitly identified (at least provisionally), and crossplots are invaluable for this phase of prognosis. In certain cases, computer analyses may produce apparently nonsensical solutions whose true meaning is most easily deduced from appropriate crossplots as a matter of diagnosis. At the most basic level, the geometrical implications of matrix algebra data provide some means of visualizing the relationships between lithofacies compositional solutions and the original log data. In situations using a variety of logs, the transfer can be somewhat mind boggling, since the geometrical realization is a crossplot in multidimensional space! However, this problem in perception can be tackled through multiple crossplots of logs taken two at a time, or by projections such as the Z-plot or *M–N* plot.

The theory of a simple matrix algebra algorithm is explored in the following example. Its bizarre nature was deliberately chosen to emphasize that the approach is one of general problem solving, rather than a technique restricted to log analysis. By implication, it follows that the method can be used with any logs that are sensitive to compositional content in any system of components regardless of their origin.

THE BASICS OF COMPOSITIONAL ANALYSIS USING MATRIX ALGEBRA

Among humankind's noblest endeavors must be included the Quest for the Holy Grail, the Search for the Lost Chord, and the Mixing of the Perfect Martini. Since its apocryphal origin in a San Francisco bar in the 1860s, the Perfect Martini has been the subject of heated and acrimonious debate. The ingredients of the classic Perfect Martini are generally agreed on; namely, dry gin, dry vermouth, and sweet vermouth. Controversy rages around the question of the relative proportions of these ingredients. Suppose that after prolonged prayer, mediation, and mixing of the three spirits, you had finally hit on the Perfect Martini. However, by a tragic oversight, you had not been keeping track of the relative quantities of gin and the two vermouths. This possibly traumatic experience would be relieved by the realization that your log analysis skills can resolve this problem. You make measurements of the alcohol and sugar contents of your Perfect Martini (corresponding to two log responses). You then measure the alcohol and sugar contents of the gin, dry vermouth, and sweet vermouth (corresponding to log responses of two minerals and a porosity fluid). The results are shown in Figure 1. What is the composition of the Perfect Martini in number of parts gin, dry vermouth, and sweet vermouth?

The problem can be solved graphically by an alcohol–sugar crossplot (Fig. 2). By linking the endmember gin and vermouth coordinates with lines, the resultant

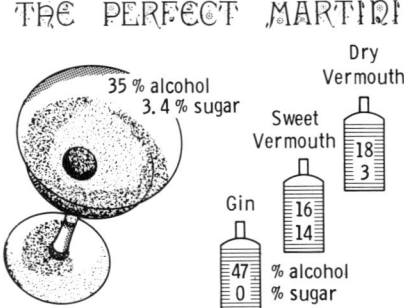

FIGURE 1. The Perfect Martini problem.

The Basics of Compositional Analysis Using Matrix Algebra

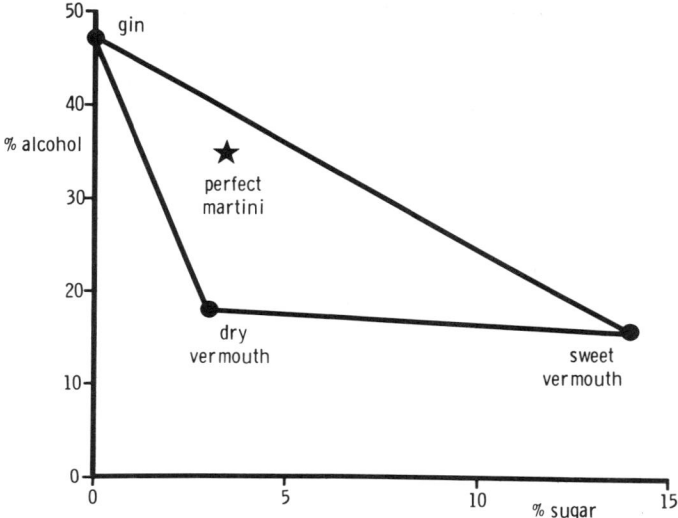

FIGURE 2. Alcohol–sugar crossplot indexed with the endmembers gin, dry vermouth, sweet vermouth, and the Perfect Martini problem "zone."

triangle bounds all possible mixtures of these three components. The composition of the Perfect Martini may then be read by reference of its alcohol–sugar content to the isolines within the triangle.

The numerical solution is contained in the three equations:

Alcohol: $47G + 18D + 16S = 35$
Sugar: $0G + 3D + 14S = 3.4$
Unity: $G + D + S = 1$

where G, D, and S are the proportions of gin, dry vermouth, and sweet vermouth, respectively. Rewritten in matrix form, the equation set becomes

$$\begin{bmatrix} 47 & 18 & 16 \\ 0 & 3 & 14 \\ 1 & 1 & 1 \end{bmatrix} \begin{bmatrix} G \\ D \\ S \end{bmatrix} = \begin{bmatrix} 35 \\ 3.4 \\ 1 \end{bmatrix}$$

which neatly segregates the known data from the unknowns. This cumbersome representation can be rewritten as

$$\mathbf{CV = L}$$

where each letter signifies an array of numbers or unknowns, rather than a single number or unknown, as in conventional algebra. The "knowns" are **C** and **L**; the "unknown" is **V**.

In simple algebra, the equation:

$$cv = l$$

is easily solved for v, since

$$v = \frac{l}{c}$$

However, the equivalent operation of division is not feasible in matrix algebra. An alternative way of solving

$$cv = l$$

is by the operations

$$\left(\frac{1}{c}\right)cv = \left(\frac{1}{c}\right)l$$

$$1 \cdot v = \left(\frac{1}{c}\right)l$$

$$v = \left(\frac{1}{c}\right)l$$

In other words, v is solved by multiplying l by the reciprocal of c.

The conceptual equivalent in matrix algebra of a reciprocal of a number is the inverse of a matrix. This inverse is itself a matrix and, if the original matrix is denoted by **C**, the inverse is written as $\mathbf{C}^{-1}$. The inverse matrix has the basic property expressed in the equation

$$\mathbf{C}^{-1}\mathbf{C} = \mathbf{I}$$

In other words, a matrix multiplied by its inverse results in an identity matrix, **I**, in which all the diagonal entries are "ones" while ths off-diagonal entries are zero.

In the Perfect Martini problem, the **C** matrix is of order 3 × 3, since it has three rows and three columns. The inverse of the **C** matrix, $\mathbf{C}^{-1}$, will also be of order 3 × 3. The identity matrix from their product will be

$$\begin{bmatrix} 1 & 0 & 0 \\ 0 & 1 & 0 \\ 0 & 0 & 1 \end{bmatrix}$$

The identity matrix fulfills the matrix role of "one," since

$$\mathbf{IV} = \mathbf{V}$$

just as

$$1 \cdot v = v$$

The operations of finding the inverse of the matrix **C** and of multiplying matrices can be worked by hand as exercises in masochism, but are far more easily done by a computer program in a few lines of code. The solution to the Perfect Martini problem is then written as

$$\mathbf{CV} = \mathbf{L}$$

$$\mathbf{C^{-1}CV} = \mathbf{C^{-1}L}$$

$$\mathbf{IV} = \mathbf{C^{-1}L}$$

$$\mathbf{V} = \mathbf{C^{-1}L}$$

and the compositional solution is

$$\mathbf{V} = \begin{bmatrix} 0.60 \\ 0.20 \\ 0.20 \end{bmatrix} \begin{array}{l} \text{Gin} \\ \text{Dry vermouth} \\ \text{Sweet vermouth} \end{array}$$

MATRIX ALGEBRA COMPOSITIONAL ANALYSIS OF LOGS

Wyllie et al. (1956) concluded that in pure lithologies with small pores, transit time measured by sonic logs is closely approximated by a linear function of porosity. This relationship is most commonly expressed by the "Wyllie time-average equation":

$$\phi = \frac{\Delta t - \Delta t_{ma}}{\Delta t_f - \Delta t_{ma}}$$

Written in expanded form, this becomes the equation set

$$\text{Sonic:} \quad \phi \cdot \Delta t_f + P_{ma} \Delta t_{ma} = \Delta t$$

$$\text{Unity:} \quad \phi + P_{ma} = 1$$

By collecting terms as a matrix formulation, this is:

$$\begin{bmatrix} \Delta t_f & \Delta t_{ma} \\ 1 & 1 \end{bmatrix} \begin{bmatrix} \phi \\ P_{ma} \end{bmatrix} = \begin{bmatrix} \Delta t \\ 1 \end{bmatrix}$$

or more simply as

$$\mathbf{CV} = \mathbf{L}$$

with obvious parallels to the matrix solution of the Perfect Martini.

In a lithology of clean but variable compositon of dolomite and calcite with small pores, an additional log, such as a density, would be required for explicit solutions. At this stage, it would be impractical to modify the time-average equation to anything manageable, but the descriptive set of equations would be

$$\text{Sonic:} \quad \phi \cdot \Delta t_f + P_D \Delta t_D + P_C \Delta t_C = \Delta t$$

$$\text{Density:} \quad \phi \rho_f + P_D \rho_D + P_C \rho_C = \rho_b$$

$$\text{Unity:} \quad \phi + P_D + P_C = 1$$

which in matrix notation can again be symbolized as

$$\mathbf{CV} = \mathbf{L}$$

although the size of the matrices are expanded. Solution of the composition of zones logged through such a sequence could either be solved graphically by recourse to a sonic–density crossplot, or numerically by a matrix algebra solution.

If the problem was further expanded to clean, cherty, dolomitic limestones for resolution by measurements from a neutron-density–sonic log suite, there would be four equations in the set. A graphic approach to the solution would require a projection of a three-dimensional situation onto two-dimensional paper which, while not impossible, would be considerably time consuming.

As an example of the matrix algebra resolution of this problem, we can analyze the Lower Viola (2500–2550 feet) in the Nemaha well, which has already been illustrated by log overlay, crossplots, and M–N plot in the last chapter (pp. 126, 129, 136). The results of the crossplots clearly suggest that the Viola is a compositional mixture of dolomite, quartz (chert), and calcite. This interpretation is readily corroborated by regional stratigraphy. Since a neutron-density–sonic log suite was run through the Viola section, it is possible to write an equation set which links the unknown component proportions multiplied by their physical properties with the log responses measured at any depth zone:

Neutron: $100.0\phi + 5.0D - 5.0Q + 0.0C = \phi_N$

Density: $1.00\phi + 2.87D + 2.65Q + 2.71C = \rho_b$

Sonic: $189.0\phi + 43.5D + 55.1Q + 47.5C = \Delta t$

Unity: $1.0\phi + 1.0D + 1.0Q + 1.0C = 1.0$

Rewritten as matrices,

$$\begin{bmatrix} 100.0 & 5.0 & -5.0 & 0.0 \\ 1.00 & 2.87 & 2.65 & 2.71 \\ 189.0 & 43.5 & 55.1 & 47.5 \\ 1.0 & 1.0 & 1.0 & 1.0 \end{bmatrix} \begin{bmatrix} \phi \\ D \\ Q \\ C \end{bmatrix} = \begin{bmatrix} \phi_N \\ \rho_b \\ \Delta t \\ 1.0 \end{bmatrix}$$

Expressed in symbols,

$$\mathbf{CV = L}$$

$$\mathbf{V = C^{-1}L}$$

On finding the inverse of the coefficient matrix, **C**, this matrix equation in expanded form is

$$\begin{bmatrix} \phi \\ D \\ Q \\ C \end{bmatrix} = \begin{bmatrix} 0.004 & -0.090 & 0.002 & 0.125 \\ 0.022 & 7.369 & 0.073 & -23.441 \\ -0.079 & 5.561 & 0.123 & -20.927 \\ 0.051 & -12.840 & -0.198 & 45.244 \end{bmatrix} \begin{bmatrix} \phi_N \\ \rho_b \\ \Delta t \\ 1.0 \end{bmatrix}$$

By means of this relationship, the proportional compositon of any zone in the sequence can be found immediately by premultiplying a column vector of the zone log readings and one by the inverse of the coefficient matrix.

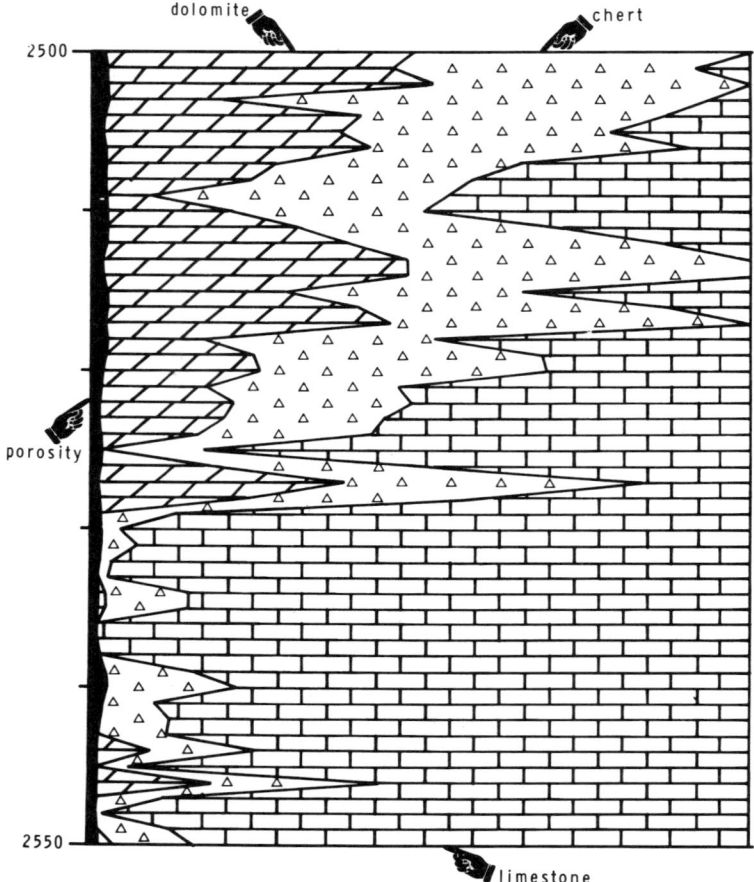

FIGURE 3. Matrix algebraic solution of zones from the lower part of the Viola Limestone in a northern Kansas well. The result is shown as a graphic lithology log of the system porosity–dolomite–chert–calcite.

It could be argued that the composition of a zone can be solved by standard algebra applied to the original simultaneous equations. This is true, but in a practical context would be a grossly inefficient procedure. When applied to the 50 zones of the Viola sequence, the simultaneous equation-solving procedure by standard algebra would have to be run 50 times, since although the left-hand side of the equations stay constant, the right-hand sides change. Using the matrix algebra approach, the computational requirements are a single inversion of the coefficient matrix (which precedes the reading of any logs), followed by one matrix multiplication operation for each zone.

A program which used the matrix algorithm outlined was applied to the solution of the zone compositions of the restricted Viola sequence. The results are presented as a graphic lithology log on the preceding page (Fig. 3) and show a low-porosity carbonate which grades upward from limestone at the base into cherty dolomites.

Computer programming languages such as FORTRAN are an ideal medium to code the numerical operations described in the preceding section. Groups of numbers can be handled easily: one-dimensional "arrays" represent vectors, two-dimensional arrays express matrices. The repetitive procedures involved in matrix algebra manipulations can be executed by simple or nested "do-loops."

The numerical and graphic solution of the Viola section composition was solved by a simple set of coding called the KIWI (Kansas Instructional Well-log Interpretation) program. The program is listed in Appendix 1, together with instructions for its operation. KIWI was designed to be general purpose in scope and can resolve rock compositions with up to six components, provided that a sufficient suite of logs is available. In all cases, n components will require $(n - 1)$ logs for an explicit solution. Typically, these logs will be drawn from either/and/or the neutron, density, and sonic logs, coupled with the gamma ray, and occasionally, the spontaneous potential or a transformed resistivity log.

CASE STUDY 1: ANALYSIS OF A CAMBRO-ORDOVICIAN DOLOMITE

The Arbuckle Limestone occurs extensively across the midcontinent and consists of Upper Cambrian and Lower Ordovician sandy and cherty dolomites which are a major exploration play for oil in Kansas. Porosity types are of all grades, ranging from intercrystalline to fractures and vugs and even cavernous developments. An Arbuckle section in a well in eastern Kansas was logged with a suite of gamma-ray, sonic, density, and neutron logs. Drill cuttings were examined and used in the preparation of a sample log and the stratigraphic subdivision of the Arbuckle into its constituent formations. Core was taken in a short interval of the upper Arbuckle and engineering measurements used as quality control to monitor log corrections, as described in the section on normalization in Chapter 8.

The service company magnetic tape of the digital logs was read into a computer file, normalized, and linked with a program similar to KIWI which incorporated the matrix algorithm described. The basic composition of the Arbuckle is a five-component system of dolomite, quartz (either chert and/or sand), shale, "primary porosity" (intercrystalline and intergranular), and "secondary porosity" (fractures and vugs). The five components are dictated by the geology as "seen" by the logs rather than other criteria, so that sand is grouped with chert as a single-quartz component, while porosity is subdivided into two separate components, due to the basic insensitivity of the sonic log to macroscopic pores as contrasted with the neutron and density logs.

In the compositional analysis of the Arbuckle section, the neutron, sonic, density and gamma-ray logs were used to provide the necessary five equations (when adding the unity equation) to solve for the five components at each successive 6-inch increment of depth. The results of this long record were averaged over 10-foot intervals for ease of comparison with the sample log description from cuttings, as shown in Figure 4. In the presentation, the computed results are drawn on the left, while the cuttings log commentary and graph are transcribed on the right, together with stratigraphic boundaries and unconformities. Overall, there is a striking concordance between these two independent sources of data, while the log analysis amplifies the compositional variation of the Arbuckle. Trends in the secondary porosity are broadly confirmed by borehole televiewer pictures over a 200-foot subdivision, which show vugs, strong horizontal, and vertical fractures, in associations which appear to be linked with the volumetric content of quartz and its identity as either sand or chert.

The subdivision of the Arbuckle into formations has been based primarily on systematic changes in the character of insoluble residues at formation boundaries (Keroher and Kirby, 1948). Early work in stratigraphic correlation found that fossils were of limited value for paleontological zonation, mainly because of their sporadic occurrence in restricted zones. The problem was further compounded in correlations based on well cuttings whose small size yielded only occasional fragmentary fossil material. Conventional lithostratigraphic methods were hampered by both the minor occurrence of distinctive shale and sandstone beds and their impersistence when traced laterally. The use of insoluble residues has proved successful in tracing correlative equivalents of the Arbuckle formations from their outcrop in the Ozark

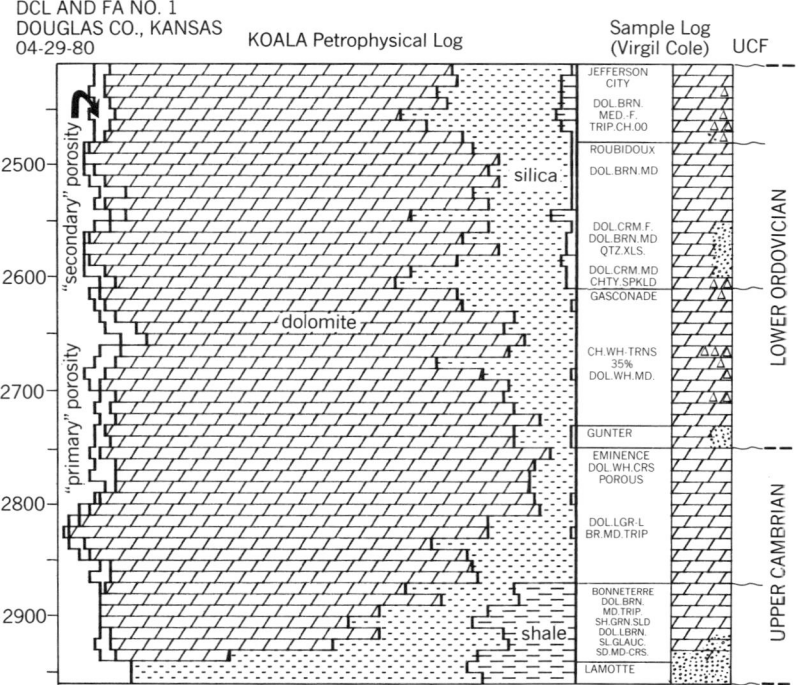

FIGURE 4. Compositional profile of an Arbuckle Limestone section derived from logs and matched with a sample log. (UCF signifies unconformity.)

Case Study 1: Analysis of a Cambro-Ordovician Dolomite

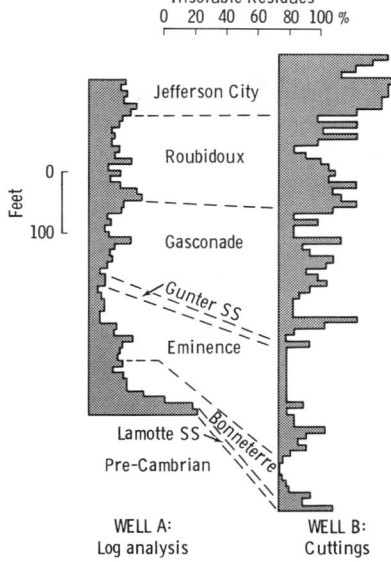

FIGURE 5. Comparison of matrix algebra log solution of silica content in the Arbuckle Limestone of well A with the insoluble residue of the Arbuckle in well B.

region of Missouri into the subsurface of both Kansas and Oklahoma (Ireland, 1944).

The procedures of insoluble residue work are time-consuming and not entirely satisfactory. Detailed analysis of rotary cuttings requires the elimination of cavings materials, selection of the hand-picked fraction in hydrochloric acid, and careful volumetric measurement. Even with these safeguards, the quantitative estimates are generally exaggerated, owing to the selective loss of part of the dolomite host as rock flour in the drilling operation. By contrast, the computer analysis of a wireline log combination in the estimation of silica content (either sand or chert) has the merits of both speed and direct measurement of the formation. The silica content can be equated with the volume of insoluble residues and applied to stratigraphic correlation problems in the Arbuckle. The major drawback of wireline logs is their failure to discriminate between chert and sand, so that their use must focus on patterns of volumetric change in silica as a function of depth.

The log analysis solution of silica content is shown in conjunction with insoluble residue estimations from a well at a distance of 25 miles to the east on Figure 5. Most of the curve features can be correlated convincingly between the two traces, when allowance is made both for the exaggerated estimates and intrinsic sampling problems associated with cuttings and variations attributed to lateral facies changes. In practice, the application of computer-processed logs to problems of zonation and correlation would be coordinated with cuttings information in much the same way that raw logs are used to aid geologists in the preparation of detailed sample logs. In the specific case of work with Arbuckle sections, the nature of the cuttings would be an essential key to allocate computed silica between components of chert and/ or sand.

CASE STUDY 2: ANALYSIS OF A PENNSYLVANIAN SANDSTONE–SHALE SEQUENCE

Elongate sandstone bodies of the Cherokee Group (Middle Pennsylvanian) in southeast Kansas have been prolific producers of oil for many years. Depositional origins of individual sands have been variously ascribed to nearshore bars, barrier islands, tidal channels, and alluvial valley-fill (Ebanks and James, 1974), as specific elements of a generalized paralic environment. The mineralogy of the sandstones consists mainly of quartz and muscovite together with minor accessory components and a dominantly kaolinitic clay content. The clay mineralogy of the interbedded shales are mixtures of illite and chlorite.

Gamma ray, neutron and density logs from a section containing the Skinner Sandstone were reviewed in pages 145–147 of Chapter 5. The use of crossplots of these logs was demonstrated in the location of a representative shale point as an aid in the qualitative interpretation of the sequence. The shale point on the crossplots specifies the log response properties of shales which can be considered to be typical in the section. The composition of the section was resolved in terms of the components quartz, mica, shale, and pore fluid as calculated by a matrix algebra solution, using the gamma-ray, neutron and density logs (Fig. 6).

Although the solution is not an unreasonable representation of lithology variation in the sequence, it can only be considered as a highly generalized result. As is always the case, the shale point typifies shales *between* the sandstones which are mineralogically different (dominated by chlorite and illite, rather than kaolinite) and morphologically different (lithostatically loaded rather than hydrostatically pressured) from shaly material *within* the Skinner Sandstone. Therefore, the estimation of shale content is only approximate in this example, in common with the vast majority of shaly sand analyses from logs.

Volumetric Solutions of Shales

More specific studies of shale compositions from logs are complicated by several factors. If a single-shale endmember is subdivided into subordinate components of individual clay minerals and a silt fraction, the additional number of unknowns requires more logs for their quantitative solution. In addition, the chemical composition of most clay mineral species shows ranges of variation and a matched variability in physical properties. The amount of absorbed water has a crucial influence on log responses. These considerations collectively make for a formidable problem in quantitative analysis. However, the economic importance of shaly sandstone oil and gas reservoirs has spurred research into refined procedures for analysis. Both the volumetric fraction and identity of the clay mineral content are important to the petroleum engineer, since both have a major bearing on the success or failure of various reservoir engineering techniques.

In summary, the volumetric analysis of sequences with a significant clay content presents special problems. Other rock-forming minerals have physical properties

Case Study 2: Analysis of a Pennsylvanian Sandstone—Shale Sequence

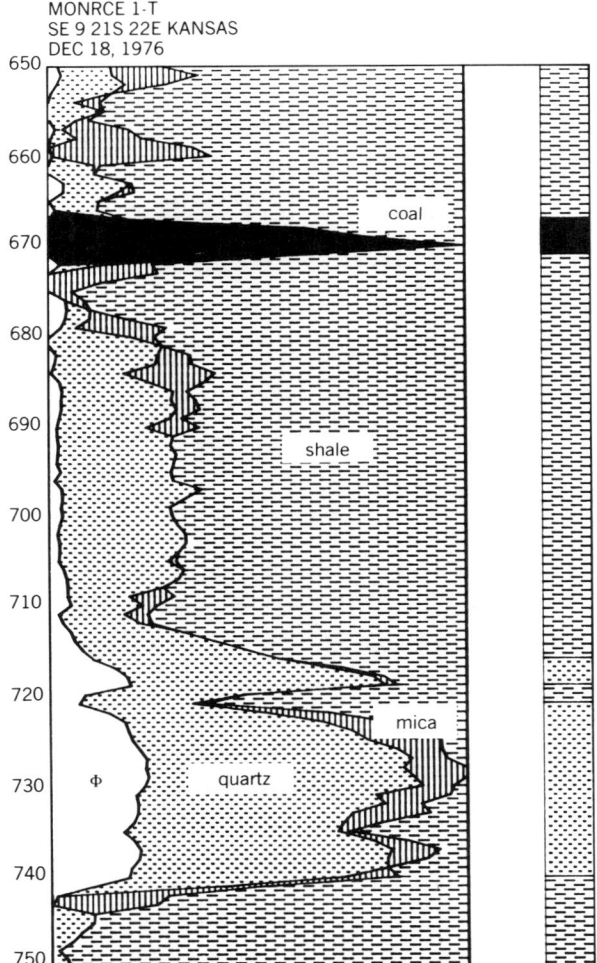

FIGURE 6. Compositional profile of a Cherokee Group (Pennsylvanian) section from southeast Kansas, based on neutron, density, and gamma-ray logs.

which are constrained within narrow ranges and are measured easily in the laboratory. The numerical analysis of a shaly sequence can be made as a "first order solution" using commonly available logs. In this approach, the properties of distinctive shales are equated with a single shale endmember. The generalized quantitative solution is satisfactory to the degree that there is low variability within the shales and there are minor differences in physical properties between clay minerals within the shales and those within the grain-supported framework of other lithologies.

A detailed "second order solution" entails an attempt to selectively differentiate quantitative variation of individual clay mineral species and requires additional log information from specialized tools. In this regard, photoelectric absorption mea-

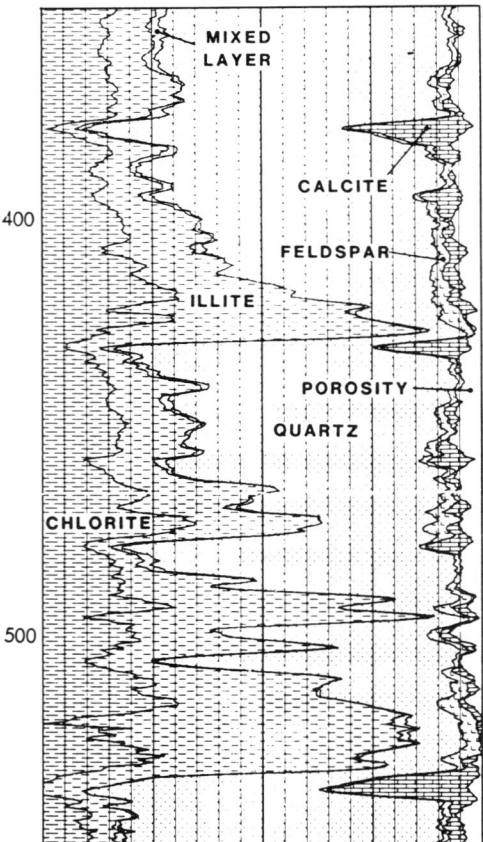

FIGURE 7. Example of mineralogical profile computed from induced gamma-ray spectrometry measurements. From Gilchrist et al. (1982).

surements from the lithodensity tool and natural gamma-ray spectral estimate partitioned between potassium, thorium, and uranium, have proved to be useful, as discussed in the previous chapter. Induced gamma-ray spectral logs show particular promise in both the identification and volumetric assays of clay minerals. Gilchrist et al. (1982) reported applications with good correlations between spectral data and X-ray diffraction analyses from core, with the result that profiles of clay mineralogy could be calculated as sound representations of actual geology (Fig. 7). The computational algorithm is necessarily more complex, both as a consequence of the nature of the spectral data and the expansion in the number of unknowns to be solved. However, the approach has much in common with simpler methods already described with its basis in an expanded set of response equations. For the foreseeable future, supporting evidence in the form of sample analyses from core or cuttings will continue to be a necessary calibration standard to check volumetric profiles computed from logs, in view of the ranges in compositional variability of clay mineral species.

CASE STUDY 3: PERMIAN EVAPORITES

In 1970, the U.S. Atomic Energy Commission drilled Test Hole No. 2 in Lyons County, central Kansas. The test hole was part of a study of the potential for storage of high-level radioactive wastes in salt beds in Kansas. In this case, the target horizon was the Hutchinson Salt (Wellington formation, Permian), which was both continuously cored and logged for stratigraphic and engineering evaluations. A 100-foot subsection of the log suite was drawn from an interval which straddles the base of the Hutchinson Salt and the anhydrite/shale beds of the lower Wellington.

The basic mineralogy of the section consists of halite, anhydrite, and a shale component, which can be resolved by two logs, the sonic and the density. The shale transit time and density were deduced from a "shale point" located on a sonic–density crossplot (Fig. 8). Results of the analysis are shown in Figure 9 and contrasted with the core description made by well-site geologists.

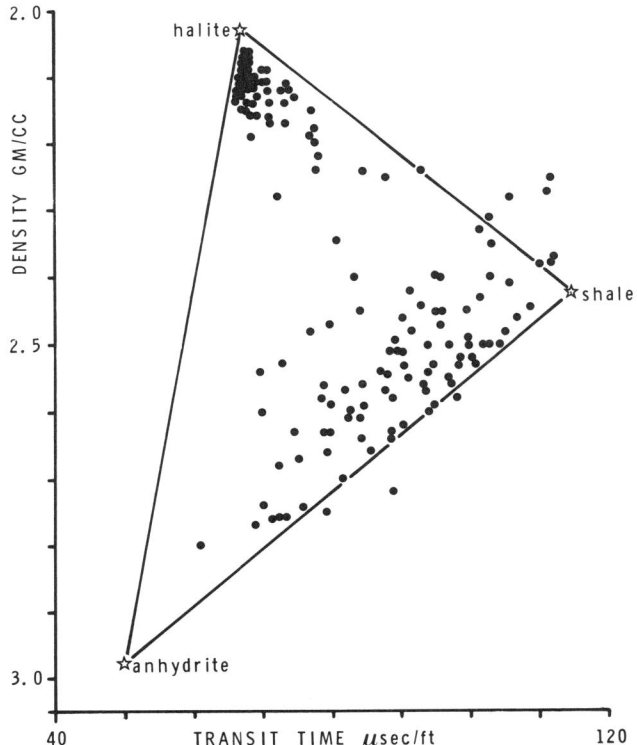

FIGURE 8. Density–sonic crossplot of part of the Wellington formation (Permian) in a central Kansas well.

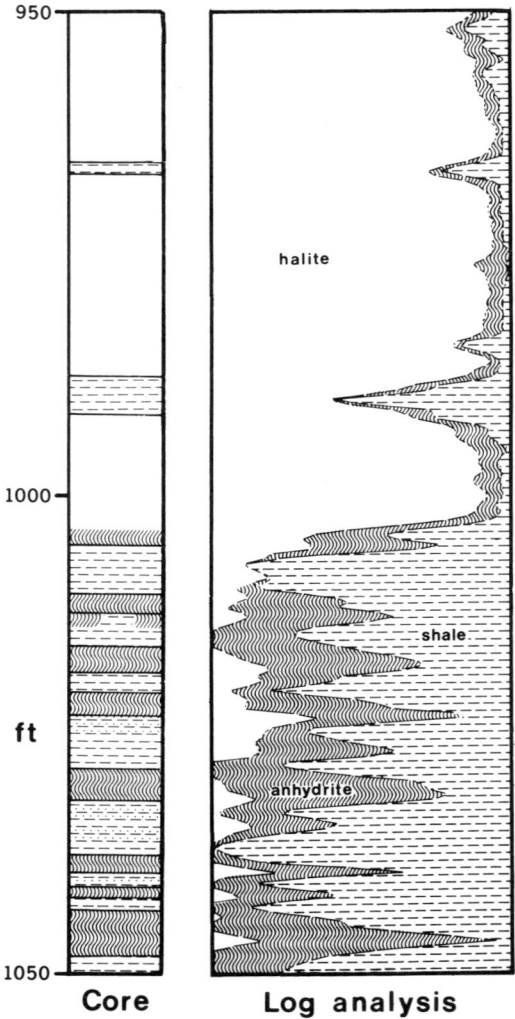

FIGURE 9. Compositional profile of part of the Wellington formation (Permian) in a central Kansas well, contrasted with a graphic core description.

How successfully does the log analysis match the core report? A common complaint by geologists is that log analysis presentations of compositional contents are unrealistic since they tend to show fluctuating variations rather than discrete beds. Obviously, there is some truth to this assertion, since the resolution of the original log data is of the order of 1 or 2 feet and will result in smoothed averaging of the actual variation by a window dimension of this order. It is possible that work with deconvolution filters could be used to remedy the situation, although filtering has its own intrinsic problems. As counterpoint, it is legitimate to ask to what degree the graphic core description expresses real compositional variation. The continuous core of this section was photographed and a graphic mosaic assembled as Figure 10. Comparison of this composite with the graphic core description in Figure 9 shows the common tendency for geologists to emphasize major components and establish discrete bedded units. The performance of log analysis solutions of compositional variation should therefore be judged with reference to real lithological character rather than simplified descriptions by geologists.

One error of this log analysis is the prediction of minor, but significant, contents of halite in the lower part of the section. The reason for this is the occurrence of some sandstone beds and sandy streaks in this interval which are rationalized as halite, since the "sandstone line" occurs within the halite–anhydrite–shale composition triangle on the sonic–density crossplot. The source of the error is therefore not one of miscalculation, but one of misidentification of the component system. When rerun with a sandstone–anhydrite–shale system this lower interval would show a more satisfactory solution. The moral follows that, while crossplots are useful in

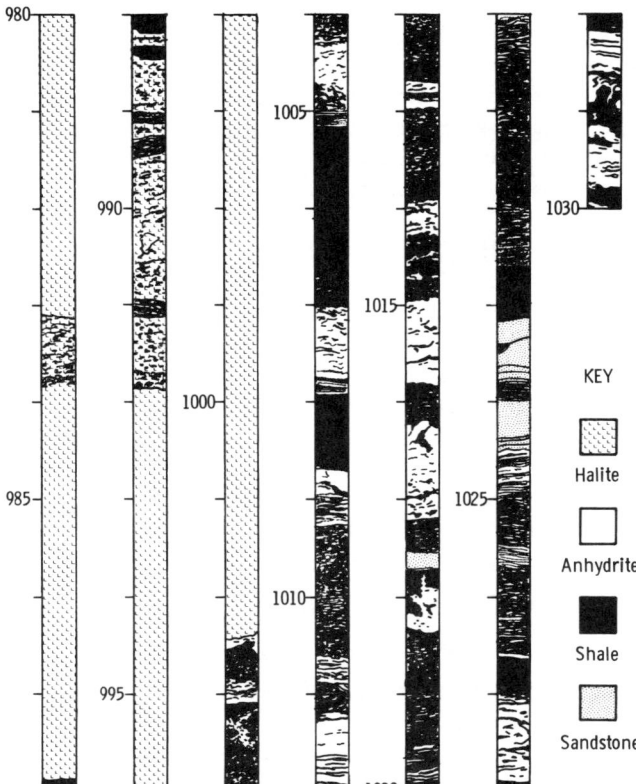

FIGURE 10. Graphic rendition of composite core photographs of a subinterval of the Wellington formation in a central Kansas well.

the identification of fundamental components, a little knowledge of the regional stratigraphy can be invaluable.

ADVANCED MATRIX ALGEBRA METHODS FOR COMPONENT ESTIMATION

The approach described in the previous section was introduced by Savre (1963), further developed by Burke et al. (1967), and by Dawson-Grove and Palmer (1968). The matrix formulation is a linear model and generally provides a satisfactory approximation for compositional solutions, as confirmed by core analyses and laboratory studies. Its basic limitation is that the number of logs should match the number of components minus one. In this situation, the equations match the unknowns and the system is said to be "determined." In reality, this condition will not always be met. In many instances, the number of logs will be insufficient to provide unique solutions of the compositions of complex lithologies ("underdetermined" systems). Less frequently, the number of logs may be more than necessary to resolve lithologies with simple mineralogies ("overdetermined" systems). However, these distressing situations are not intractable, but can be analyzed using methods which are an extension of the matrix algebra already discussed.

Solutions of compositions in the underdetermined, determined, and overdetermined systems will be reviewed in the following sections. A component system of calcite–dolomite–porosity is used as a medium for graphic illustration of the geometrical implications of each mathematical procedure.

The Underdetermined System

When the number of logs available is restrictive and such that the number of log response equations plus unity equation does not match the number of components, the system is underdetermined and no unique solution is immediately available. Under this stricture, the situation is one of uncertainty and predictions concerning the composition of any zone must be phrased inevitably in terms of "least error" rather than "most correct." McCammon (1970, 1972) pioneered the research in this area by applying concepts from information theory to the problem.

The tenets of information theory dictate that the least biased solution corresponds to that which maximizes the entropy quantity given by

$$H = - \sum v_i \log v_i$$

where v_i is the proportion of component i, subject to the constraints of the information supplied by the available logs.

An alternative expression of variation that closely parallels entropy is that of proportional variance given by the equation

$$p = \frac{n}{n-1} \sum v_i(1 - v_i)$$

Advanced Matrix Algebra Methods for Component Estimation

A component estimation that matches the maximum proportional variance rather than maximum entropy is more easily implemented as a simple matrix algorithm. Because the expression for proportional variance can be rewritten as

$$p = \frac{n}{n-1}\left(1 - \sum v_i^2\right)$$

the goal of maximizing the proportional variance therefore is equivalent to minimizing the quantity $\sum v_i^2$. It can be shown that this solution corresponds to

$$\mathbf{V} = \mathbf{C}^T(\mathbf{C}\mathbf{C}^T)^{-1}\mathbf{L}$$

where $\mathbf{L}$ is the vector of log responses. (The symbol $\mathbf{C}^T$ signifies the transpose of the $\mathbf{C}$ matrix, which simply means a matrix in which the rows and columns have been interchanged.) Inspection of the matrix formula shows that in the extreme limiting situation of no log information, use of the unity equation gives the component vector estimate as

$$\mathbf{V} = \left[\frac{1}{n}\right]$$

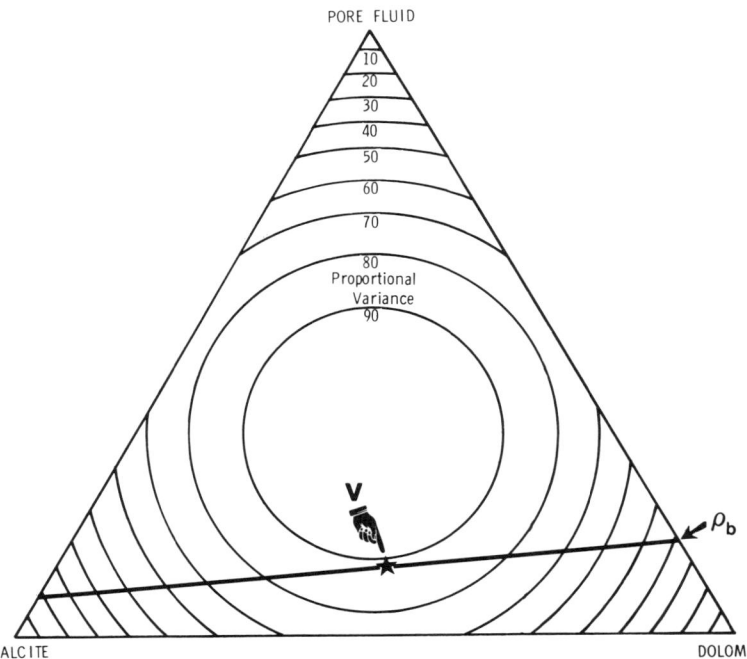

FIGURE 11. The maximum variance solution of the underdetermined system for a hypothetical zone in a dolomitic limestone sequence.

If several logs (but an insufficient number) are available, the equation will generate compositions which are the nearest feasible solutions to a homogeneous mixture of the components. As an example, a single-density log reading of 2.6 g/cm^3 is sketched as a line on a calcite–dolomite–porosity composition triangle (Fig. 11) and shows all the possible solutions in the system which satisfy this reading. The location of the maximum variance solution is indicated by a star. While this composition is not necessarily correct, it is the single solution which is likely to have the least error when related to the real composition.

The method has the major drawback that it assumes that the entire range of possible compositions are equally likely. This is generally not the case for real rocks, whose composition tends to be dominated by relatively few components whose proportions are distributed in a highly unequal manner. In the case of the dolomite–limestone range, for example, analytical data suggest that intermediate dolomitic limestones and limy dolomites occur significantly less frequently than the extreme endmembers (Fig. 12). In a typical composition triangle, the most common compositions are usually located close to the vertices, secondarily along the sides, and less frequently in the interior of the triangle. The maximum variance solution therefore provides the least biased solution, although it must be viewed generally as an intermediate of the set of realistic alternatives. If proportional variance of the solution is low, the solution is likely to be a close estimate; if high, the solution faithfully indicates that there is a high degree of uncertainty regarding an appropriate choice from a set of rather disparate alternatives.

In problem sections (high proportional variance solutions), additional geological information is clearly needed and this can be incorporated within the matrix algebra. The maximum variance approach can be considered as an attempt to minimize proportional variance with respect to a homogeneous mixture. The method can be

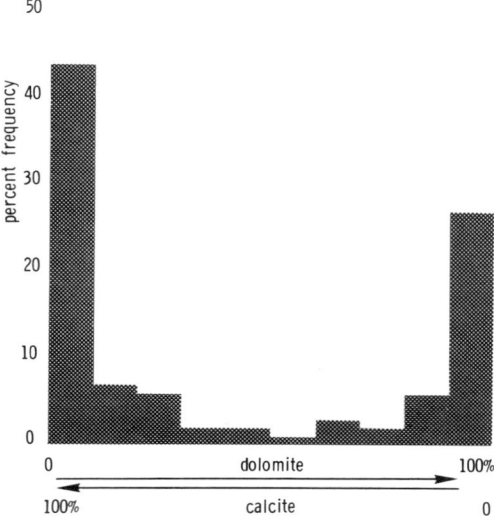

FIGURE 12. Histogram of calcite–dolomite proportions from a sample of 1148 analyses of North American carbonate rocks. Redrawn after Steidtmann (1917).

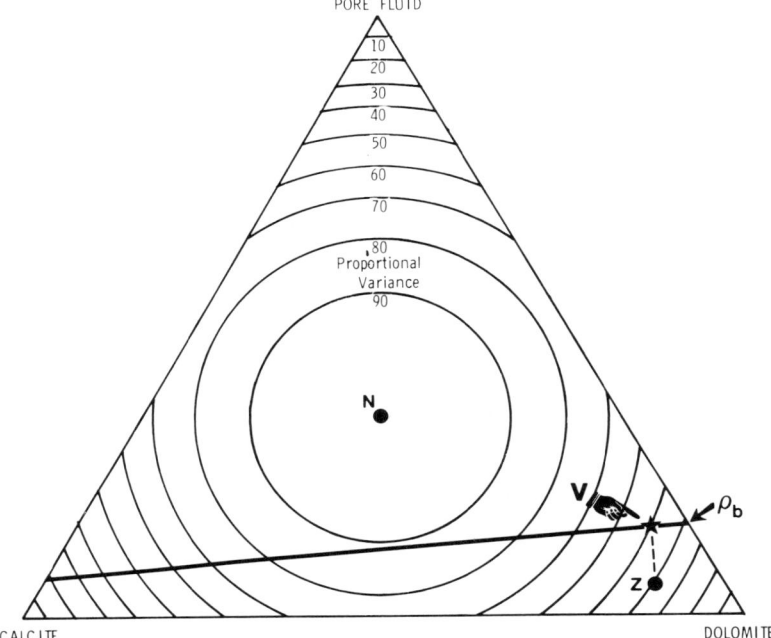

FIGURE 13. The guided minimum variance solution of the underdetermined system for a hypothetical zone in a dolomitic limestone sequence.

modified to minimize variance related to a new reference point which is more likely to represent the real composition of the section. A "guidepoint" composition can generally be estimated, based on regional stratigraphic considerations or studies of nearby outcrop or cores from neighboring wells. The modified "guided minimum variance" solution is then

$$V = C^T(CC^T)^{-1}(L + CD) - D$$

$$D = N - Z$$

where N is a vector of proportions, $1/n$, and Z is a vector of the guide-point proportions. The geometrical implications of this solution, as applied to a single zone, are shown in the composition triangle of Figure 13.

The method has useful "feedback" information. If the computed compositions differ radically from the guide-point composition, either the log data are poor in quality or the guide point is illusory and not representative of the section analysed.

The Uniquely Determined System

When the number of logs is less than the number of components by one, the equations satisfy the unknowns and the appropriate matrix algorithm is a conventional simultaneous equation solution procedure. Because,

$$CV = L$$

then

$$V = C^{-1}L$$

Figure 14 illustrates a unique composition solution for a hypothetical zone in the calcite–dolomite–porosity system using two porosity log readings.

This is the most common matrix method that is used for the calculation of porosity and mineral-component proportions. Clearly, its application in practice requires a felicitous coincidence that matches the number of components with the logs available. In general, it is impossible to resolve all the components that actually occur by the limited log suites which are run typically through a section. However, the selection of the major components for solution usually will reduce errors in estimation to low levels introduced by minor contaminating minerals.

The influence of unaccounted components may be detected by a feature of the algorithm which provides no constraint for the proportions to be positive in value. Whereas the proportions collectively sum to unity, individual negative proportions

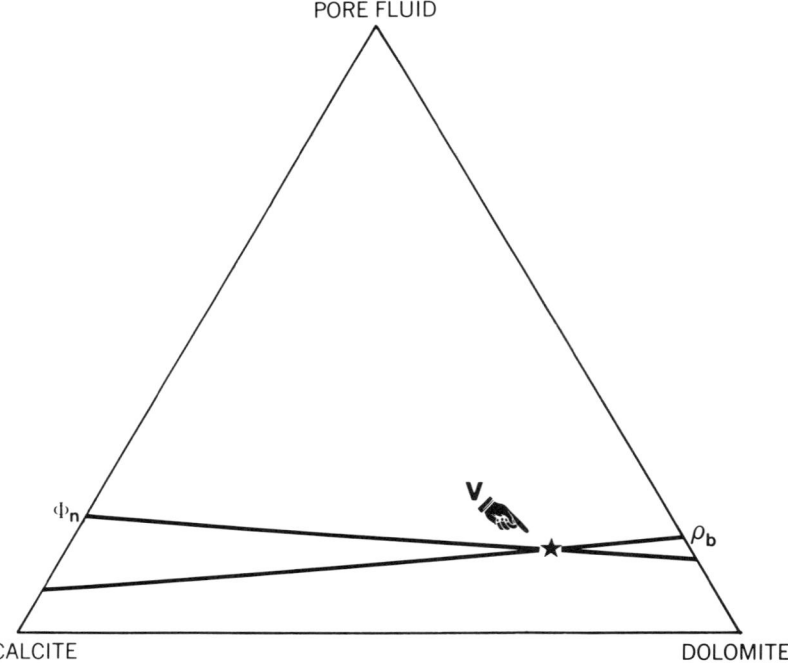

FIGURE 14. Unique solution of the determined system for a hypothetical zone in a dolomitic limestone sequence.

can occur, and designate zones which plot outside the fundamental composition space linking the specified component endmembers.

Errors of this type can be monitored for each zone by an "absolute sum error," a, where

$$a = \left(\sum |v_i|\right) - 1$$

Small values in absolute sum error may be attributed to minor instrument recording errors or imprecise depth calibration of the logs. Larger values reflect either gross logging errors or the systematic influence of additional mineralogical components.

The problem introduced by significantly negative components was recognized by both Burke, Curtis, and Cox (1967) and Dawson-Grove and Palmer (1968). Their responses were a series of maneuvers which seek to determine the most feasible set of components whose number is prescribed by the log suite to give a unique solution. However, the "negative components" in any zone give important clues regarding the physical properties of an unresolved component and more detailed analysis of the location of anomalous zones on log crossplots is an additional aid in the recognition of a more feasible component system.

The Overdetermined System

If a full suite of logs is run, certain lithologies may occur where the major components are severely restricted in number and the system becomes overdetermined. In this situation, the number of equations provided by the logs exceeds the number of components.

An appropriate estimate of zone composition can be drawn from a least-squares model in which the error is minimized between the log response and its corresponding value predicted by the solution. Because the logging measurements are recorded in radically different units, a common reference scale is required to avoid undue weighting of the solution with respect to any of the logs. A natural choice for standardization in most applications is a porosity scale referenced to a reservoir rock mineral such as calcite.

It must be noted that in this model, the log-response equations carry an associated error and are contrasted with the unity equation which is a constraint. In a pragmatic accommodation, an arbitrarily high weighting constant may be applied to the unity equation to ensure that the constraint is satisfied to a desired level of accuracy. (A more sophisticated resolution of a regression/constraint equation set is the minimization of a quadratic form by the method of Lagrange multipliers.)

The matrix solution of the least-squares model then becomes

$$\mathbf{V} = (\mathbf{C}^T\mathbf{C})^{-1}\mathbf{C}^T\mathbf{L}$$

A least-squares solution in the system calcite–dolomite–porosity is shown for a hypothetical zone with three log readings in Figure 15.

The vector of errors, **E**, associated with the solution for each zone may be computed readily as

$$\mathbf{E} = (-\mathbf{L} : \mathbf{C}) \begin{bmatrix} 1 \\ \vdots \\ v \end{bmatrix}$$

and a mean absolute deviation, e, is given by

$$e = \frac{\sum\limits_{j}^{m} |e_j|}{m}$$

where m is the number of logs.

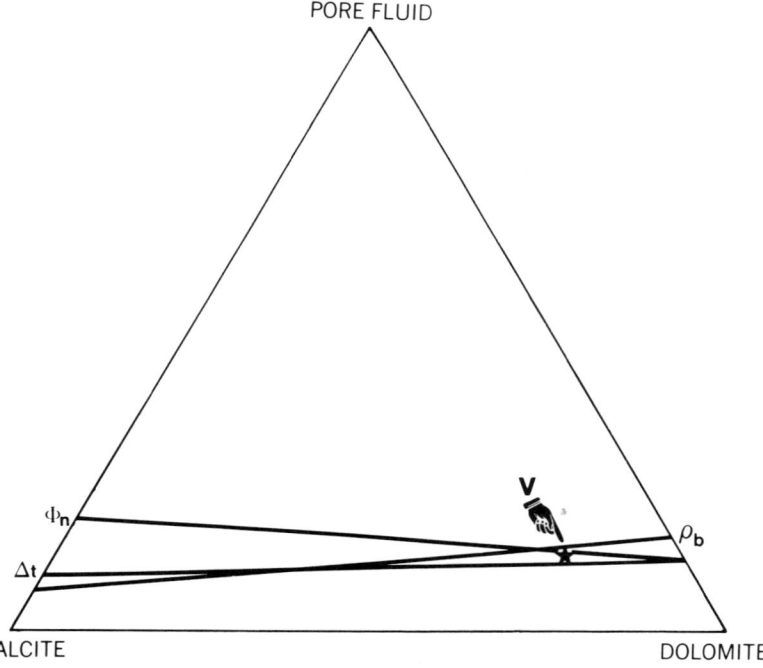

FIGURE 15. Least-squares solution of the overdetermined system for a hypothetical zone in a dolomitic limestone sequence.

TABLE 1
Matrix Algorithms for Different Contingencies Relating Number of Component Unknowns with Number of Available Logs[a]

Number of Components	Number of Logs	System	Solution	Algorithm	Error Diagnosis
n	$< n - 1$	Underdetermined	Maximum variance	$V = C^T(CC^T)^{-1}L$	p, a
n	$< n - 1$	Underdetermined	Guided min. variance	$V = C^T(CC^T)^{-1}X - D$	p, a
n	$n - 1$	Determined	Unique	$V = C^{-1}L$	p, a
n	$> n - 1$	Overdetermined	Least squares	$V = (C^TC)^{-1}C^TL$	p, a, e

[a] p = proportional variance; a = absolute sum error; e = mean absolute log deviation.

The ideal of a zero error is impractical due to logging measurement errors, nonlinearity effects, variable borehole conditions, and the inevitable contribution of minor components. In practice, a small error value indicates a feasible solution for the major components which is compromised mildly by the factors mentioned. Large error values are diagnostic of either significant environmental effects or systematic perturbation by unaccounted major components. Because the system is overdetermined, there is adequate flexibility for the introduction of additional mineral candidates and the reapplication of an appropriate algorithm for the new system.

A Coordinated Solution Procedure

The matrix algorithms described form a coherent hierarchy of equivalent operations in the solution of porosity and mineral components in lithologies from the analysis of log suites. Their main features are summarized in Table 1.

In each situation, the appropriate procedure follows a similar computational sequence. A system of components is selected on the basis of prior geological knowledge or from inspection of log crossplots of the zones. A matrix of component log responses is specified a priori for the designated components from available laboratory measurements of their physical properties. The match of the number of components with the number of logs specifies the degree of determinancy of the solution system. The appropriate matrix operations of transposition, multiplication, and inversion then are applied to the coefficient matrix as a single execution which precedes the processing of the logs. Component proportion solutions for any zone may be obtained immediately by the postmultiplication of the vector of log responses and unity.

When implemented in an interactive environment of a minicomputer and linked graphic display terminal (Fig. 16), the simplicity of the algebra allows rapid processing of large volumes of log data. As a consequence of the solutions and their associated

FIGURE 16. Example of a computer hardware system appropriate for compositional analysis from logs: the equipment used to run the KOALA package at the Kansas Geological Survey.

diagnostics, the user may respond immediately with modifications at all levels (components, coefficients, stratigraphic intervals, logs) in an interactive strategy to isolate solutions which are most consistent with the available data.

As an example of this type of analysis, the sonic, density and neutron log traces from the entire Viola in the Nemaha well were processed by matrix algebra techniques. When directed to porosity–dolomite–chert–calcite, the system is determined and solved by simple matrix inversion. The analysis is satisfactory in the lower part, but is problematical in the upper section, as shown in Figure 17. At this level, high-sum deviation errors (SUM DEV) are generated which signify the calculation of negative proportions of calcite in some zones and dolomite and chert in others. Since both borehole and log quality are reasonably good, the indications are that the analysis is responding to components other than those specified.

An M–N–Z plot of gamma-ray values in this problem interval (Fig. 18) clarifies the situation. The systematic increase in gamma-ray values toward the base of the plot clearly indicates the influence of a "shale point" which would lie in this direction. Zones that are located below the triangle would generate negative calcite compositions and nonzero sum deviations, which can be attributed to the neglect of shale as a component in the initial solution. Zones that lie above the triangle may be interpreted as reflecting the influence of "secondary porosity" (vugs and fractures) and would be resolved as negative dolomite and chert proportions.

A comprehensive solution of the upper Viola should therefore be solved in terms of six components: porosity, secondary porosity, dolomite, chert, calcite, and shale. The situation is now underdetermined and efforts should be directed to generating the most feasible solution which is consistent with the available data. By incorporating the gamma-ray log to aid in the estimation of the shale component and applying the guided minimum variance algorithm, the upper section analysis was revised to a new solution. The final, composite analysis of the entire Viola interval in this

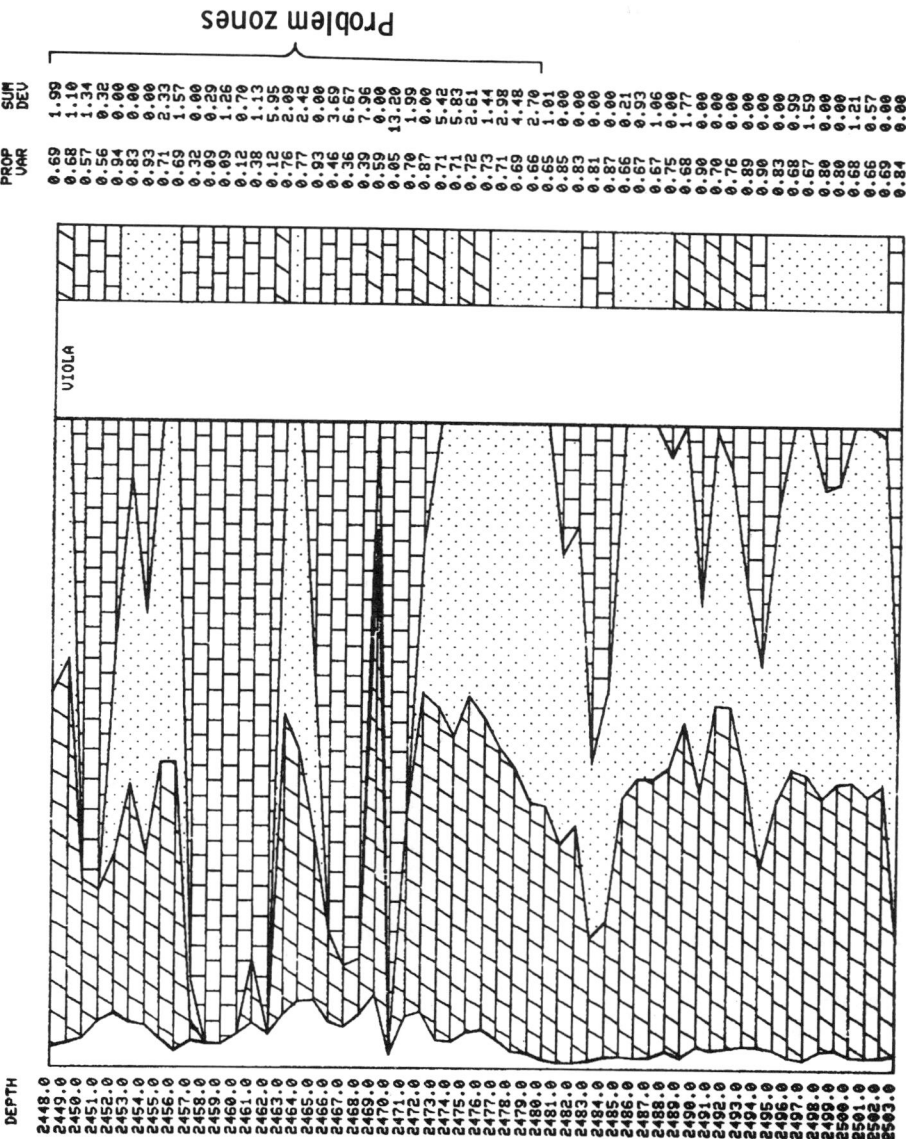

FIGURE 17. Problems in compositional analysis of the upper part of the Viola Limestone indicated by the diagnostic of high sum deviation values.

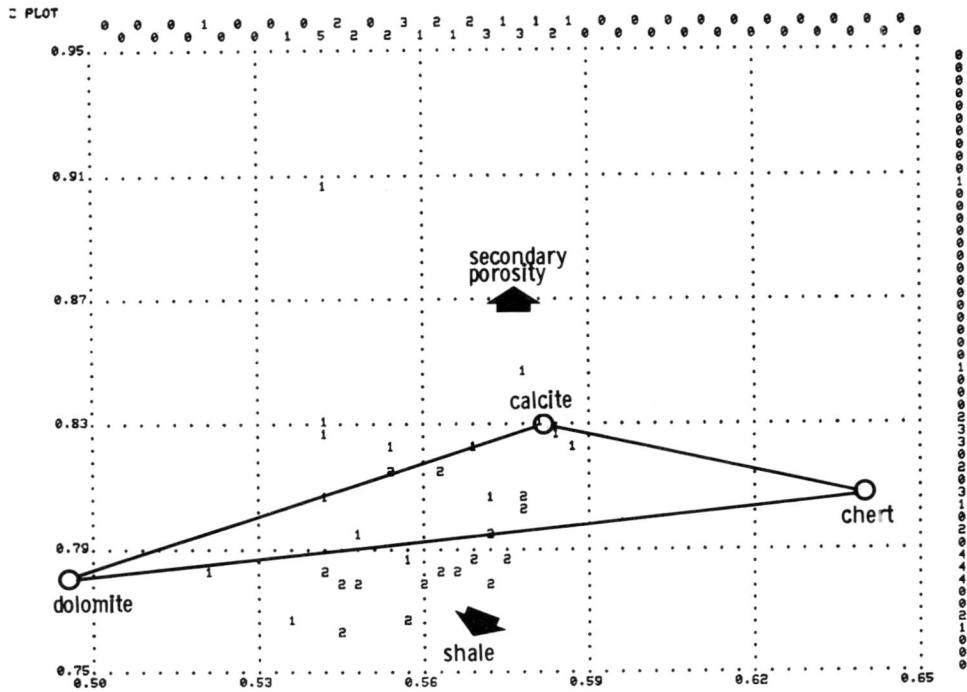

FIGURE 18. *M–N–Z* gamma-ray crossplot of problematical zones in the Upper Viola.

well is shown in Figure 19. This view is smoothed by the coarse resolution of the log combination, although much of the fluctuation of components, such as chert, is a reflection of true varied compositions. It is theoretically possible that the resolution could be sharpened to finer detail using deconvolution theory. Such work would involve the design of inverse filters to unravel the convolution of real variation by logging tools. Good filter design requires a subtle blend of art and science, particularly since the enhancement of the underlying signal will also amplify any associated noise. However, the theoretical groundwork has been developed and it remains to be seen whether research in this area can overcome the practical limitations of tool performance and the problems of the borehole environment.

NORMATIVE LOG SOLUTIONS AND MODAL PETROGRAPHY

The estimation of mineral compositions from log combinations is obviously subject to a variety of constraints. These include effects introduced by the vagaries of the borehole environment and measurement errors associated with the logging equipment (both reviewed in Chapter 8). The mathematics used for solution conform to a linear model and are approximations of the nonlinear structure of reality. Mineral properties are idealized as constants rather than ranges of variation. The number of logs available is generally limited and restricts analysis to the major mineral components.

Any serious work with logs must take all these factors into consideration. Consequently, log analysis results inevitably differ to some degree from real compositions. Petrographic and geochemical analyses of core are measurements of *actual* composition, although they are subject to measurement error and sample size estimate variation. Computations of mineral compositions from wireline logs are the results of a mathematical *model* which links physical measurements with the properties of minerals in an interpretative scheme of analysis. Stated in this manner, it can be seen that there is no inherent conflict between the two types of data. In common with all models, the aim of log analysis is to approximate reality to the closest degree possible, although exact duplication is not a feasible goal. However, it is appropriate to make the conceptual distinction between a log analysis estimation of mineralogy and the actual rock composition.

A useful analogy can be found in igneous petrology. Mineral compositions are computed both as normative and modal assemblages. Cross et al. (1902) introduced

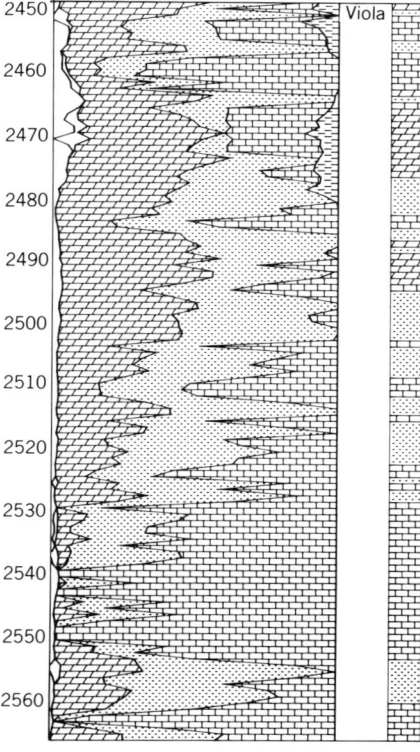

FIGURE 19. Final compositional solution of the Viola Limestone in a northern Kansas well. The components are primary and secondary porosity, dolomite, chert, calcite, and shale.

a system to recast chemical analyses of igneous rocks from oxide compositions to a set of hypothetical minerals. Their aim was the design of a classification scheme, but the classification has been abandoned for decades. However, their procedure is retained as the CIPW norm calculation which is still used widely today. Chemical distinctions between texturally and mineralogically different igneous rocks are contained within oxide analysis data. Petrologists generally find the distinctions more obvious when the oxides are transformed into suites of hypothetical norm minerals (Bickel, 1979). The logic of normative allocation of oxides to different minerals attempts to conform with the genetic mechanisms responsible for igneous mineral formation. Alternative normative schemes have been developed for other rock associations. Barth (1959) introduced the mesonorm as a scheme appropriate for mineral calculations in metamorphic rocks. McGetchin and Smyth (1978) describe a method for the calculation of standard mineral suites from chemical oxides as an aid in modeling assemblages in the upper mantle of Mars. In all cases, the normative method is a model that generates hypothetical mineral assemblages that are not unreasonable simulations of actual mineral compositions. However, they are not confused with the natural mineral assemblage, assessed by petrographic methods as a *modal* composition.

There are parallels in the methodology of log analysis. Rather than work with chemical oxide analyses, the physical properties of rocks measured by logging tools are recast in the form of theoretical mineral assemblages. The transformation results in data that are more easily assimilated by the geologist than the raw log responses. As is the case with the igneous norm, the assemblages are computed to be a reasonable match with natural associations, but are bound by the limitations of the log analysis model.

Log analysis interpretations of mineral compositions can therefore be regarded as normative solutions, whose results both compare and contrast with modal estimations from core and drill cuttings. The concept has useful properties that can expand the role of logs. As discussed earlier, sequences with shale as a significant component pose special problems because of the wide variability of clay mineral properties. "Normative" solutions of clay mineral proportions from log analysis could be made based on a consensus on the physical properties of a "standard" set of clay minerals. While these estimates would deviate to varying degrees from modal compositions, they would be valuable diagnostic tools in themselves, when viewed in the same sympathetic light accorded to norms by igneous petrologists.

REFERENCES

Barth, T. F. W., 1959, Principles of classification and norm calculation of metamorphic rocks, *J. Geol.*, Vol. 67, No. 2, pp. 135–152.

Bickel, C. E., 1979, The CIPW normative calculation, *J. Geol. Education*, Vol. 27, pp. 80–82.

Burke, J. A., Curtis, M. R., and Cox, J. T., 1967, Computer processing of log data enables better production in Chaveroo field, *J. Petrol. Tech.*, Vol. 19, pp. 889–895.

Cross, W., Iddings, J. P., Pirsson, L. V., and Washington, H. S., 1902, A quantitative chemico-mineralogical classification and nomenclature of igneous rocks, *J. Geol.*, Vol. 10, pp. 555–690.

References

Dawson-Grove, G. E., and Palmer, K. R., 1968, A practical approach to analysis of logs by computer, *Trans. 2nd Form. Eval. Symp., Can. Well Logging Soc.*, pp. 1–12.

Ebanks, W. J., Jr., and James, G. W., 1974, Heavy-crude oil bearing sandstones of the Cherokee Group (Desmoinesian) in southeastern Kansas, *Canadian Soc. Petroleum Geologists, Memoir 3*, pp. 19–34.

Gilchrist, W. A., Jr., Quirein, J. A., Boutemy, Y. L., and Tabanou, J. R., 1982, Application of gamma ray spectroscopy to formation evaluation, *Trans. Soc. Prof. Well Log Analysts 23rd Ann. Logging Symp.*, Paper B, 28 pp.

Ireland, H. A., 1944, Correlation and subdivision of subsurface Lower Ordovician and Upper Cambrian formations in northeastern Oklahoma, *U.S. Geol. Survey, Oil and Gas Investigations, Prelim. Chart No. 5.*

Keroher, R. P., and Kirby, J. J., 1948, Upper Cambrian and Lower Ordovician rocks in Kansas, *Kansas Geol. Survey Bull.*, Vol. 72, 140 pp.

McCammon, R. B., 1970, Component estimation under uncertainty, in *Geostatistics, A Colloquium* (D. F. Merriam, Ed.), Plenum, New York, pp. 45–61.

McCammon, R. B., 1972, Estimating lithologic components in stratigraphic sequences under uncertainty, *Geol. Soc. America Special Paper 146*, pp. 11–24.

McGetchin, T. R., and Smyth, J. R., 1978, The mantle of Mars: Some possible geological implications of its high density, *Icarus*, Vol. 34, pp. 512–536.

Savre, W. C., 1963, Determination of a more accurate porosity and mineral composition in complex lithologies with the use of the sonic, neutron and density surveys, *J. Petrol. Tech.*, Vol. 15, pp. 945–959.

Steidtmann, E., 1917, Origin of dolomite as disclosed by stains and other methods, *Geol. Soc. Amer. Bull.*, Vol. 28, pp. 431–450.

Wyllie, M. R. J., Gregory, A. R., and Gardner, G. H. F., 1956, Elastic wave velocities in heterogeneous and porous media, *Geophysics*, Vol. 21, No. 1, pp. 42–59.

CHAPTER SEVEN

MATHEMATICAL ANALYSIS OF LOG TRENDS AND PATTERNS

Compositional analysis from logs or conventional geological sources, such as cuttings, core, or outcrop, is not an end in itself, but a means to an end. The scope of a geological investigation may include questions concerning stratigraphy, bedding sequence patterns, depositional facies, diagenetic changes, and other factors. Collectively, these aspects are interrelated elements of an evolution from the original sediment through a history of compaction, diagenesis, and tectonic events. Useful interpretations may be made from visual examination of raw logs, log crossplots, and processed compositional profiles, when these are coordinated intelligently with available core and cuttings. However, the numerical form of log data makes logs an ideal medium for quantitative analysis.

The intrinsic numerical character of logs contrasts with most "orthodox" geological information which is generally highly qualitative. Many rock properties are observed either to be present or absent; rock types are allocated to broad subdivisions of ranges between lithologic endmembers. While numerical data are available from petrographic analyses, they are usually subordinate to the bulk of standard geological information. This is the inevitable consequence of the high cost of detailed petrographic work in terms of money and time. Log data are both numerical and numerous.

The field of "mathematical geology" has drawn on theory from statistics, signal theory, pattern recognition, and other areas in the adaptation of methods for the numerical analysis of geological variation. In this chapter, we examine a selection of mathematical techniques applied to logging data as aids in the interpretation of geological patterns. Texts such as Davis (1973) provide a more comprehensive treatment of both theory and algorithms of a broader range of methods.

PATTERNS IN COMPOSITIONAL ASSOCIATIONS

The transformation of wireline log suites to estimates of mineral properties was reviewed in the last chapter. In each case, the product is a record of compositional variation as a function of depth. The isolation of systematic relationships between individual components may give insight into facies associations and diagenetic mechanisms. Wireline log data from a Viola Limestone section in southern Kansas are used to illustrate some useful techniques to extract patttems and interrelationships.

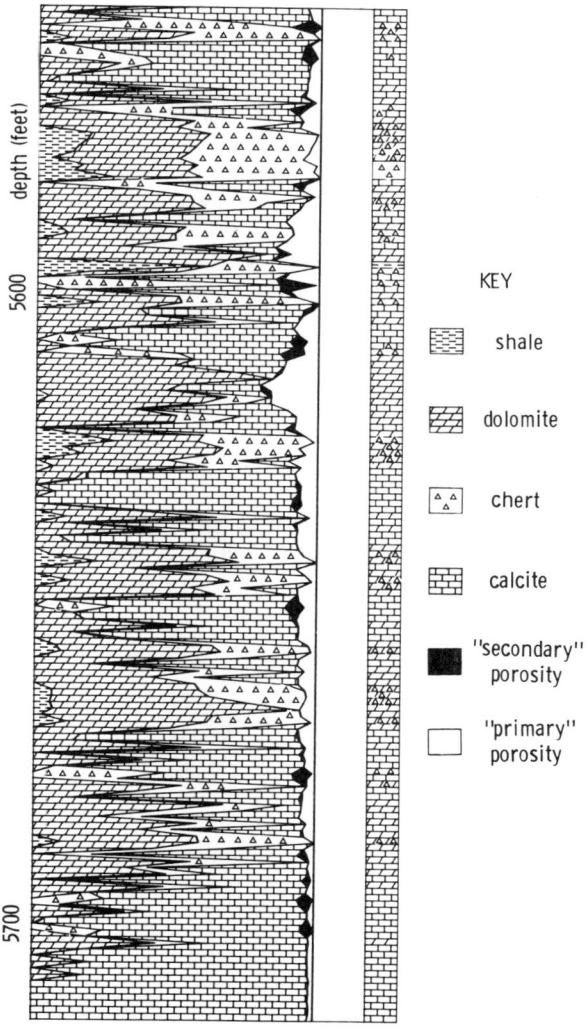

FIGURE 1. Compositional profile of the Viola Limestone in the Belcher A-1 well computed from logs using matrix algebra.

Sonic, density, neutron, and gamma-ray logs were transformed by computer processing to a depth profile of mineralogy and porosity variation. The result (Fig. 1) is expressed in terms of shale, dolomite, chert, calcite, "primary" porosity (intercrystalline and intergranular grades), and "secondary" porosity (vugs and fractures). The lithological associations indicate an interbedded sequence of limestones and cherty dolomites, with a minor shale phase.

This well is located in an area where the Viola was studied by St. Clair (1981) in a detailed petrographic examination of available core. She concluded that the Viola was composed primarily of low-porosity limestones of crinoidal packstones and grainstones which are interbedded with moderately porous cherty dolomitic limestones, representing dolomitized wackestones and mudstones. The two endmembers could be traced as correlative subdivisions across the area on both cuttings profiles and wireline log traces.

This petrographic information is useful in the geological interpretation of the log compositional profile of Figure 1. A ternary diagram of zone compositions in the Viola from this well in terms of the endmembers calcite, dolomite, and chert is shown in Figure 2. The calculation of a compositional trend to summarize the basic pattern of variation is complicated by the fact that the mineral properties collectively describe a closed system. If the three minerals are referenced to orthogonal axes, then all possible compositions are confined to a triangular plane (Fig. 3). Consequently, the three apparent dimensions of the system are really described by two axes in a plane. Coordinates of composition mixtures on this plane may be calculated by a Helmert transformation given by

$$z_1 = \frac{y_2 - y_1}{\sqrt{2}} \quad \text{and} \quad z_2 = \frac{2y_3 - y_1 - y_2}{\sqrt{6}}$$

where y_1, y_2, y_3 are proportions of calcite, dolomite, and chert and z_1, z_2 are the

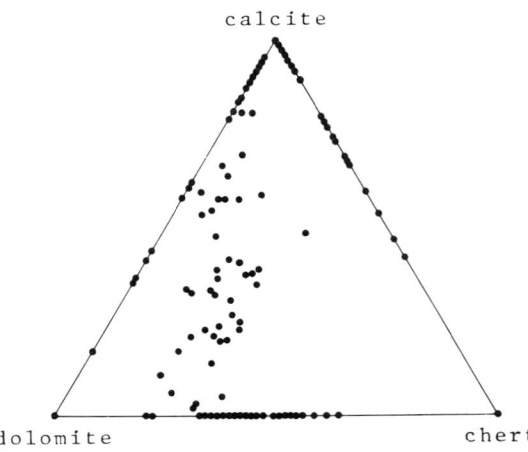

FIGURE 2. Composition triangle of zones from the Viola of the Belcher A-1 well.

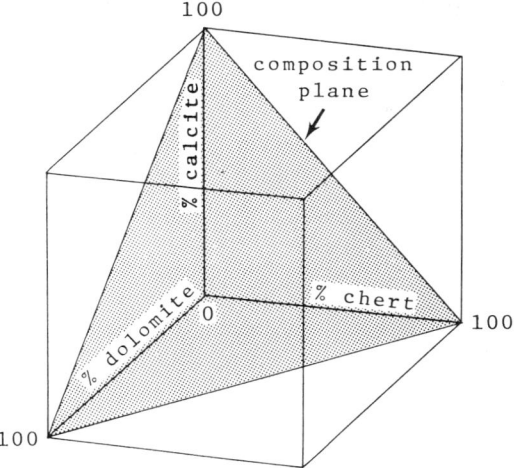

FIGURE 3. Graphic demonstration of system closure for compositional variables which shows that all possible mixtures of their endmembers are confined to a two-dimensional composition plane.

transformed planar coordinates. A reduced major axis was fitted to the transformed compositional data as a best fit line (Fig. 4). The reduced major axis is a line whose slope is given by the ratio of the standard deviations of two variables (in this case the reference coordinate axes, z_1 and z_2) and is constrained to pass through the centroid of the data. The fitted line shows a basic compositional trend ranging between two implicit endmembers of limestone and cherty dolomite. The trend matches observed data from core and cuttings reported earlier, which contrast limestone and cherty dolomitic limestone facies.

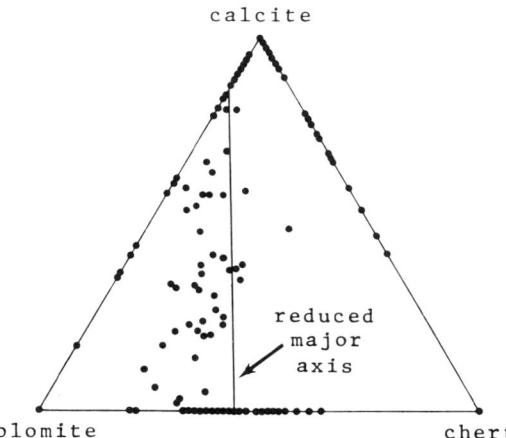

FIGURE 4. Reduced major axis computed as a best-fit trend of compositional variation of the Viola Limestone zones in the Belcher A-1 well.

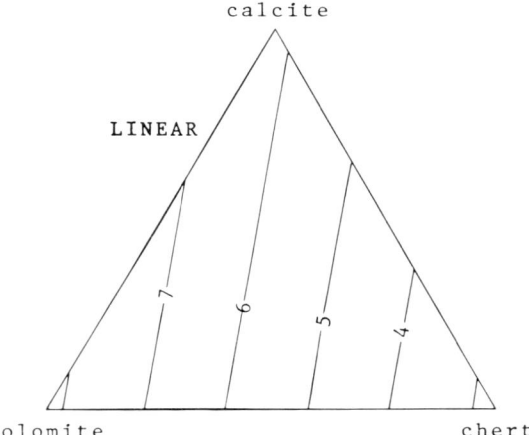

FIGURE 5. Linear trend surface of porosity variation in the Viola Limestone of the Belcher A-1 well, computed as a function of calcite, dolomite, and chert content.

Trends in porosity variation with respect to compositional variation were mapped by regression analysis. The equation for a linear trend is given by

$$\hat{p} = a_0 + a_1 z_1 + a_2 z_2$$

and describes a best-fit plane of porosity trend ($\hat{p}$) which crosses the compositional triangle as determined by the constants a_0, a_1, and a_2. This linear response surface (Fig. 5) has a fit of only 4% to the raw porosity data. However, it is statistically significant at the 5% α level and shows a generalized increase in porosity with increasing dolomite content. The equation for a quadratic (or second order) trend is

$$\hat{p} = a_0 + a_1 z_2 + a_2 z_2 + a_3 z_1 z_2 + a_4 z_1^2 + a_5 z_2^2$$

and provides a markedly improved estimation of variation in porosity with a fit of 31% (Fig. 6). Low-porosity limestones show a gradation of increasing porosity toward the chert endmember probably caused, in part, by vugs and fractures in the cherty limestones. The relatively low porosities of highly cherty dolomites may register the influence of small amounts of argillaceous material in this facies which reduces overall porosity. The marked increase in porosity moving from the limestone to dolomite compositional field is readily explained in terms of original depositional facies and subsequent diagenesis. In the core analysis study reported earlier, limestones were identified as representing crinoidal packstones and grainstones in which pore volumes had been partially occluded by syntaxial overgrowths of calcite. These were contrasted with dolomites which appeared to be matched with wackestones whose high porosities were a causative factor in their selective dolomitization.

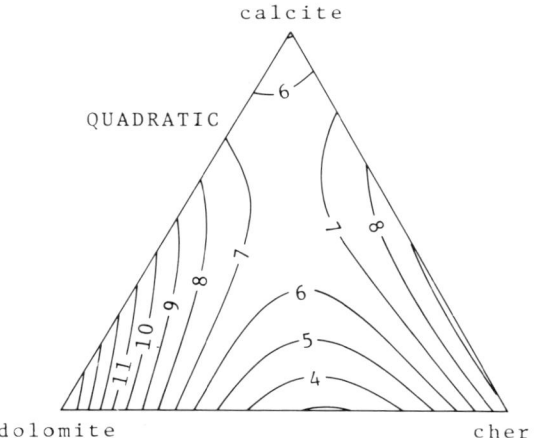

FIGURE 6. Quadratic trend surface of porosity variation in the Viola Limestone of the Belcher A-1 well, computed as a function of calcite, dolomite, and chert content.

The Pearson product–moment correlation coefficient for two variables from an open system is given by the equation

$$r_{12} = \frac{\text{cov}(x_1 x_2)}{s(x_1) s(x_2)}$$

where r_{12} is the correlation coefficient between variables x_1 and x_2, $\text{cov}(x_1 x_2)$ is their covariance, and $s(x_1)$, $s(x_2)$ are their standard deviations. The correlation coefficient can range between one (perfect correlation) and minus one (perfect inverse correlation), while zero reflects no correlation. A correlation coefficient matrix is shown in Table 1 for the components solved by log analysis in the Viola section example. The values represent the association between the components taken a pair at a time. The collective sense of these correlations gives some indications

TABLE 1
Correlation Coefficient Matrix for Viola Components

	Calcite	Dolomite	Chert	Shale	Primary Porosity	Secondary Porosity
Calcite	1.00	−0.82	−0.75	−0.44	−0.09	0.15
Dolomite	−0.82	1.00	0.27	0.11	0.30	−0.47
Chert	−0.75	0.27	1.00	0.39	−0.31	0.29
Shale	−0.44	0.11	0.39	1.00	−0.26	−0.17
Primary porosity	−0.09	0.30	−0.31	−0.26	1.00	−0.13
Secondary porosity	0.15	−0.47	0.29	−0.17	−0.13	1.00

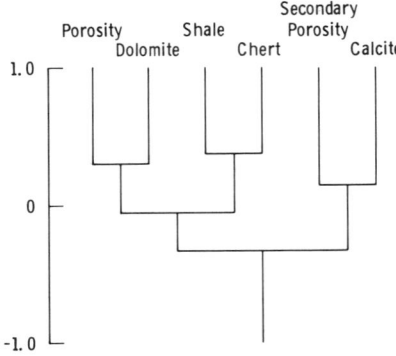

FIGURE 7. Dendrogram of relationships between Viola section mineral and porosity components resulting from cluster analysis of their correlation matrix.

of natural groupings matched with lithology types. This intrinsic classification can be brought out by the use of cluster analysis, which condenses the complex network of associations into a simple pattern. The correlation matrix was transformed by a simple clustering algorithm (Davis, 1973, pp. 458–460) to a dendrogram (Fig. 7). The dendrogram distinguishes the two endmember associations of limestones and porous cherty dolomites while showing more detailed interrelationships. All dendrograms are limited as to the amount of information that they can display without distortion, since their presentation is fixed in two dimensions. This effect can be seen on Figure 7, where secondary porosity is linked with calcite, although secondary porosity shows a maximum correlation with chert. The dendrogram should therefore be viewed as an averaged expression of component associations.

TIME-SERIES ANALYSIS OF LOG DATA

A limitation of the preceding methods of compositional analysis is that their application takes no account of depth as a contributory variable. The depth axis of any log is some monotonic function of geological time in sequences which are not faulted or overturned. As a result, most log traces are a natural expression of a time series of geological variation. Strict equivalence with an orthodox time series is complicated by variations in sedimentation rates and discontinuities caused by erosional breaks and periods of nonsedimentation. Ager (1973) even considers the "gap" to be more important than the record and the lithological succession possibly a phenomenon of quantum catastrophic sedimentation. While none of these considerations invalidate a mathematical approach, they do suggest the importance of the sedimentation model as the touchstone for the interpretation of results of methods successful in economic forecasting and electronic signal theory.

Polynomial Analysis of Long-Term Trends in Log Traces

Time series of fluctuating data, such as economic indices or meteorological variables, often are fitted by polynomial functions of time to distinguish long-term trends from local fluctuations. The application of the methodology can be illustrated in the analysis of calcite variation as a function of depth in the example Viola section. Any systematic trends that are extracted should reflect major stratigraphic patterns in the disposition of limestone and cherty dolomite facies.

A first order polynomial takes the form

$$\hat{c} = a_0 + a_1 d$$

where $\hat{c}$ is the trend value of calcite content and d is the depth, in an equation which defines a straight line. The second-order polynomial is

$$\hat{c} = a_0 + a_1 d + a_2 d^2$$

and describes a parabola, a curve with a single maximum or minimum. The general equation of an nth-order polynomial is

$$\hat{c} = a_0 + a_1 d + \cdots + a_n d^n$$

which will have $(n - 1)$ maxima or minima and $(n - 2)$ inflection points, expressing peaks/troughs and boundaries of subdivisions, respectively.

Each equation is fitted by the principle of least-squares deviation, which minimizes the variation about the trend. The degree of fit improves with increasing complexity of the polynomial equation and can be measured by the proportion of raw variation satisfied by the polynomial. A graph of polynomial fit against polynomial order is sensitive to any major "natural" subdivision of the calcite profile (Fig. 8). A fundamental subdivision appears to be located by a fourth order polynomial, since

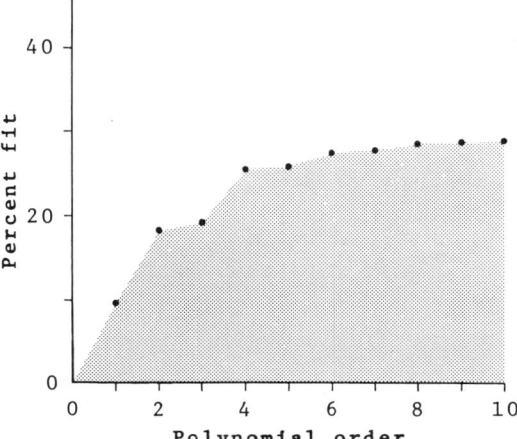

FIGURE 8. Graph of percent fit versus order of polynomial curve functions of calcite content regressed on depth in the Viola of the Belcher A-1 well.

there is both a marked decline in the rate of improvement by higher orders and an analysis of variance isolates the fourth order as the statistically significant major trend.

This graphic polynomial trend is superimposed on the Viola raw calcite variation in Figure 9 and defines five basic stratigraphic subdivisions. The lower four are readily identified with the informal units recognized from core studies in the east of the area: the Basal Limestone, the Lower Cherty Dolomitic Limestone, the Upper Limestone, and the Upper Cherty Dolomitic Limestone. Since the well is located to the west in a thicker development of the Viola, it appears that the uppermost unit represents a "third limestone" which overlies the Upper Cherty Dolomite Limestone.

Conventional methods of zonation used in basic log analysis are easily duplicated in the subdivision of polynomial trends. Individual log zones are usually identified

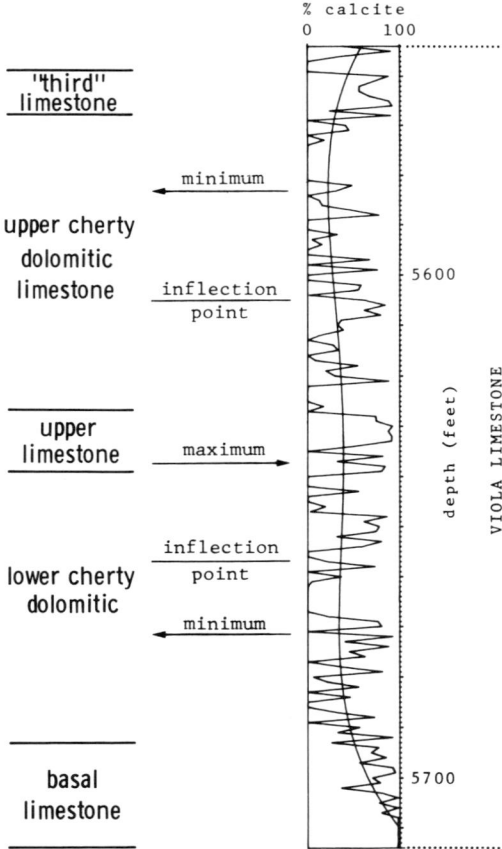

FIGURE 9. Calcite variation of the Viola Limestone in the Belcher A-1 well overlain by a fourth-order polynomial regression curve.

with distinctive peak or trough features on the log trace. Boundaries between adjacent zones are located at the position of inflection points on the curve shoulder which links the two zone extremes. A polynomial equation may be solved for depths matched with maxima, minima, and inflection points by simple methods of calculus.

In the case of the fourth order polynomial fitted to calcite variation in the Viola section, the descriptive equation can be expressed as

$$\hat{c} = a_0 + a_1 d + a_2 d^2 + a_3 d^3 + a_4 d^4$$

When this equation is differentiated with respect to depth, the result is

$$\frac{d\hat{c}}{dd} = a_1 + 2a_2 d + 3a_3 d^2 + 4a_4 d^3$$

The formula is an expression of the slope of the polynomial trend in calcite at any given depth. The slope will be zero at depths matched with the maximum and minimum extremes on the trend. When the slope equation is set to zero, the formula is a cubic expression with three roots. These are the depths of one maximum and two minima whose positions are marked on Figure 9. They represent a trend interpretation of the maximum and minimum developments of limestone within the section. If the slope equation is itself differentiated, the result is

$$\frac{d^2\hat{c}}{dd^2} = 2a_2 + 6a_3 d + 12a_4 d^2$$

and gives the rate of change of slope at any depth. When equated with zero and solved for depth, the two quadratic roots locate the depths of inflection points on the trend curve (Fig. 9).

The use of calculus on a polynomial trend function is therefore a simple method for stratigraphic subdivision, where inflection points locate boundaries, while extremes mark the zenith or nadir of the fitted variable within each zone. The criteria of this "polynomial stratigraphy" differs philosophically from conventional stratigraphic methods. Instead of matching stratigraphic horizons with discrete marker beds or distinctive breaks, the trend curve geometry accommodates the total variation of the unit. In the Viola section example, the polynomial trend subdivision represents stratigraphy based on the aggregate pattern of limestone development. As such, it responds to relatively long-term changes in character, but is less sensitive to isolated thin marker beds. Both the trend and raw variation of estimated calcite content in the Viola section give a valuable insight of the character of vertical (and, by implication, lateral) changes in lithofacies which supplement the restricted and qualitative information from core and cuttings. Boundaries between the fundamental facies units show gradual transitions which appear to reflect long-term waxing and wanings of calcite packstones/grainstones and dolomitized wackestones in the lateral migration of facies across the area. At a smaller scale, the boundaries between individual beds of these two facies appears to be fairly sharp.

Polynomial Filtering of Short-Term Trends

The polynomial functions described in the last section are a useful device to extract major trends of variation which reflect broad changes in lithofacies as perceived by logs. Since the curve-fitting procedure utilizes all the data in the interval of interest, it is sometimes termed a "global" technique. By contrast, "local" methods are useful in the isolation of systematic intermediate and short-term trends through the analysis of interval subdivisions. Their operation results effectively in a smoothing of the raw log to a trace which is sensitive to localized trends.

The simplest method to smooth fluctuating data is to compute an equally weighted moving average. The method is easy to program and requires only a specification of a window length. If a log is averaged by a window of 5 feet, the smoothed result will screen out variations whose thickness is less than about half of this dimension, while retaining variation whose scale is greater. The operation is shown diagrammatically on Figure 10. Here, a column of digitized raw log data is transformed by a moving average operator to a sequence of smoothed log values by successively sliding the operator past the log at incremental step positions. At each step, the operator elements are cross multiplied with corresponding elements on the matched log segment and the results summed. This value is then divided by a normalizing factor, which in this case is the sum of the operator elements, to obtain the smoothed estimate at this position. The arithmetic procedure of cross multiplication and summation of products is known as "convolution." The moving average operator is an example of a "filter," whose elements are one within the window and zero outside the window limits. The simple shape of this operator gives rise to its informal name as a "box-car filter." In summary, the raw log variation is convolved by a filter to a smoothed trace.

The moving average is only one example of a convolution filter which can be used in the transformation of data sequences. Since its form is extremely crude, it

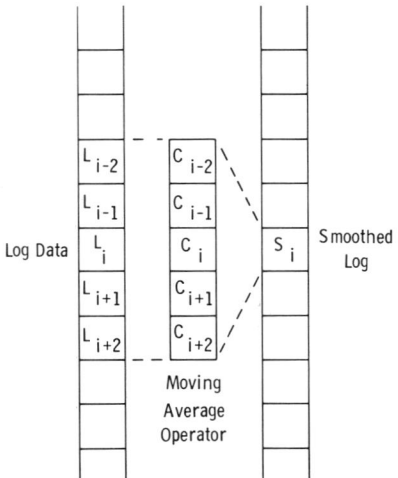

FIGURE 10. Diagrammatic representation of the operation of a moving average filter.

does not extract as much information as more sophisticated filters. A better smoothing operation would be one which caused a best-fit curve to be drawn through a data segment scanned by the window. If the criterion of fit is set by the principle of least squares and the curve represented by a polynomial function, then an approach similar to that of the preceding section can be applied.

The generalized method of polynomial curve fitting involves the calculation of sums of powers and cross products of the two descriptive variables in a matrix format, with matrix operations of inversion and multiplication. This procedure would be extremely cumbersome to apply at each successive step in the migration of the window down the length of the raw log trace. However, an equivalent computation can be achieved through the design of a convolution filter which takes advantage of the fact that successive values of depth are separated by a fixed increment on a digitized record.

When a filter is applied to fit a curve through a data segment, the value of the curve is computed at the center of the segment. The window is then moved by an incremental step and the convolution repeated to obtain the adjacent smoothed point. When all the smoothed points are linked together, the result is a continuous curve whose character is controlled by the window length and the complexity of the fitted polynomial. The values of the appropriate filter weights are therefore keyed to the requirement of the computation of the polynomial curve at the midpoint of the segment. Other filters can be designed which calculate the first derivative (slope) and second derivative (rate of change of slope) at the midpoint of the curve. The mathematical derivation of filter weights for various window lengths and a hierarchy of polynomial functions are reviewed by Savitzky and Golay (1964), who also list tables of values.

These general remarks can be clarified by a specific application to the calcite profile of the Viola section example. Let us select a window length of 9 feet and a cubic polynomial as a curve-fitting function. The cubic polynomial is defined by the equation

$$\hat{c} = a_0 + a_1 d + a_2 d^2 + a_3 d^3$$

which describes a curve with both a possible maximum and minimum. The equation is generally adequate to fit peaks, troughs, and intervening shoulders on the scale of the window span. In using this local polynomial, variations whose wavelength are less than 9 feet will be screened out effectively, but features of longer duration will be retained. The filter coefficients for a nine-point cubic function are listed in Table 2, together with the normalizing factor. The result of convolution of the calcite variation with this filter is shown in Figure 11 as a generalized curve which accentuates the bulk calcite variation.

The calcite trace was also convolved with filters which extract the first and second derivations of the cubic function at the window midpoint, using coefficients listed in Table 2. The first and second derivative logs are shown in Figure 11 and graph the slope and rate of change in slope of the cubic smoothed curve of calcite. The interpretation of these logs follows the same concepts used for the global

TABLE 2
Coefficients of Convolution Filters for (a) Smoothed Estimate, (b) First Derivative, (c) Second Derivative of a Nine-Point Cubic Polynomial

Filter Element	(a)	(b)	(c)
1	−21	86	28
2	14	−142	7
3	39	−193	−8
4	54	−126	−17
5	59	0	−20
6	54	129	−17
7	39	193	−8
8	14	142	7
9	−21	−86	28
Normalizing Factor	231	1188	462

polynomial function. Depths at which the first derivative is zero locate positions of smoothed peaks and troughs. Zero crossings of the second derivative match the positions of inflection points which mark smoothed zone boundaries separating limestone beds and cherty dolomites. The net result is a method of stratigraphic subdivision keyed to the perception of a 9-foot moving window.

The operation of smoothing filters of this type is most easily understood in terms of frequency. The nine-point filter used in the example is known as a "low-pass filter," since it screens out high-frequency variation but passes low-frequency trends. Components with a wavelength of less than nine points have been suppressed. By the selection of an appropriate window length, the user can eliminate high-frequency components beyond any specified breakpoint in the frequency spectrum. Davenport (1979) also noted that the method could be extended in the design of a "band pass" adaptation. The operation is achieved in two phases. First, the data are filtered with a breakpoint at the desired high-frequency cutoff. On a second pass, the data are filtered using a lower frequency limit. The subtraction of the second pass from the first generates a profile which restricts frequency components to those between the two frequency limits.

In practice, the choice of polynomial filters and window sizes is a decision for the user in the extraction of components deemed to be diagnostic of systematic local changes. The range of possibilities allowed by these techniques makes for a powerful and flexible methodology in the dissection of trend components over a wide spectrum of frequencies.

Fourier Analysis of Cyclic Components

The occurrence of cyclic, rhythmic, and repetitive sequences in the vertical ordering of lithologies has been discussed in the geological literature for decades. An expansive historical summary is provided by Duff et al. (1967). The isolation of a basic motif of lithological ordering simplifies the interpretation of mechanisms of deposition and diagenesis. Where such patterns exist, they can be expected to be reflected in cyclic characteristics of wireline logs and their transforms.

Signal decomposition for cyclic components is commonly resolved by Fourier analysis, which translates the variation in the "time domain" to a spectral representation in the "frequency domain." The operation of the mathematical process is analogous to the manner in which a prism breaks white light into its colored components as a spectrum of different frequencies. One of the conditions of Fourier analysis is that the input data should be stationary (no long-term drift or trend). The global fourth order polynomial fitted to the calcite data of the Viola section is an expression of long-term variation, while the residuals retain information on short-term fluctuations and are effectively stationary.

The subtraction of the polynomial trend from the raw data isolates the residual variation which is keyed with relatively short-term facies changes. It is interesting to speculate whether the pattern of these residuals is essentially random or if it is

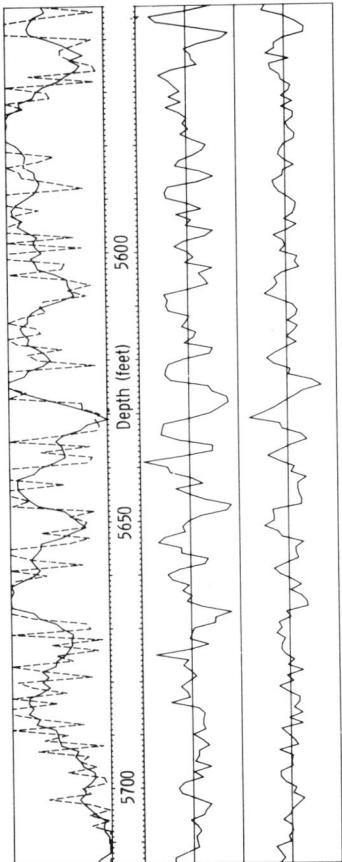

FIGURE 11. Calcite variation (dashed line) in the Viola section convolved by a 9-foot cubic polynomial (solid line), together with first- and second-derivative logs.

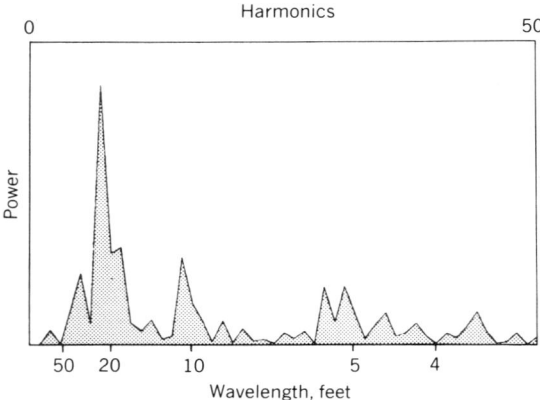

FIGURE 12. Power spectrum of residuals of a fourth-order polynomial regression fitted to calcite variation in the Viola Limestone of the Belcher A-1.

described by some type of periodic component. A random character would imply that individual limestone and dolomite beds represented chance local oscillations of facies which were totally subordinate to the major migrations of facies belts recognized in the informal stratigraphic subdivisions of the Viola. Alternatively, a systematic periodic component would suggest a cyclic mechanism of depositional facies control.

The power spectrum of the calcite residuals (Fig. 12) is marked by a pronounced peak at approximately 20 feet, with harmonics at 10- and 5-foot wavelengths. The most common explanation for harmonic generation is that it is the result of a nonsinusoidal character of the fundamental frequency as suggested by Mackay (1973). This effect would be caused by distinctive interbedding of limestones and cherty dolomites, as calcite variation would more closely follow a square wave rather than a sine-wave pattern.

The power spectrum clearly indicates a periodic structure in the alternation of facies units within the informal stratigraphic subdivisions. The subdivisions probably represent the regional migration of facies belts as a consequent of some mechanism such as eustatic changes in sea level or platform subsidence. The presence of an internal cyclic pattern indicates that the process was not abruptly episodic, but continuous and oscillatory. From this interpretation, it follows that divisions that can be correlated over the region represent trends in either a major transgressive or regressive phase which are perturbed by minor reversals, and that the whole process has a distinctive cyclic character.

Cyclic variations in lithofacies and faunal distributions within the Ordovician are most commonly attributed to eustatic changes in sea level. So, for example, Sweet (1979) traced inflection points on relative-abundance diagrams of interpreted shallow and deep-water conodont taxa in an attempt to define correlative sea level cycles. In Late Ordovician (post-Viola) times, lithofacies variations in the Appalachian Queenston Delta have been linked directly with a glacial control of sea level (Dennison,

1976). Extensive Late Ordovician glaciation has been recorded in northern Africa, southern Spain, Argentine, Peru, and Bolivia. The influence of polar ice-cap fluctuations at an earlier time is more speculative, but has been invoked to explain sedimentary cycles in the Middle Ordovician Trenton Limestone of southern Ontario by Brookfield (1982) and is the approximate time equivalent of the Viola.

Short-Term Spectral Analysis

Fourier analysis of a lengthy sequence can be considered as a "global" application, since the resultant power spectrum summarizes the frequency content of the trace variation taken as a whole. In many formations, localized trends in bedding character are to be expected and this property will be reflected in a drift of frequency composition with depth. Short-term spectral methods are used to provide a three-dimensional view of the frequency content of data as a function of time or depth. The result is conceptually similar to sonagrams or voice prints, which are analogue transforms of sound signals broken down into their frequency composition over a time range (Fig. 13).

A short-term spectral profile of a sequence can be made through the calculation of a sequence of Fourier transforms. Each transform is computed for a limited segment of the data sequence defined by a window. The window is moved progressively by incremental steps, with the repetition of the transformation at each step and conversion to a power spectrum. Each power spectrum graphs the square of the amplitude of component frequencies versus their wavelength. A composite plot of the sequence of power spectra is presented as a map of squared amplitude referenced to the two scales of time (or depth) and frequency. The size of the window is a matter for the judgment of the user. A broad window will encompass a wide spectrum of frequency content, but be less sensitive to local variations. A narrow window will be restricted as to the frequency content that it can perceive. The computations involved are formidable, but are considerably simplified by use of the fast Fourier transform introduced by Cooley and Tukey (1965).

A Fourier series represents the transformation of a signal into component frequencies matched with sine and cosine waveforms. The components are mutually independent

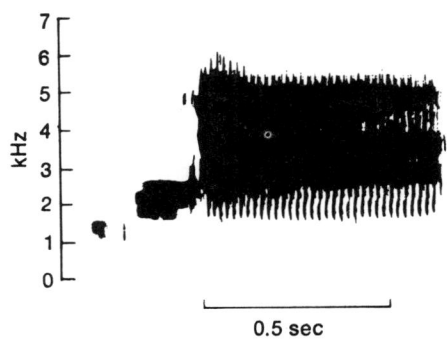

FIGURE 13. Sonagram of a song type from the repertoire of an adult male redwing. From Searcy and Yasukawa (1983).

and are calculated as a set of orthonormal trigonometric functions. Other orthonormal function sets are possible, including one proposed by Walsh (1923). Walsh functions are an orthogonal set of square waves which assume the value of either one or minus one. The first eight Walsh functions are shown in Figure 14, as a hierarchy of square waves with increasing complexity. The functions are also classified in terms of their similarity with equivalents in the Fourier series as CAL or SAL types (hybridized from "Walsh" with "sine" or "cosine"). Oscillation character is defined by the number of zero crossings. The frequency of sine and cosine waves is generally measured in cps (cycles per second). The corresponding units of "sequency" for Walsh functions are expressed in terms of zps (half the number of zero crossings per second).

The major advantage of the Walsh transform over the Fourier transform is the great reduction in computer resources required for their calculation. However, an important additional consideration is the nature of the signal to be analyzed. If the signal takes the form of a sequence of block pulses, the Walsh transform has a considerably simpler structure than the equivalent Fourier transform. This is caused by the multiple sinusoidal harmonics which are necessary to accommodate the discontinuous form of the trace. Interbedded sequences of lithologies with sharp boundaries and relatively homogeneous character conform quite closely with a model of a complex square-wave signal. When logged by wireline tools, the discontinuities will be smoothed to some degree by the averaging process of the tool over an interval set by its vertical resolution. However, log traces still remain much of the sharp boundary features in a logged sequence. Some evidence from this is suggested by the generation of multiple harmonics in the Fourier power spectrum of Viola calcite variation (Fig. 12). The use of Walsh transforms for the generation of short-term spectrograms of logs therefore has the merit of computational speed but is also compatible with the pattern of lithological variation.

The mechanics of Walsh function computation are outlined in the following simple example. Consider a segment of a signal whose values take the sequence

$$(2 \quad -1 \quad 2 \quad 1)$$

This four-point segment is completely described by the first four Walsh functions whose coefficients and waveforms are shown in Figure 15. The corresponding Walsh transform is the set of quantities

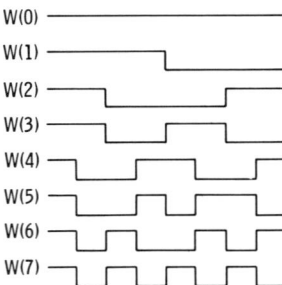

FIGURE 14. Square-wave characteristics of the first eight Walsh functions.

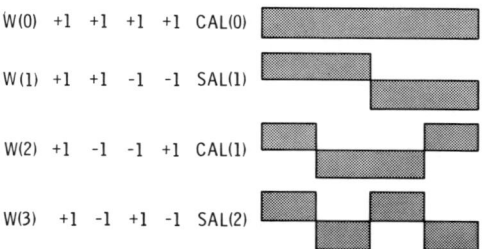

FIGURE 15. Coefficients and waveforms of the first four Walsh functions for a four-point segment.

$$[W(0) \quad W(1) \quad W(2) \quad W(3)]$$

Each transform value is computed by the convolution of the data segment with the coefficients of the appropriate Walsh function, modified by a division by the normalizing factor which is the number of segment points. The Walsh transform of the data segment is

$$(1.0 \quad -0.5 \quad 0.5 \quad 1.0)$$

and expresses the relative amplitudes of the fitted Walsh function series. If the Walsh transform is itself convolved with the Walsh functions, the result is the original data segment. The segment and the transform are a transform pair.

The computation of a power spectrum follows similar rules to those used in Fourier analysis. Power is a measure of the energy or variance of the signal at a given frequency and is equal to the squares of matched couplets of CAL and SAL functions which have the same sequency. Consequently,

$$P(0) = CAL^2(0) = W^2(0)$$

$$P(1) = SAL^2(1) + CAL^2(1) = W^2(1) + W^2(2)$$

$$P(2) = SAL^2(2) = W^2(3)$$

where $P(i)$ is the ith spectral point associated with the sequency i. The values for the power spectrum of the data segment are

$$(1.0 \quad 0.5 \quad 1.0)$$

The first element is the square of the mean value in the segment, which corresponds to the square of the amplitude of a wavelet of infinite wavelength and zero sequency. The other two elements measure the power of one and two wavelength components. The dominance of the last element reflects the alternating character of the numbers in the data segment.

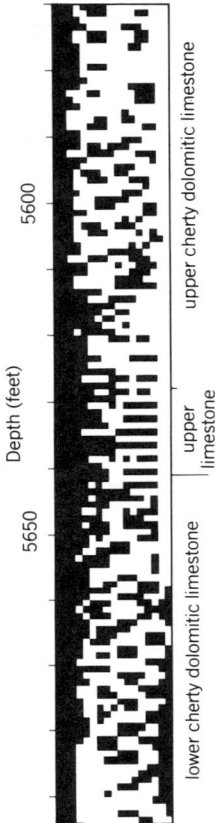

FIGURE 16. Depth versus sequency graphic spectrogram of calcite variation in the Viola section generated by a 32-point Walsh function window.

The elementary calculations of this example are indicative of the simplicity of a computer program written to process extended sequences of data such as logs. The size of the window is set at some integer power of two. Computation of any individual element of the Walsh transform involves the convolution of a matched data segment with a filter whose elements are either one or minus one. The ordering of values is prescribed by the Walsh function associated with the element. After each convolution, the result is divided by the normalization factor of the window length. The convolution is repeated at successive steps as the window is moved past the digitized sequence of log values. This process generates a moving Walsh transform of the entire sequence. When sequency-matched CAL and SAL transform elements are squared and summed, the final result is a sequency–depth plot of local power spectra.

The Viola section calcite variation was processed by Walsh functions using a 32-point window. The computer program used is listed and described by Ahmed et al. (1976). Interpretation of the results is aided by a graphic output, where higher powers are assigned darker symbols. The graphic spectrogram is shown in Figure

16 indexed against the profile of calcite variation. The spectrogram accentuates trends in sequency differentiation which reflect changes in the rate of interbedding of the limestone phase. The Lower Cherty Dolomitic Limestone unit shows a marked drift from high-sequency elements at the base, through an intermediate- and low-sequency range toward its contact with the Upper Limestone. By contrast, the Upper Cherty Dolomitic Limestone has a more diffuse character in its sequency distribution.

Because of the sensitivity of Walsh transforms to discontinuities, Lanning and Johnson (1983) have suggested that they can be usefully integrated into a computer program designed for the automatic location of bed boundaries on a log. In their paper, they demonstrate the application of the concept to the zonation of several log examples. By low-pass filtering the logs in the Walsh domain, changes in signal level are accentuated and result in a stepped version of the original data. The low computer requirements of Walsh transform algorithms is particularly valuable in allowing fast interaction and a flexibility in the implementation of different window lengths to resolve problem sequences.

PATTERN RECOGNITION METHODS

The techniques discussed earlier in this chapter have been applied to compositional variations deduced from logs. The mathematical processing of the original logs has been a two-step procedure: first, an explicit matrix algebra solution of mineral and porosity content, followed by a statistical analysis of intervariation and trends with time. As an alternative, mathematical methods may be used directly with raw log data to isolate distinctive patterns of geological variation.

This second approach is particularly useful in the distinction of lithofacies which have similar mineralogies. As an example, a limestone sequence may show little variation in composition, but reveal a textural subdivision between packstones, grainstones, and wackestones which are obvious in core. However, interbeddings of these facies are often reflected in changes in log character, caused by differences in porosity, pore geometry, and cementation character.

The cumulative responses of a log suite are a multivariate data set whose discrimination of facies can take subtle and complex forms. Each individual log may contribute useful information, but the simultaneous analysis of all logs will bring out additional diagnostic characters contained in their mutual variaton. Consequently, methods are needed which can handle the multidimensional structure of the data in a logical and practical manner.

Serra and Abbott (1980) introduced the term "electrofacies" to describe collective associations of log responses which appeared to typify certain zones and differentiate them from others. Conventional geological facies (sedimentary, faunal, metamorphic) are defined by sets of attributed which characterize certain rocks formed under equivalent conditions. "Electrofacies" differ from conventional facies in that they are empirical associations of log responses, rather than ensembles of properties thought to be of genetic significance. However, the disposition of sedimentary rock facies can be expected to be duplicated to some degree by electrofacies, since many of the diagnostic sedimentary properties have some petrophysical expression.

Principal Component Analysis

Readings of zones from a digitized log suite may be thought of as points located in a multidimensional space, whose reference axes are the logs. The gross distribution of points can be summarized by some basic statistics. A vector of the mean values of each log are the coordinates of the center of the data cloud. A matrix of the variances of each log and the covariances of each pair of logs expresses both the relative inflation of the cloud along each reference axis and the dispersion in the space between the axes. Obviously, these relatively simple descriptors will not contain all the nuances of what will generally be a complex cloud pattern of points. Instead, they model the cloud as a hyperellipsoid, with its centroid located by the log means and cross-sectional profiles specified by the variances and covariances.

These basic ideas can be clarified in a simple example. Consider a hypothetical crossplot of density and neutron porosity readings from a sequence of logged zones (Fig. 17a). Since the two logs are recorded in radically different units, it is appropriate to rescale them in a standardized form. This step will ensure that the difference in measurement units neither obscures nor distorts the underlying pattern of mutual variation. Standardization is achieved by relocating the cloud centroid to a new origin (equivalent to transforming both log means to a measurement of zero), and

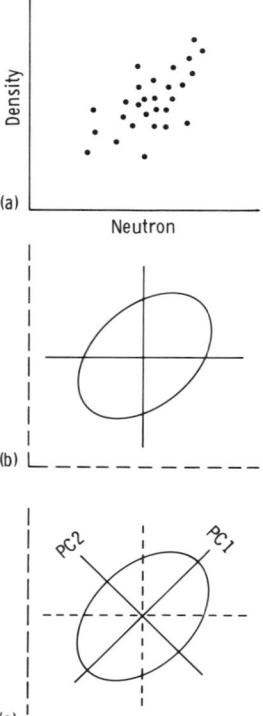

FIGURE 17. Schematic diagram sequence of the geometry involved in principal component analysis. (a) Hypothetical neutron–density crossplot; (b) correlation matrix ellipse superimposed on standardized neutron–density axes; (c) location of ellipse principal axes by eigenvectors.

making units of variation on each log axis equal to one standard deviation. The standardization converts the variance–covariance matrix to a matrix of correlation coefficients. The correlation matrix is described geometrically by an ellipse (Fig. 17b). The eigenvectors of the correlation matrix locate the major and minor axis of the ellipse. The eigenvalue associated with each eigenvector gives the relative length associated with each axis. These eigenvectors are called the "principal components" of the correlation matrix and define a rotated set of axes which are keyed with the principal sources of variation in the raw data (Fig. 17c).

In this example, data points projected onto the first principal component are scaled in terms of the major source of common variation between the neutron and density logs. Their relative locations on this axis are calculated as principal component "scores." The residual variation is contained on the second principal component. If the density and neutron porosity log readings were perfectly correlated, the data cloud would form an alignment, the descriptive ellipse would be constricted to a line, and only one principal component would account for all variation. Put another way, the apparent two-dimensional structure of the data would collapse to only one dimension of information, since a second dimension is redundant.

When this operation is applied to a larger set of logs, the mathematics remains unchanged, although the geometry becomes difficult to visualize. The eigenvectors of the correlation matrix locate the axes of a hyperellipsoid which represents the data cloud, while the eigenvalues measure their relative elongation. By ordering the eigenvalues from largest to smallest, the associated eigenvectors are ordered in their importance in absorbing the total variation of the data cloud. The proportion of the total variation satisfied by any principal component is given by its eigenvalue divided by the sum of all the eigenvalues. For standardized data, this eigenvalue sum is equal to the number of logs used in the analysis.

Mutual similarities in response by different logs gives a result in which the first few principal components absorb almost all of the entire variation of the log suite. This is an important property, since it means that the original, poorly tractable structure of many dimensions may be condensed to a few descriptive axes with little loss in information. The information content of commonly used logs in terms of lithological variation can be examined in the analysis of an Ordovician sequence from northern Kansas. This same sequence was used to illustrate log overlay interpretation in Chapter 5 (Figs. 2–4). The contained lithologies are those most common in the stratigraphic record, namely shales, sandstones, limestones, dolomites, and cherty carbonates. Six logs were used in the analysis: spontaneous potential, gamma-ray, root conductivity, sonic, density, and neutron porosity. The square root of the conductivity was chosen in preference to the raw resistivity trace in order to approximately linearize its variation, as suggested by the Archie equation.

The first two principal components of these six log variables account for 98% of the total variation. A crossplot of zone projections on the two axes, given by their principal component scores, is shown on Figure 18. Different symbols were allocated to the Ordovician stratigraphic subdivisions, and boundary lines drawn to link endmembers of minerals and pore fluid. The extremely high-information content of the first two principal components is clear evidence that the bulk of the

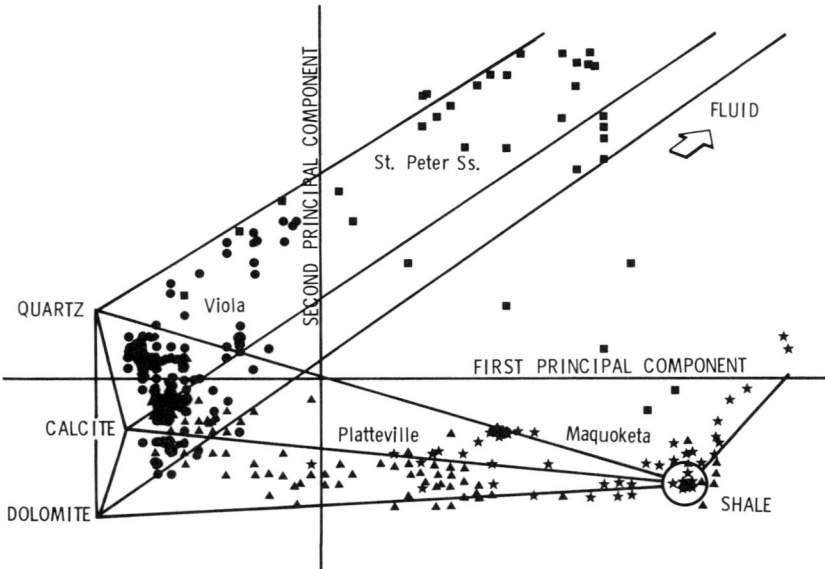

FIGURE 18. Crossplot of the two major principal components of six logs run through a Kansas Ordovician section. Key: squares, St. Peter Sandstone; triangles, Platteville formation; circles, Viola Limestone; stars, Maquoketa Shale.

data cloud is constrained closely to a two-dimensional plane which obliquely crosses the six-dimensional space of the log measurements. This is not a surprising result, since the logs are collectively most sensitive to variations in shale content and pore volume. The principal component analysis has automatically extracted these two intrinsic factors as major descriptive trends.

The condensation of the readings from a large suite of logs to a smaller set of composite variables simplifies considerably the task of electrofacies recognition. Wolff and Pelissier-Combescure (1982) used principal component analysis as an initial data condensation phase applied to an extensive suite of logs. In addition to the common logs, they included shear character and amplitude ratios derived from sonic waveform measurements, as well as correlation curve activities and correlation frequencies processed from high-density dipmeter runs. The expanded data set included information which is sensitive to textural variations of the logged lithologies. The distribution of zone data points could then be examined in the reduced dimensional space matched with the first few important principal components.

Zones with similar gross petrophysical aspect will tend to be clustered as localized cloud concentrations of data points. The clouds will have different shapes and sizes, as dictated by the internal variability of each electrofacies and the effects of log errors. Wolff and Pelissier-Combescure (1982) applied a combination of clustering algorithms to locate and aggregate similar zones into a range of mutually distinct electrofacies types. Core data and geological experience are used to help in decisions concerning the appropriate level to terminate the clustering process.

The geological meaning of each electrofacies can be deduced from cored sections of a well. Zones assigned to a common electrofacies are compared with depth-equivalent core samples to isolate descriptive themes of mineralogy and texture. If successful, the method can be extended to the prediction of equivalent lithofacies in uncored intervals. The basic idea of this approach has great potential as a means to integrate logging data with petrographic observation in sedimentological interpretations of log suites.

Discriminant Function Analysis

As an alternative method, log response associations with lithofacies observed in cored intervals may be analyzed directly by discriminant function analysis. The approach differs philosophically from the use of principal components and clustering, where electrofacies are defined implicitly from the data structure and then matched with lithofacies. Instead, lithofacies are identified a priori on the core and logged zones allocated between them. A classification function can be developed that attempts to distinguish separate lithofacies, based on the log responses of their contained zones. If the function appears to be effective, then zones in intervals which are not cored can be assigned to one or other of the lithofacies on the basis of their log characters.

Discriminant function analysis has been applied successfully in many geological studies (e.g., Middleton, 1962; Chayes, 1964). Mathieu and Rice (1969) distinguished sections dominated by sandstone from those dominated by shale in a discrimination between seismic events based on their amplitude characterization. Heseldin (1976) developed a discriminant function to differentiate dry holes close to oil production from those farther away, from an analysis of a set of log responses.

Discriminant function analysis is both a means to distinguish samples from predefined populations in terms of measurement variables and a classification tool to type individuals of unknown affinities. The simplest (and most common) application is directed to the linear discrimination of two populations. Implementation of the method is prefaced by the assignment of a test group of samples into two classes (each corresponding to a distinctive population) on the basis of prior geological knowledge. If k variables are measured for each sample individual, the samples may be plotted as two data point clouds in k-dimensional space. The location and shapes of the clouds are specified by their multivariate means, variances, and covariances.

It is possible to locate an axis on which the distance between each class cluster is maximized while, at the same time, the dispersion within each cloud is minimized. The axis is specified by a linear discriminant function, or equation which is derived by a computational treatment of the multivariate means, variances, and covariances of the two clusters. A detailed mathematical background on the method is given by Davis (1973). The equation takes the form

$$Z = d_1 X_1 + d_2 X_2 + \cdots + d_k X_k$$

where d_i is the discriminant coefficient associated with the variable X_i, and Z is the "discriminant score" which locates the position of any individual on the axis according to its variable measurements. Substitution of the midpoint between the sample means locates a "discriminant index" value. The index defines a partition point on the discriminant function which marks a boundary between the two classes. Discriminant scores may be used to classify individuals with one or other of the classes, depending on which side of the discriminant index they lie. In practice, the index may be displaced to an alternative position which causes more effective discrimination of the classes from the test group.

A simple pictorial example is illustrated in Figure 19, based on two hypothetical variables. The two group classes are not successfully distinguished when related to a single variable. Plotted in two dimensions, a distinctive separation becomes apparent, and a discriminant function discriminates the two classes. The example is idealized, in the sense that a perfect separation is illustrated, unlike realistic situations where there is commonly an overlap of distribution clouds across the discriminant index. In these cases, assignation of unknown individuals to one or other of the classes is framed in terms of likelihood. The performance of discriminant function analysis applied to real data suggests that it is one of the most robust and powerful tools available for multivariate analysis.

An application of discriminant function analysis to log differentiation of lithofacies is described in the following example. A Mississippian section in western Kansas

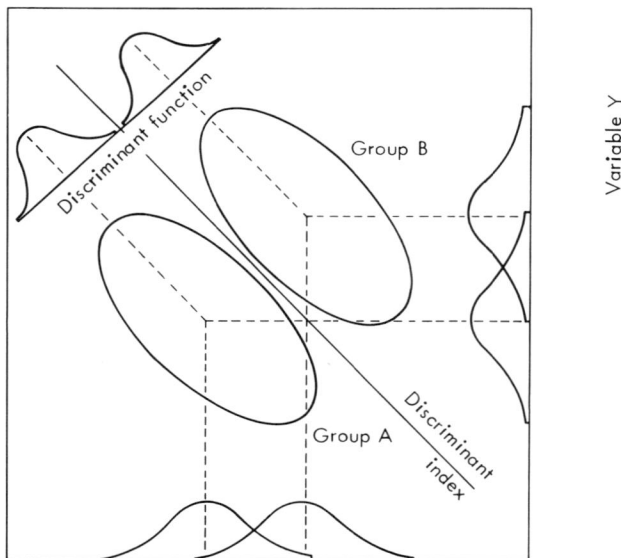

Variable X

FIGURE 19. Pictorial representation of discriminant function analysis applied to a simple bivariate example. Modified after Davis (1973).

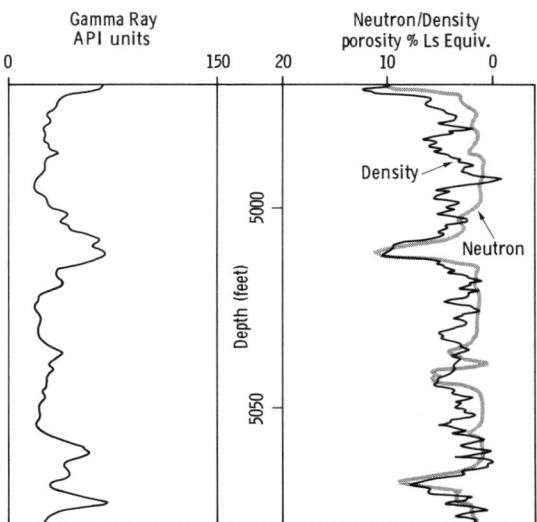

FIGURE 20. Gamma-ray, density, and neutron logs of a Mississippian section from western Kansas.

was both cored and logged over an extensive interval. Core examination showed a sequence dominated by limestones and sandy limestones, with subordinate shaly sandstones and siltstones. Many of the limestone beds are fine grained, with fossil contents of sponges, algae, and occasional brachiopods. At various horizons, the limestones are oolitic and sometimes show distinctive cross-bedding structures. The sequence is interpreted in the context of a depositional model of a carbonate shelf with high-energy oolite shoals grading to wackestones and mudstones of an inner shelf lagoon. Interfingered shaly sandstones and siltstones represent prograding clastics from a probable western land source.

Core samples which were marked oolitic were distinguished as one class of lithofacies and differentiated from the remainder as a second class. Discriminant function analysis was made between the two classes, based on the depth-equivalent zone readings from the gamma-ray, density, and neutron logs (Fig. 20). The computed discriminant function took the form

$$Z = -0.153G + 45.4D + 1.77B$$

where G, D, N are the gamma-ray, density, and neutron responses. The radically different scales of the three logs mask their relative contribution to the composite discrimination in this equation. However, the discrimination is dominated by the neutron (78%) and gamma-ray (16%) logs, with a minor influence of the density log (6%). The role of the gamma-ray log is clear, in the distinction of low-radioactive, high-energy oolites from the shaly phase present in both inner shelf wackestones and mudstones and the prograding clastics. The composite loadings of the two porosity logs appear to be a more complex function which makes a distinction in terms of both porosity and mineralogy between the two classes.

More detailed interpretation is considerably simplified through a comparison of a "discriminant score log" with a mineralogy/porosity profile of the sequence computed by matrix algebra processing (Fig. 21). Zones classified as oolites can be seen to coincide broadly with segments of high calcite content in a sequence marked by long-term fluctuations in mineral composition. The processed logs are readily interpreted in terms of lateral facies migrations between oolite shoal, inner shelf lagoon, and near-shore clastics. The common theme of the compositional profile and the discriminant score log is demonstrated by crossplot of discriminant scores against estimated calcite content (Fig. 22). This plot demonstrates the sensitivity of the discriminant function as a generalized measure of current energy, both in the trend to high-energy oolites at the top of the plot, and a misclassification of some sandstones at the base.

As applied in this example, discriminant function analysis can be appreciated as more than simply a classification tool to distinguish two discrete classes of facies. If samples from two extreme endmember classes can be identified, observations from the remainder of a sequence can be converted to scores which reflect their relative position on the intervening continuum. An interpretation of the analysis in terms of process sedimentology gives the necessary insight to the meaning of a discriminant log. This synthetic log has the valuable property that it condenses the pattern of variation with time from its multiple log source to a single profile.

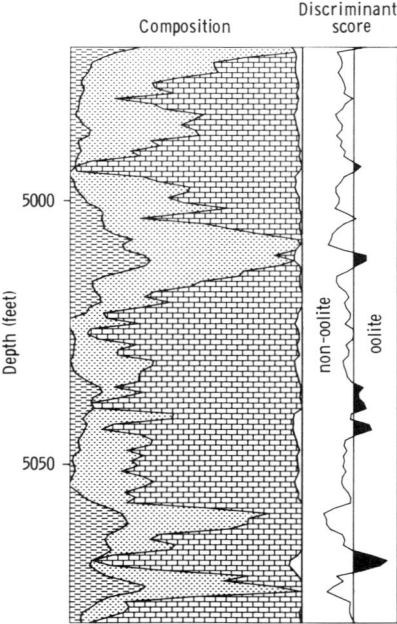

FIGURE 21. Compositional profile and discriminant score log of the Mississippian section.

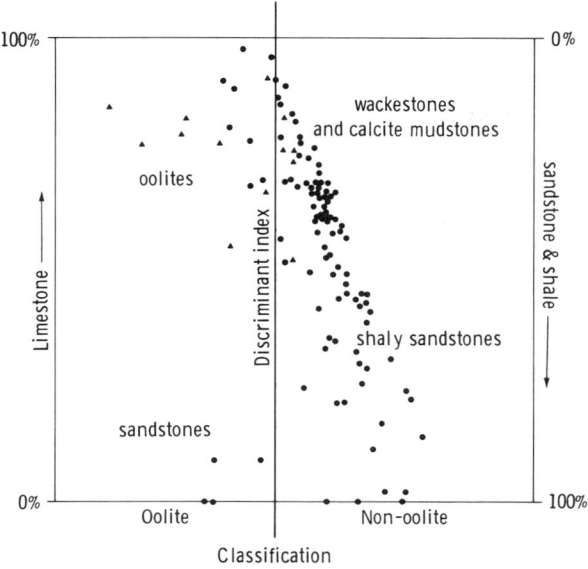

FIGURE 22. Crossplot of discriminant scores versus calcite content in the Mississippian section.

Discriminant function analysis has the basic restriction that it is limited to the distinction and classification of two classes of observations. Many studies can be structured in terms of this bipartite differentiation, either in the recognition of two classes or two endmembers of a continuum. In other cases, a larger number of classes is a more appropriate description of variations in compositional attributes and processes of origin. These types of problems can be treated through the use of canonical analysis (alternatively known as multiple discriminants). The mathematics of canonical analysis is a multivariate extension of discriminant function analysis. Multivariate means, variances, and covariances are computed for variables measured on individuals assigned to each of several classes. The axes are ordered in their effectiveness in total discrimination. Each has a descriptive equation which allows the computation of a score of any individual set of observations as a projection onto the axis. Discriminant function analysis can be seen to be the simplest limiting case of canonical analysis in the computation of a single axis of differentiation.

The distinction of a larger set of classes necessarily complicates the problem of differentiation and interpretation. However, the methodology and application is simply an amplification of the basic approach of discriminant function analysis. In this more complex case, a suite of discriminant score logs will be the final result of computation. Each log is keyed to different aspects of class distinction whose geological meaning can be interpreted from consideration of the variable loadings on its associated axis and the use of crossplots. In most cases, the high correlation between many of the common logs will result in a drastic condensation of a large number of input logs to a minor set of discriminants which contain the major source of differentiation between the facies classes.

REFERENCES

Ager, D. V., 1973, *The Nature of the Stratigraphical Record*, John Wiley & Sons, Inc., New York, 114 pp.

Ahmed, N., Natarajan, T., and Rainbolt, H. R., 1976, On generating Walsh Spectrograms, *IEEE Trans. on Electromagnetic Compatibility*, Vol. EMC-18, No. 4, pp. 198–200.

Brookfield, M. E., 1982 Glacio-eustatic sedimentary cycles in Trenton limestone (Middle Ordovician) of southern Ontario (abs.), *Bull. Am. Assoc. Petrol. Geolog.*, Vol. 66, No. 8, p. 1165.

Chayes, F., 1964, A petrographic distinction between Cenozoic volcanics in and around the ocean, *J. Geophys. Res.*, Vol. 69, pp. 1573–1588.

Cooley, J. W., and Tukey, J. W., 1965, An algorithm for the machine computation of complex Fourier series, *Math. Computation*, Vol. 19, pp. 297–301.

Davenport, C. M., 1979, Digital Filtering Techniques for Use in Minicomputers, U.S. Department of Energy Report Y-2184, 22 pp.

Davis, J. C., 1973, *Statistics and Data Analysis in Geology*, John Wiley & Sons, Inc., New York, 550 pp.

Dennison, J. M., 1976, Appalachian Queenston delta related to eustatic sea-level drop accompanying late Ordovician glaciation centered in Africa, in *The Ordovician System* (M. G. Bassett, Ed.), University of Wales Press, Cardiff, pp. 107–120.

Duff, P. Mcl. D., Hallam, A., and Walton, E. K., 1967, *Cyclic Sedimentation*, Elsevier, Amsterdam, 280 pp.

Heseldin, G. M., 1976, Discriminant analysis in petrophysics, *Trans. Soc. Prof. Well Log Analysts 17th. Ann. Logging Symp.*, Paper OO, 10 pp.

Lanning, E. N., and Johnson, D. M., 1983, Automated identification of rock boundaries: An application of the Walsh transform to geophysical well-log analysis, *Geophysics*, Vol. 48, No. 2, pp. 197–205.

Mackay, D. B., 1973, A spectral analysis of the frequency of supermarket visits, *J. Marketing Research*, Vol. 10, pp. 84–90.

Mathieu, P. G., and Rice, G. W., 1969, Multivariate analysis used in the detection of stratigraphic anomalies from seismic data, *Geophysics*, Vol. 34, No. 4, pp. 507–515.

Middleton, G. V., 1962, A multivariate statistical technique applied to the study of sandstone composition, *Trans. Royal Soc. Canada*, Vol. 56, Ser. 3, Sec. 3, pp. 119–126.

St. Clair, P. N., 1981, Depositional History and Diagenesis of the Viola Limestone in South-Central Kansas, unpublished M.S. Thesis, University of Kansas, 66 pp.

Savitzky, A., and Golay, M. J. E., 1964, Smoothing and differentiation of data by simplified least-squares procedures, *Analytical Chemistry*, Vol. 36, No. 8, pp. 1627–1639.

Searcy, W. A., and Yasukawa, K., 1983, Sexual selection and Red-winged blackbirds, *American Scientist*, Vol. 71, pp. 166–174.

Serra, O., and Abbott, H. T., 1980, The contribution of logging data to sedimentology and stratigraphy, *Society of Petroleum Engineers Preprint 9270*, 19 pp.

Sweet, W. C., 1979, Late Ordovician conodonts and biostratigraphy of the western Midcontinent province, *Brigham Young Univ. Geol. Studies*, Vol. 26, Part 3, pp. 45–85.

Walsh, J. L., 1923, A closed set of normal orthogonal functions, *American J. Math.*, Vol. 45, pp. 5–24.

Wolff, M., and Pelissier-Combescure, J., 1982, Faciolog-automatic electrofacies determinations, *Trans. Soc. Prof. Well Log Analysts 23rd. Ann. Logging Symp.*, Paper FF, 23 pp.

CHAPTER EIGHT

REMEDIAL CORRECTION OF LOGS

All physical measurements are subject to error. Although logging service company standards are generally high and tools are carefully calibrated in test pits and field operations, small deviations from true values are to be expected routinely. Systematic errors are introduced by environmental effects of borehole rugosity, variable mudcake thickness, and degree of formation invasion by mud filtrate. Deviations from an ideal situation of a straight, in-gauge hole with minimal invasion can be remedied through the use of "environmental corrections." Test pit simulation run by logging service companies provides the necessary engineering data for log corrections over ranges of variation in the borehole environment. These accommodate changes in borehole diameter, mud weight and resistivity, mudcake thickness, and other factors. The corrective measures can either be drawn from graphs published in service company chartbooks, or applied as corrective equations in computer processing.

As part of the continuous improvement in tool design, many of the basic environmental corrections are incorporated automatically as a modification to the raw sonde readings at the time of the logging run. The use of dual detector systems on modern sonic, density, and neutron tools provides two sets of measurements whose analytical comparison is used to screen out effects induced by borehole variability. At the same time, the caliper record of the borehole diameter monitors the rugosity of the borehole and is used to compensate for the influence on log measurements.

The careful planning and execution involved in the drilling program of most modern wells results generally in a well-conditioned borehole. Under these conditions, environmental corrections to log traces are relatively minor in character. However, locally unfavorable geological sequences may cause significant degradations in hole quality. The occurrence of such sections is usually obvious both on the caliper log and anomalous behavior of other logs. Corrective measures are then essential to eliminate spurious effects and rectify raw log measurements to closer estimates of their true values.

Even after the most comprehensive environmental corrections, the log trace still includes a distinctive source of error which is compounded from tool malfunctions, shop and field miscalibrations, and operator error. Far from being a libel of service company standards, this error is a fact of life which applies to all physical measurements, whether made in the subsurface or the laboratory. Procedures that attempt to eliminate logging measurement error are usually referred to as "normalization" or "quality control." Neinast and Knox (1973) and Patchett and Coalson (1979) have reviewed a variety of approaches to the art of normalization and have discussed the merits and drawbacks of each technique in a number of illustrative examples. The most commonly used methods are reviewed in the following sections.

SINGLE-WELL METHODS FOR NORMALIZATION

Use of Calibration Units

The well bore may penetrate certain stratigraphic horizons that can be used for internal calibration of the well logs. The most obvious example is an anhydrite bed, whose physical properties are distinctive and allow the logs to be standardized using the anhydrite as a reference. More commonly, a more pragmatic decision must be made concerning a suitable calibration unit. Certain shales or relatively tight carbonate horizons are often selected on the basis that they can be correlated over a wide area and their physical properties appear to vary only to a minor degree; thus, while not precisely constant, they are at least "consistent."

The notion of consistency has intrinsic problems. All lithologies show some degree of variation in composition and texture. As a result, there is no unique value that is representative of the log response of even a well-chosen calibration unit, but (hopefully) there will be a narrowly restricted range of variation. Changes in calibration unit properties will be linked with geography, reflecting a combination of depositional facies variation, diagenesis, and compactional effects linked with structural position. This composite factor results in a geographic drift of the calibration standard, with a selective bias in the calibration of any well log according to its location.

To a certain extent, the systematic geographic effect can be compensated by the procedure suggested by Neinast and Knox (1973) who stated that, "daisy chain circles are used from a base map to cross-check the possibility of normalizing gradual formation changes instead of tool variation."

Application of Crossplots

Subsurface units with simple mineralogies (clean sandstone, pure limestones, and dolomites) and logged by two or three porosity logs can be crossplotted and their trends compared with theoretical sandstone, limestone, and dolomite lines to determine the necessary shifts to be applied to the logs. *M–N* plots may also be used for this purpose. Resistivity logs can be corrected through computations of theoretical porosities in water-saturated units and comparison with porosity logs on a dual-porosity crossplot.

Comparison with Core Analyses

If cores are taken from an interval within the well, engineering data of porosity and grain density may be used in the correction of the porosity logs. A comparison of the mean porosities from core and a porosity log allow the computation of the necessary shift to be applied to the log. It should always be remembered that core measurements are themselves subject to error, and also that different methods of porosity measurement (using Boyle's law for gas expansion or the summation of fluids) may yield systematically different results. A particularly important consideration is the difference in sample size when relating core with log measurements. A vertical profile of porosity measurements from small plugs will generally show more variation than a porosity log, since the log is smoothed by the coarser vertical resolution of the tool. However, the means of the two sources of porosity data for the same interval should be similar, because they are estimates of the same parameter.

As a preparatory step to normalization, the various logs which are sensitive to porosity must be shifted to common depth registration with one another and with the cored interval. The same problem occurs in the processing of dipmeter logs, where the relative vertical shifts in the separate microresistivity traces must be determined. A simple mathematical method for this operation is that of cross correlation, in which correlation coefficients between traces are computed for a sequence of match positions. At match position, k, the correlation coefficient is

$$r_k = \frac{\text{covariance (trace 1} \cdot \text{trace 2)}}{\text{s.d. trace 1} \times \text{s.d. trace 2}}$$

where s.d. signifies standard deviation. By progressively migrating the traces past each other at incremental shifts or lags, the computed correlation coefficients can be graphed against lag to determine the position of best match. The procedure is illustrated in the following example.

A test well was drilled through the Arbuckle of northern Kansas and logged by a variety of tools. Engineering measurements were made at 1-foot intervals of a 24-foot cored section of cherty dolomite. Neutron, density, sonic and laterolog resistivity logs of the general interval were each cross correlated with the core sequence of porosities to determine the best match. These match positions dictated the necessary vertical shifts of the logs to converge on a common depth registration. The plots of cross correlation versus lag are shown in Figure 1. (Prior to computation, the laterolog readings were transformed to root conductivity values to approximately linearize variation with porosity.)

Once common depth registration is satisfied, the core and log intervals are positioned for normalization operations. If grain densities and porosities are measured on core, the average bulk density of the cored interval is the appropriate criterion to normalize the density log. Transit times can be measured on core samples, and

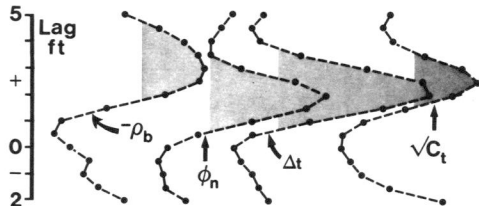

FIGURE 1. Cross correlation of logs with core data graphed against lag shift in a section of Arbuckle cherty dolomite.

their average would be suitable to calibrate the sonic log. However, this information is not commonly available on routine core analysis reports. Instead, apparent transit times may be estimated if the lithology is simple, based on mineralogical compositions deduced from the grain density. Following a similar logic, neutron porosities can be estimated from core measurements of grain density and porosity. The averages of these estimates of transit time and neutron porosity over the interval are useful (if not precise) standards to compare with corresponding sonic and neutron log averages. In the Arbuckle well example, the comparative figures from core and logs were

	Core	Log
Average bulk density	2.64	2.64
Estimated average transit time	60.3	57.0
Estimated average neutron porosity	14.1	14.3

These data show extremely close agreement between core and log estimates of average bulk density and neutron porosity, with no necessity for correction. The discrepancy between the respective transit times is easily explained. Vugs are reported in several of the core samples and would generate slower transit times in the computation of apparent transit time based on total porosity. The conclusion from this specific example is that log quality in this well was extremely good. However, in many other cases, systematic deviations between core and log data would form the basis for log normalization.

DUAL-WELL METHODS

Overlays

In neighboring wells with similar stratigraphy, an overlay of the same log from two wells, correlated to match equivalent units, gives an immediate indication of the relative shift between their separate calibrations. The method does not directly indicate whether one or both of the logs requires correction, or indeed the magnitude of the correction(s). Results from methods described for single wells must be incorporated for these decisions.

Histograms

If the same log in different wells is digitized for a stratigraphically equivalent interval, histograms of the relative frequency of different readings may be compared to determine a relative shift to match the histogram peaks and troughs in moving between the wells. Again, as with the overlay technique, some external criteria are needed to provide explicit information on the necessary corrections.

MULTIWELL METHODS FOR NORMALIZATION

Simple Mapping

The average log response in a correlateable interval from a set of wells can be mapped areally as a continuous surface. Localized anomalies in the resulting contour map indicate possible problem wells whose calibration appears to deviate significantly from the surrounding well control. The method is most successful when applied to an interval whose log response is likely to show only gentle regional changes. Under this condition, local anomalies are more likely to reflect measurement error rather than representing genuine local variation.

Trend Surface Analysis

The empirical method of simple mapping may be formalized in a statistical model through the use of trend surface analysis. The theory is drawn directly from regression, which partitions observational data between a regional systematic component and a spatially uncorrelated "random" element. The regional component is estimated by a regression of the mapped variable on polynomials of the geographic coordinates of the well control. The regional surface is fitted to the data in such a way that the sum of the squares of their deviations from the surface is the minimum possible (Fig. 2).

A hierarchy of trend surfaces matches polynomial equations of differing degrees of complexity. A linear (first order) surface is a plane; a quadratic (second order) surface is a paraboloid which may describe a basin or a positive feature. Higher order trend equations will represent successively more complex surfaces. As a regression procedure, trend surface analysis functions as an analysis of variance, in which the total variation of the data is partitioned between variation accounted

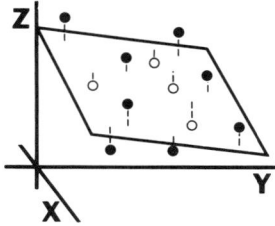

FIGURE 2. Hypothetical linear trend surface (a plane) fitted to a variable, Z, measured at locations with coordinates X and Y.

for by the surface and the remainder as the sum of the squared deviations from the surface. When the residuals show no spatial correlation, the associated surface contains an estimate of the systematic spatial structure of the data variation.

This model can be applied to the mapping of a log response in a "calibration unit." A calibration unit is defined as a geological horizon which is judged to show only minor variation across the region of interest. The definition implies that a log response measured at any location can be subdivided into two components: a systematic regional element, and a random error associated with the miscalibration of the tool. The theory is best illustrated in a practical example of the method described by Doveton and Bornemann (1981).

A neutron log of the Viola Limestone is shown from a well in south central Kansas (Fig. 3). At the base of the Viola is a distinctive "low porosity zone," which can be correlated in all wells of the area. The Lower Limestone is essentially a pure limestone with only minor amounts of chert and dolomite, and appears to have a very restricted range of porosity on the order of 3%. Its simple mineralogy, moderately uniform porosity, and absence of hydrocarbons suggest that it is a reasonable choice for a log calibration unit. Analytical data were drawn from 254 wells which penetrated the Viola and were logged by at least one porosity tool. The porosity logs were digitized over the range of the Viola and the total sample consisted of 194 neutron, 62 sonic, and 36 density logs. In each well, average neutron porosity, transit time, and density readings were calculated for the Lower Limestone, with an exclusion of the upper and lower 2 feet of the unit to minimize

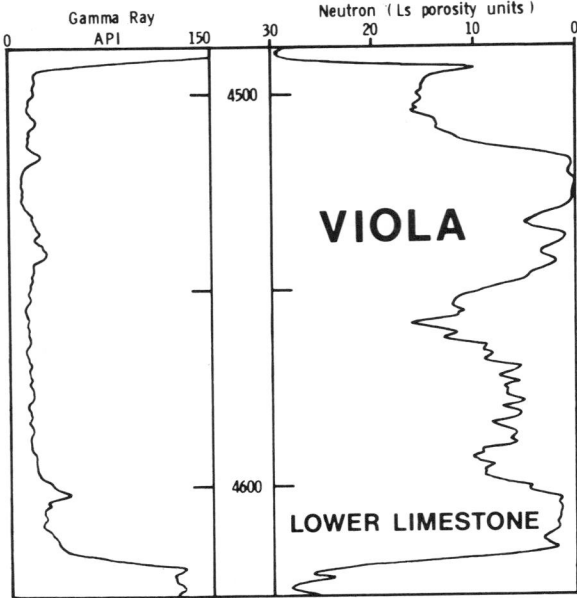

FIGURE 3. Representative neutron log of the Viola Limestone from a well in south-central Kansas.

TABLE 1
Analysis-of-Variance Table of Trend Surfaces Fit to the Lower Limestone Neutron Data[a]

Source of Variation	Sum of Squares	Degree of Fit	Mean Squares	F Ratio
Linear regression	27.22	2	13.61	5.73*
Linear deviation	377.62	159	2.37	
Quadratic-linear regression	21.41	3	7.14	3.13*
Quadratic deviation	356.22	156	2.28	
Cubic-quadratic regression	5.99	4	1.50	0.65
Cubic deviation	350.23	152	2.30	
Total variation	404.85	161		

[a] Asterisks indicate F ratios significant at the 5% level.

the effect of adjacent beds. The three sets of neutron, sonic, and density readings were made independent subjects for trend surface analysis.

Linear, quadratic, and cubic trend surfaces calculated for the Lower Limestone neutron data had fits of 6.72, 12.01, and 13.49% of the total variation. At first glance, these figures appear to be depressingly low, but they are an indication that the Lower Limestone is a good choice as a calibration unit. If the fits were high, the trend surfaces would imply that there was a major drift in the calibration standard across the area.

In moving from a linear to a quadratic surface, the degree of fit almost doubles, but the cubic is only a minor improvement on the quadratic. The quadratic surface appears to be the most "natural" expression of the systematic regional variation of the Lower Limestone neutron response. This interpretation was checked by a simple analysis of variance procedure which tests whether successively higher orders of surface make a statistically significant improvement over lower orders (Table 1). The linear surface does a significantly better job of predicting the true neutron response of the Lower Limestone at any location, when contrasted with use of the average neutron response in the data set. This demonstrates that the trend surface analysis has detected a systematic geographic component of variation in the calibration unit. The quadratic surface makes a significant improvement in fit over the linear, but the cubic fails to add significantly to the quadratic.

The configuration of the quadratic surface (Fig. 4) shows a broad central area of relatively high porosity flanked by lower porosities to the west and southeast. It is worthwhile to relate log response trend surfaces with regional geology patterns in order to assign meaning to the trends and to cross-check their validity. The quadratic surface shows a striking concordance with the axis of the Pratt anticline and suggests a genetic relationship. The neutron trend surface indicates that the Pratt anticline was an active positive feature contemporaneous with the deposition of the Lower Limestone. The trend in porosity may well reflect a regional pattern of the low-porosity, grain-supported crinoidal packstones and grainstones on the margins of the Pratt anticline, which grade toward the axis as a slightly more micritic and higher porosity facies equivalent.

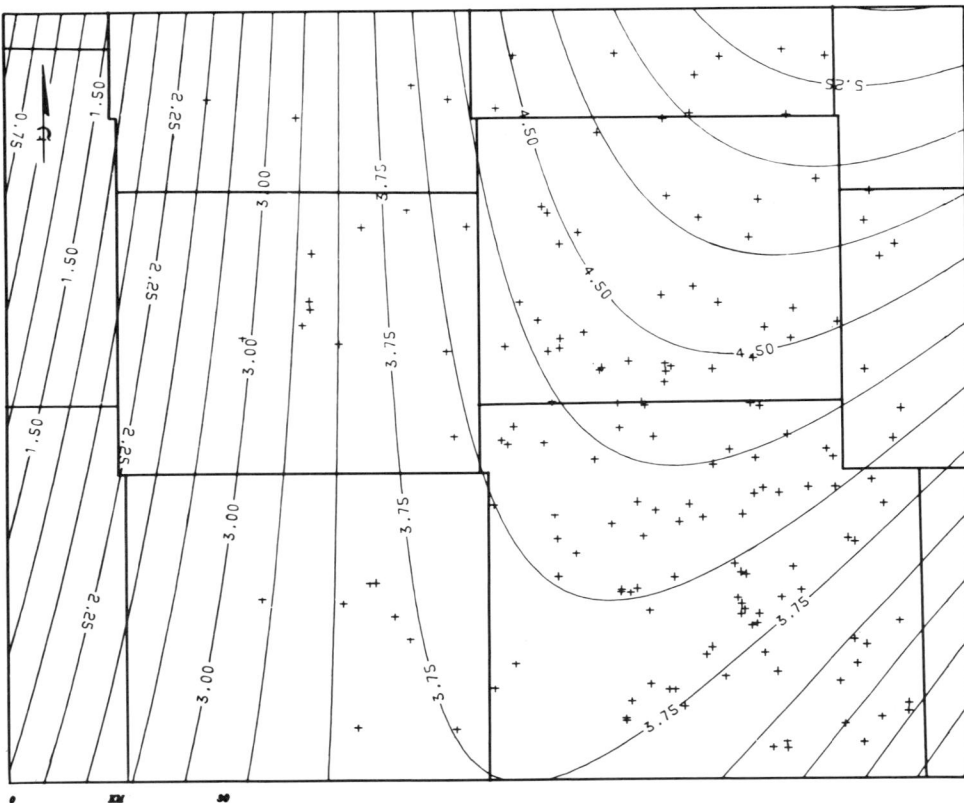

FIGURE 4. Quadratic trend surface map of neutron porosity variation of the Viola "Lower Limestone" zone. Units are in percent limestone equivalent porosity.

Trend surfaces were computed for the transit times and densities of the Lower Limestone and, in both cases, the linear surface was found to be most significant in portraying the regional variation of the data. A contour map of the transit time linear surface (Fig. 5) shows a simple decline in porosity, moving from the northeast to the southwest, which is only a general approximation of the trend in the neutron data. A major reason for this difference is that the sonic log sample is less than one-third the size of the neutron sample. As a result, the restricted sample size is not large enough to pick up any small systematic improvement by the quadratic surface.

The same argument applies to the density linear surface where the sample is further reduced to 36 wells. However, the density linear surface (Fig. 6) shows a decline in porosity to the east which seems to contradict the trends observed in both neutron and sonic surfaces. The apparent discrepancy is explained by the distribution of the density well control, which is almost entirely restricted to the eastern half of the area. The density surface is therefore primarily sensitive to the fall in porosity on the eastern flank of the regional structure. This characteristic makes the important point that trend surface predictions of calibration unit response must be restricted to areas of moderate well control and not extrapolated beyond their limits.

While the trend surfaces of the Lower Limestone porosity values are estimates of the systematic regional variation, the deviations of the raw values from the surface are attributed to a contribution of tool error and any systematic local variation. If the residuals are controlled by tool error alone, they should be approximately normally distributed with a mode located at zero. Frequency polygons of the neutron, sonic, and density residuals are shown in Figure 7, plotted on a compatible scale of limestone porosity units. In all cases, the distributions show a slight positive skew with a displacement of the modal peak to approximately minus one porosity

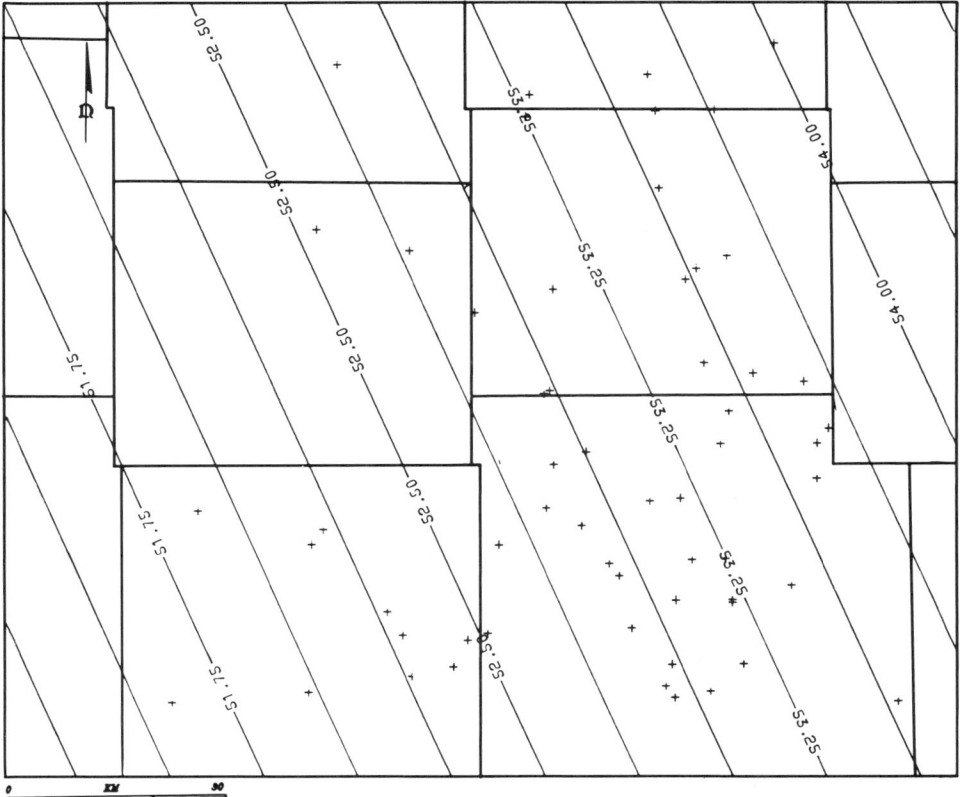

FIGURE 5. Linear trend surface map of sonic transit time variation of the Viola "Lower Limestone" zone. Units are in microseconds per foot.

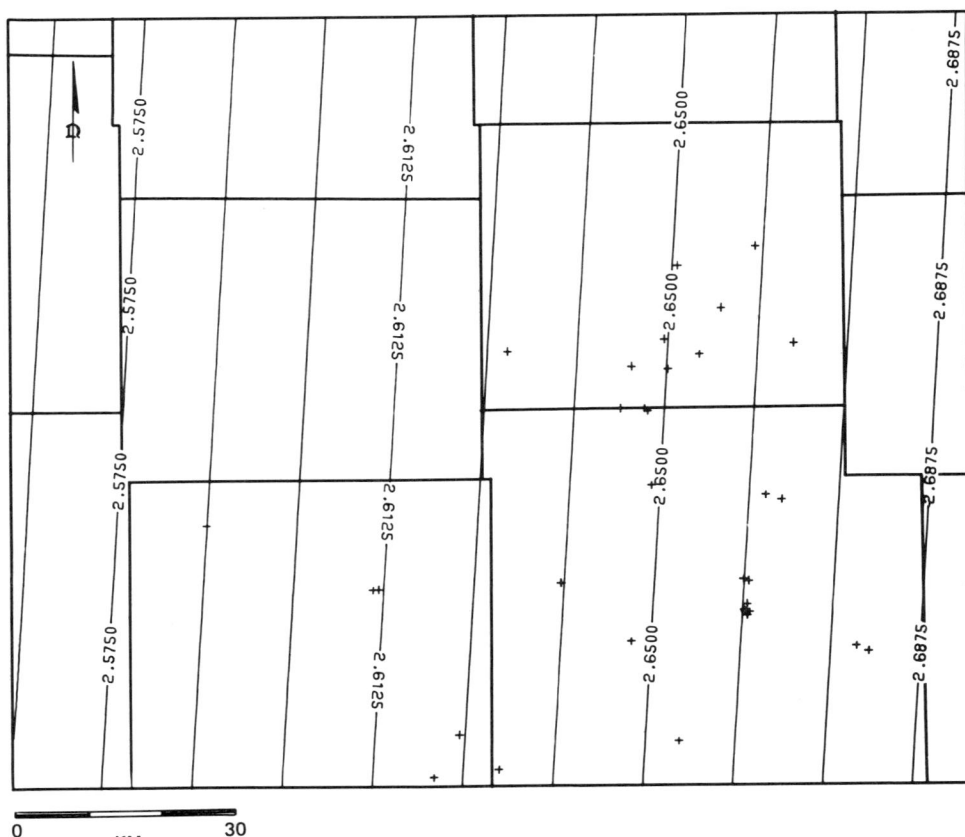

FIGURE 6. Linear trend surface map of density variation of the Viola "Lower Limestone" zone. Units are in grams per cubic centimeter.

units. This common feature implies that the true variation of the Lower Limestone closely follows the form of the trend surfaces, but is broken in local areas by the development of either enhanced porosity zones or increased shale content. As a result, the trend surface estimates will tend to be slightly optimistic over most of the area by about one porosity unit, but locally pessimistic by approximately two porosity units. Following conventional practice, each trend surface equation can be modified by subtracting the displacement of the mode from zero and inserting this quantity as a correction factor. The modified equation is now geared to the most typical values for the area, rather than the grand average. Potential problem areas of enhanced porosity can be examined through contour maps of the residuals, although it will often be difficult to distinguish spurious features caused by positive tool errors.

The analysis of both the trend surfaces and their associated residuals therefore provides useful feedback concerning the performance of the Lower Limestone as satisfactory calibration unit, as well as numerical relationships for prediction and control. The trend surface equations are used in prognosis, while the residual distributions are useful for diagnosis. Simple descriptive statistics and inferential tests may be applied in a systematic dissection of log quality in the area if the residuals can be reasonably approximated by normal distributions. The area under the curve of a normal distribution is a function of a scale of standard deviations about the mean and dictates the proportion of tool errors to be expected within specified ranges. So, for example, if an accuracy of plus or minus one porosity unit is considered an acceptable tolerance level, 49% of the neutron logs will require correction, since the standard deviation of the neutron is 1.51 porosity units. This figure compares with 69% of both sonic and density logs.

The trend surface normalization method can be expanded to a generalized strategy of analysis designed for routine applications. The major steps are:

1. Define an area of interest in which well logs are to be normalized.

2. Select zones from the regional stratigraphy which are both easily correlated and appear to show only minor changes in logging character across the area. These zones are provisional calibration units whose suitability as normalization standards is to be examined in the first phase of analysis.

3. Compile data sets for each provisional unit to include geographic coordinates and log responses. For most logs, the raw response is appropriate. However, resistivity values should be transformed to the square root of the conductivity in order to

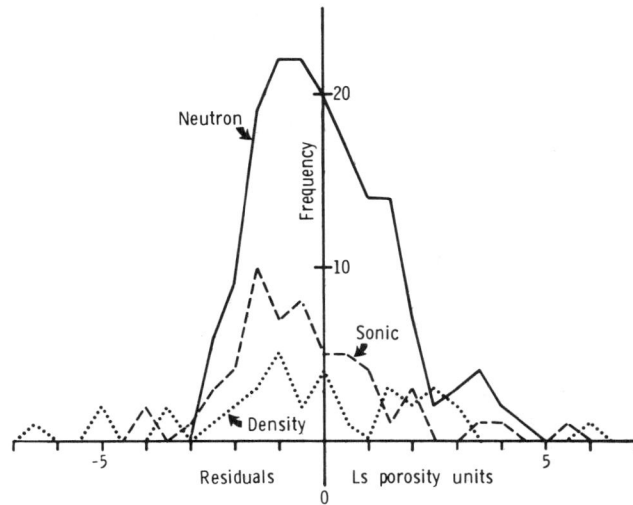

FIGURE 7. Frequency polygons of the Lower limestone neutron, sonic, and density trend surface residuals.

approximately linearize the variation. Any necessary environmental corrections should be applied.

4. For each provisional unit, compute trend surfaces for the separate log responses. Extract a specific surface from each trend surface series which appears to match the regional variation, as judged from inspection of the degrees of fit or by analysis of variance. Compare maps of the selected trend surfaces with regional geological expectations of depositional facies variations and compactional trends as an external check on their validity. The residuals of an ideal calibration unit should be approximately normally distributed with a mode located at zero. Displacement of the mode and skewness of the residual distribution or the development of distinctive secondary modes all indicate localized areas of systematic variation which has been compounded with tool error. However, some allowance for deviations from the ideal should be made when dealing with small sample sizes. Provisional units with satisfactory trend surface and residual properties may be adopted as calibration units for prediction and control.

5. Compute prognoses of calibration unit normalized responses at existing well control or prospective drilling sites by insertion of the geographic coordinates into the trend surface equations. In practice, it is wise to run several calibration units as a cross-check on consistent normalization corrections and to highlight possible changes in tool error as a function of depth. Prognosis should not be made outside the area of well control, and the overall reliability of any prognosis will be controlled by the density of neighboring wells.

6. For more detailed studies, analyze the residual distributions in assessments of error character contrasted between tools and between service companies in terms of relative precision and bias.

7. Update calibration unit data sets with log responses and geographic coordinates from newer control as it becomes available. Periodically, recompute trend surfaces as successively more refined estimates of regional variation.

The procedure outlined is extremely straightforward as a practical program for log normalization. Programs for trend surface analysis are widely available since they are used by exploration geologists in structural studies. These programs are cheap to run and their output routinely lists trend surface equations, residuals, and analysis of variance data, as well as generating simple line-printer maps or files for the plotter production of maps.

REFERENCES

Doveton, J. H., and Bornemann, E., 1981, Log normalization by trend surface analysis, *The Log Analyst*, Vol. XXII, No. 4, pp. 3–8.

Neinast, G. S., and Knox, C. C., 1973, Normalization of well log data: *Trans. Soc. Prof. Well Log Analysts 14th Ann. Logging Symp.*, Paper I, 19 pp.

Patchett, J. G., and Coalson, E. B., 1979, The determination of porosity in sandstones and shaly sandstones, Part 1—Quality control, *The Log Analyst*, Vol. XX, No. 6, pp. 3–12.

CHAPTER NINE

LITHOFACIES MAPPING FROM LOGS

Up to this point, the methods described in this book have been restricted mainly to the analysis of log variation along one dimension—the depth (or time) axis of the well bore. The mapping of compositional or lithofacies changes across geographic space introduces two extra dimensions. However, this additional complexity does not require a substantive change in the techniques suitable for single wells, but modifications to accommodate the expanded reference framework.

A particular advantage of wireline logs is their intrinsic numerical character which makes them amenable to bulk processing by computer. Copmputer contouring programs are most commonly used in subsurface geological studies to solve conventional topological problems—the mapping of structural surfaces and their differences (isopachs), based on a sample of well control. Their use is equally applicable to the contouring of hypothetical surfaces of transformed log data, such as variations in porosity, water resistivity, or mineral composition estimates.

Most regional lithofacies studies have drawn heavily on cuttings and core descriptions. Both of these traditional sources of data have practical limitations. Cores provide the most comprehensive information, but are generally severely restricted in both areal coverage and vertical sampling because of the expense of core recovery operations. Cuttings are the most common source of information, but are dogged by a variety of problems. Skillful work is required to identify contamination by caved materials and the effect of selective bias in description by different geologists. Allowance must be made for errors caused by the lag times of the cuttings in their journey to the surface and by differential flotation of components within the mud column. Estimation procedures are, at best, semiquantitative and often distorted by the selective comminution of more friable components by the drillbit.

By contrast, log readings are direct measurements of physical properties of subsurface formations which can be transformed to quantitative estimates of geological attributes. The wireline logging operation has its own intrinsic problems of errors introduced by deviations of the borehole environment from its ideal and the statistical variability of physical measurements. Useful remedial methods were discussed in the previous chapter.

In a literal sense, wireline logs are remotely sensed data from the subsurface. Interpolation between a network of well control will generate a spatial construct whose form is a reflection of geological variability. However, just as satellite image analysis requires a certain element of "ground truth" as a necessary ingredient for intelligent work, subsurface rock samples are the appropriate guides for maps based on log data. In a detailed study, core and cuttings serve two crucial functions: validation and interpretation. Log analysis of lithofacies must first be consistent with the geology of the formation found by the drill bit. At another level, the less tangible rock characteristics of palentology, sedimentology, and diagenesis are the necessary keys to provide genetic meanings to patterns derived from logs.

In the following sections, the mapping of lithofacies from logs is subdivided into two categories—simple and complex. In simple mapping, a single log is used as the data source to resolve a basic model of facies variation between two component endmembers. Methods are described which allow an extension beyond geographic space to a three-dimensional representation incorporating the third axis of depth. In the resolution of complex mineral associations, several logs are combined in an approach which makes use of the theory developed for composition analyses of stratigraphic zones within a single well.

SIMPLE LITHOFACIES MAPPING FROM A SINGLE LOG

Most of the commonly available logs are primarily sensitive to variations in the volume of pore fluid and/or the fractional content of shale, when run through typical sedimentary sequences of limestones, dolomites, sandstones, and shales. The mapping of porosity trends is obviously not a direct expression of lithology. However, several examples in previous chapters have shown that there is often an implicit relationship between porosity and lithofacies as a cumulative product of sedimentological and diagenetic processes. The simplest application of lithofacies mapping from a single log is the evaluation of lateral shale content across an area. The obvious choice for this task is the gamma-ray log, because it is most insensitive to porosity variation.

As reviewed in Chapter 3, gamma-ray log values may be transformed to profiles of estimated shale content by interpolation between limits that appear to typify shales and sandstones (or carbonates) as the scale boundaries. The rescaled gamma-ray trace expresses shale content and its vertical variation as a fluctuating distribution which must be condensed in some manner to descriptors useful for mapping. The most basic measure is the mean shale proportion, which reflects the aggregate composition. The values can be considered to be normalized, since they are individually calibrated, well by well. Interpolation of mean shale values of a formation between wells results in the creation of a continuous surface of shale variation across geographic space. The form of the surface reflects lateral changes in the gross aspects of shale

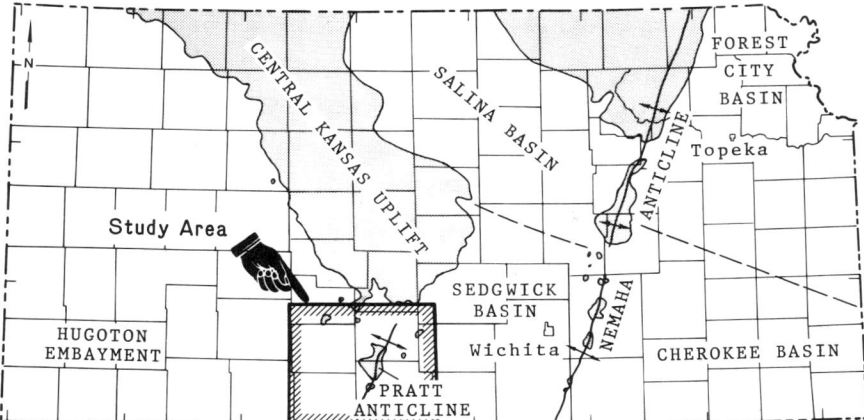

FIGURE 1. Location of study area in south-central Kansas for lithofacies mapping of the Simpson Group and Viola Limestone.

geometry. The method and interpretation of its results are reviewed in the following case study described by Doveton et al. (1984).

The Simpson Group of southern Kansas is a complex of sandstones and shales, with subordinate local occurrences of sandy limestones and dolomites. These deposits were formed as a diachronous sequence of clastic sediments in the major transgression of Middle Ordovician seas northward across the midcontinent. The unit is present throughout a four-county region of southern Kansas (Fig. 1) which straddles the Pratt Anticline. Small oilfields are located in the Simpson Group within the eastern half of the area.

A representative gamma-ray log from a Simpson Group section is shown in Figure 2 and expresses the major pattern of lithological variation. The Simpson Group is separated from the dolomites of the underlying Arbuckle Limestone by a major unconformity. The basal shale is succeeded by a sandstone capped by a thin sandy dolomite, and separated from the overlying Viola Limestone by a shale. Lateral correlation between wells suggests that the geometry of the sandstone unit consists of both lenses and sheetlike bodies.

The positions of the top and base of the Simpson Group, together with its mean shale content are indexed on the reference gamma-ray log on Figure 2. These same descriptors were estimated from gamma-ray logs run in 148 wells in the area. An isopach map of the Simpson Group (Fig. 3) shows a complex pattern of thickness variation across the northern part of the area, but a marked hinge line in the south, with rapid thickening toward the Oklahoma border. The mean shale ratio map (Fig. 4) contains broad facies belts arranged in a general north–south orientation. Low-shale regions to the east and west are broken by a high-shale trend centered over the Pratt Anticline.

In common with all conventional lithofacies maps, the shale ratio map is strictly limited as to the information it presents. Shale composition is aggregated at all

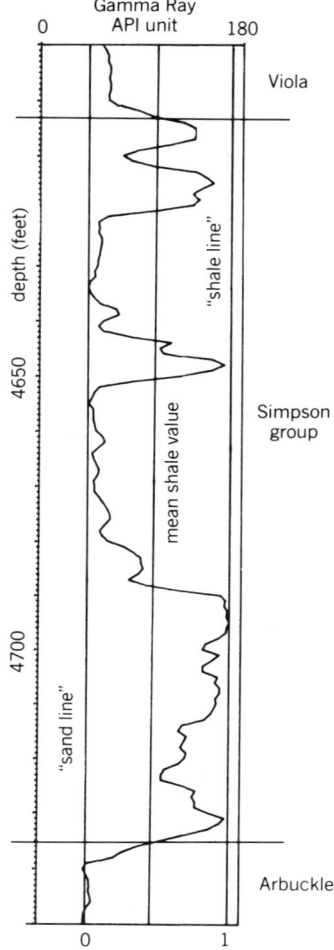

FIGURE 2. Representative gamma-ray log from a Simpson Group Section indexed with sand line, shale line, and mean shale content.

geographic locations so that variations are confined to expressions of lateral changes and preclude any assessment of fluctuations with depth. This fundamental limitation has been recognized for many years and various methods suggested to provide supplemental maps keyed to vertical variation. Krumbein and Libby (1957) termed these characterizations "vertical variability maps" and reviewed earlier attempts. They ranged from simple statistics listed by Busch (1950), including number of sandstone beds and average bed thickness, to ratios of sand compositions between unit subdivisions (Krumbein and Sloss, 1963; Forgotson, 1954).

Krumbein and Libby (1957) introduced a new approach to the problem through the computation of the moments of lithology distribution as a function of depth, and their mapping in geographic space. The method of moments offered important new advantages. Moments are fundamental measures of variability, which have a simple meaning in both statistics and mechanics, are easy to compute, and are an advance on the pragmatism of earlier methods. Although there is a hierarchy of moments, Krumbein and Libby (1957) restricted their consideration to the first two moments. These correspond to the center of gravity and radius of gyration of a lithology distribution in the vertical dimension. When coupled with a conventional facies presentation, maps of these two moments were shown to give a basic insight of gross compositional changes in three dimensions.

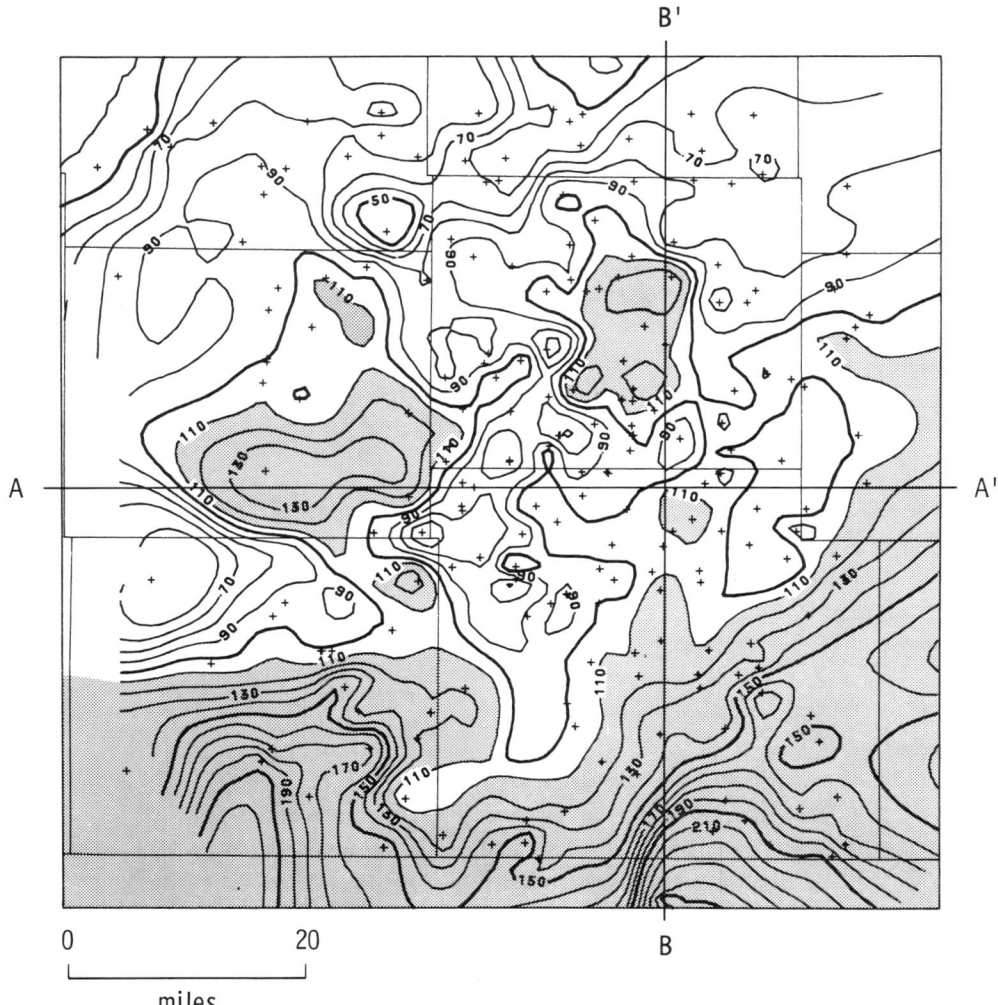

FIGURE 3. Isopach map of the Simpson Group (in feet). Contour levels greater than 110 feet are shaded. Lines A–A' and B–B' locate cross sections used in Figures 10, 12, and 13.

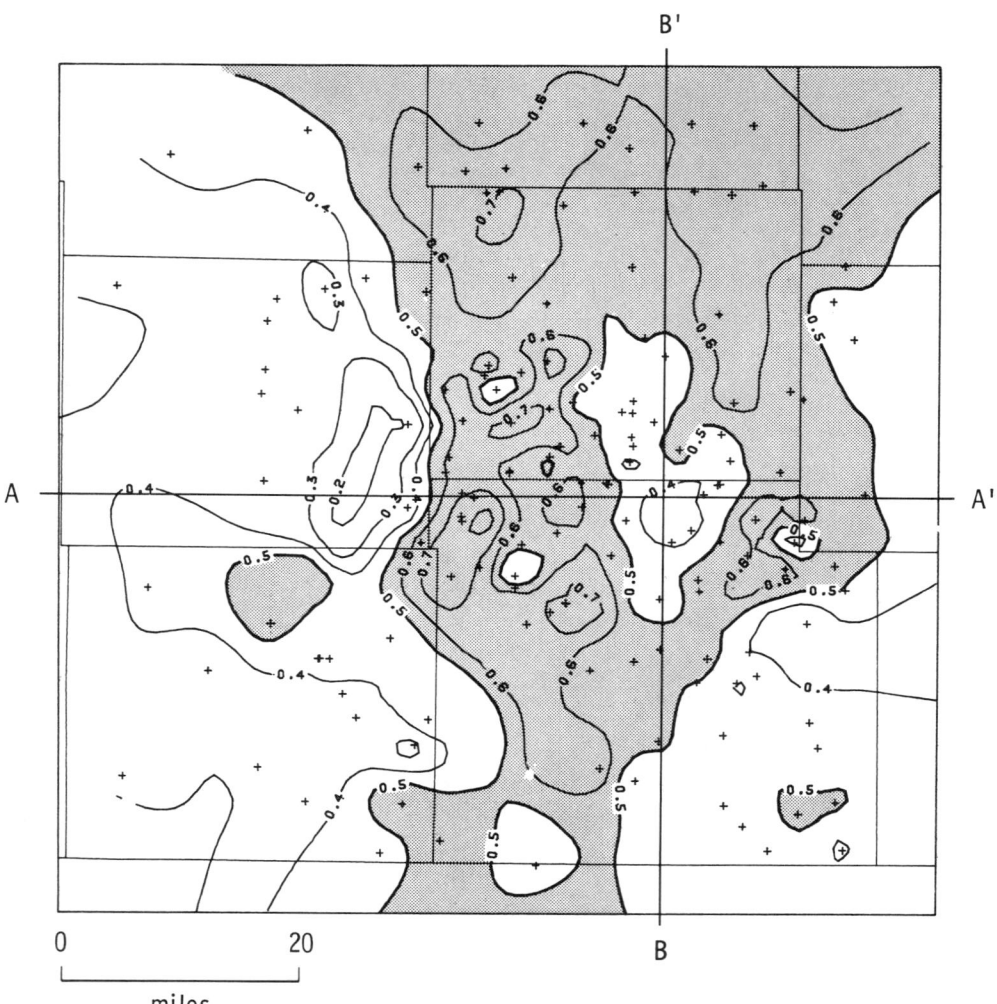

FIGURE 4. Mean shale ratio map of the Simpson Group. Ratios higher than 0.5 are shaded. Lines A–A' and B–B' locate cross section of Figures 10, 12, and 13.

The concept and mathematics of moment derivation are explained with reference to the gamma-ray log used earlier, redrawn in Figure 5. The selection of gamma-ray readings that set shales and sandstones (or carbonates) as scale boundaries, result in a transformation of the log into a "shale profile." This new scale is keyed to estimated proportional shale content rather than arbitrary radioactivity count units. Through use of the scale, the average shale composition can be computed in the manner already described. If the log is digitized at an appropriate depth interval (conventionally two readings per foot), then the mean shale proportion is given by

$$\overline{S} = \sum_{i=1}^{n} \frac{S_i}{n}$$

where S_i is the shale proportion at depth i and the sequence contains n digitized values. The center of gravity has both a statistical and mechanical meaning and is computed from the equation

$$v_1 = \frac{\Sigma S_i d_i}{\Sigma S_i}$$

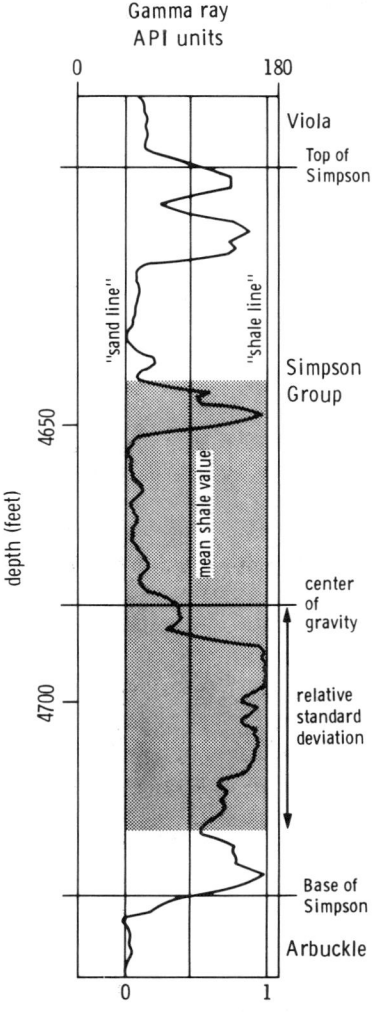

FIGURE 5. Reference gamma-ray log indexed with center of gravity and standard deviation range.

where d_i is the depth of the ith value. Depths are referred to a measurement origin which is usually selected as the top of the stratigraphic unit. The center of gravity is the first moment and locates the balance point of the shale distribution on the depth axis if it was modeled as an object of mass.

The second moment of the shale variation is given by

$$V_2 = \frac{\Sigma S_i d_i^2}{\Sigma S_i}$$

which, in terms of mechanics, represents the moment of inertia divided by the total mass. Its statistical function is to give an expression of the relative degree of dispersion of the distribution along the depth axis. The basic formula is referenced to the depth scale origin at the top of the unit. Generally, the second moment is modified to a measure of dispersion about the center of gravity by the equation

$$m_2 = \sqrt{V_2 - V_1^2}$$

The quantity m_2 is termed the "relative standard deviation" and corresponds to the radius of gyration of the distribution rotated about the center of gravity.

A progressive series of moments may be calculated from the general formula

$$v_m = \frac{\Sigma S_i d_i^m}{\Sigma S_i}$$

where V_m is the mth moment. The third moment is a measure of skewness which summarizes both the sense and the degree of asymmetry of a distribution. The fourth moment assesses kurtosis, and reflects the relative "peakedness" of a distribution shape. It is difficult to assign simple descriptive meanings to moments higher than the fourth, although these moments are legitimate, but more subtle, measures of distribution shape. In addition, the higher powers associated with the more complex moments can generate excessively large numbers, which can lead to unstable estimates caused by data precision and computer round-off problems.

Centers of gravity and relative standard deviations were computed for gamma-ray logs of the Simpson Group in all the wells of the study area. Maps of these separate moment characterizations were made in order to assess the variation of the contained shale in both the vertical dimension as well as the two lateral dimensions of geographic space. Center of gravity values from each log were divided by total thickness to yield proportional measures of their location within each section. The center of gravity map (Fig. 6) is typified by relatively high values which imply that the bulk of the shale is in the lower part of the Simpson Group. The proportional standard deviation map (Fig. 7) suggests that the shale distribution is more highly dispersed in sections with greater shale concentrations.

Interpretations from the isopach and three shale statistic maps can be amplified by more detailed comparison of the separate maps or examination as overlaid sets. Essentially, the process involves exercising a nimble mind to collate several two-

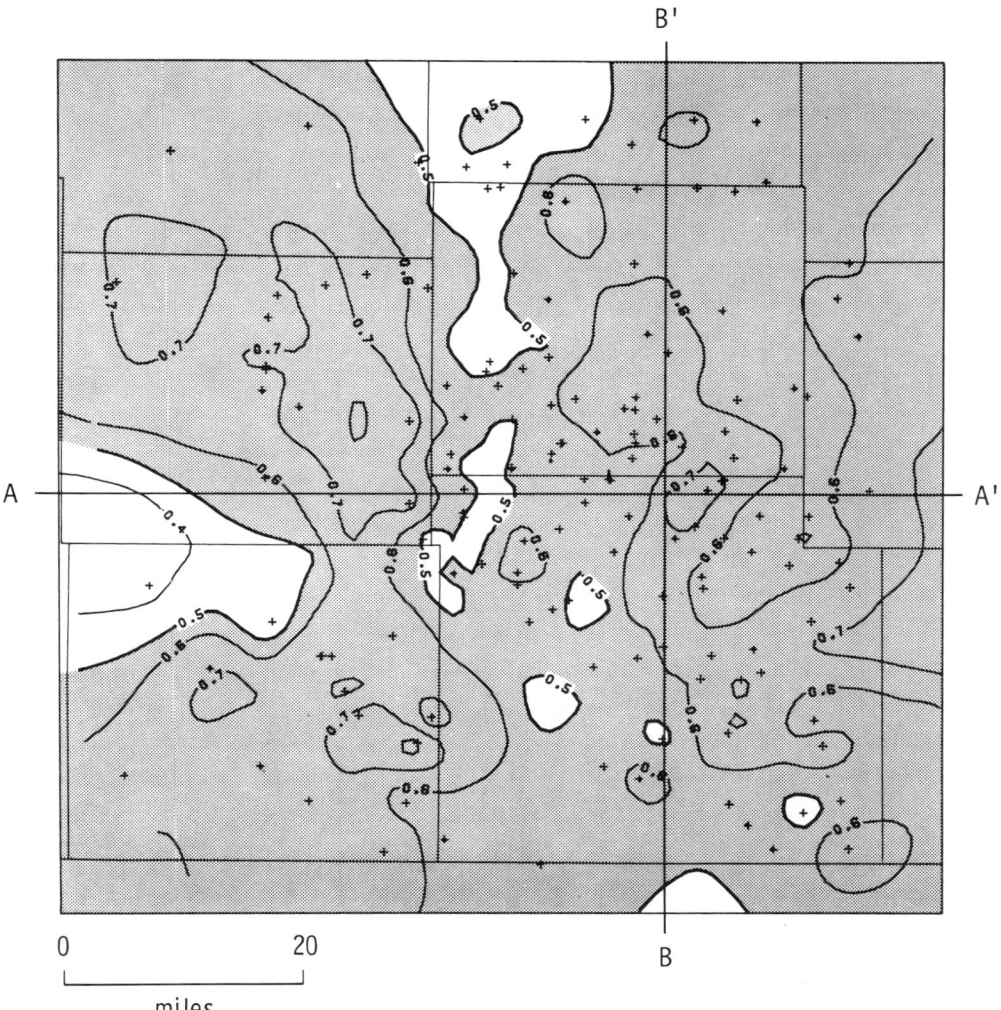

FIGURE 6. Center of gravity map of shale distribution in the Simpson Group. Lines A–A' and B–B' locate cross sections of Figures 10, 12, and 13.

dimensional summaries of simple variation in the vertical dimension and arrive at a three-dimensional visualization. However, because moments are statistical parameters, their values specify unique shapes of vertical variation. Therefore, a set of moment maps *defines* three-dimensional variation in a precise quantitative manner. It follows that moment maps and their qualitative interpretation are only an intermediate step in the quantitative generation of a three-dimensional model. The mathematical operations that achieve this objective exploit the fact that moments are the fundamental parameters of polynomial regression curves as described in the next section.

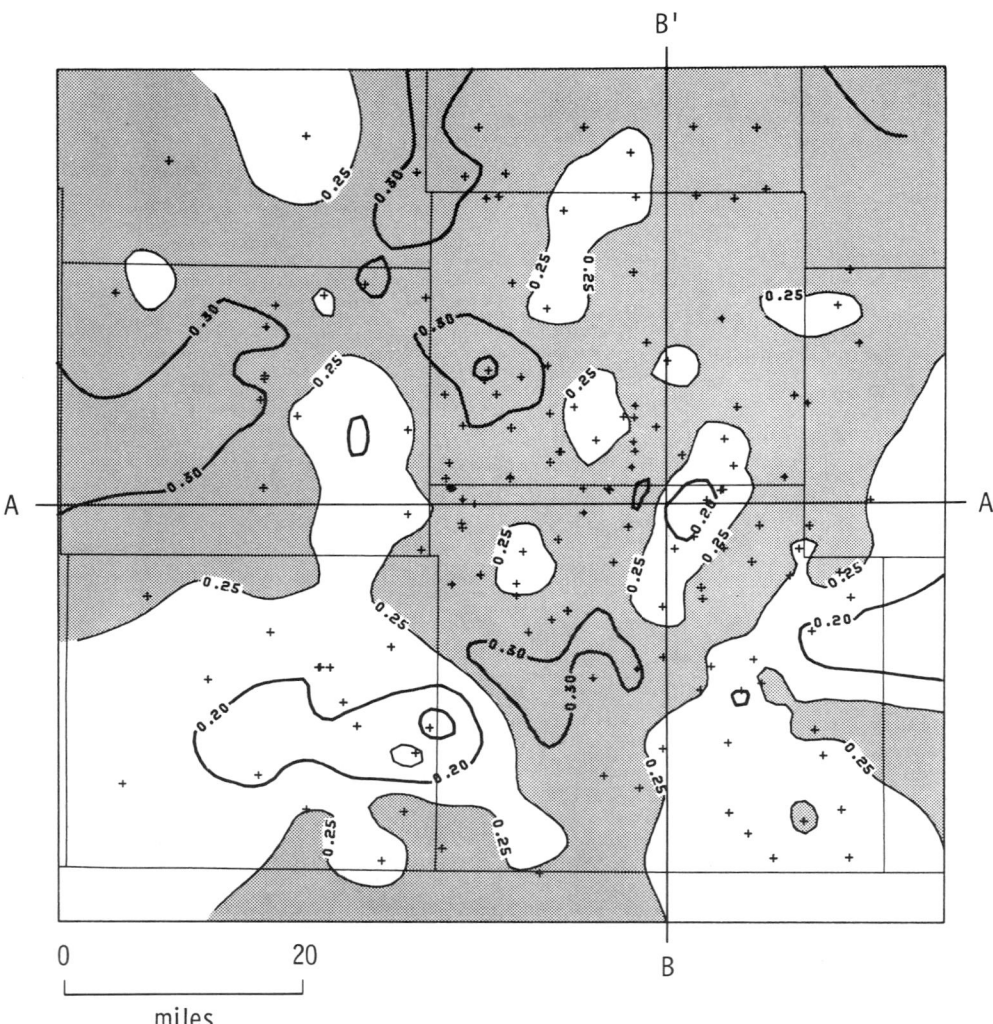

FIGURE 7. Relative standard deviation map of shale distribution in the Simpson Group. Lines *A–A'* and *B–B'* locate cross sections of Figures 10, 12, and 13.

Moments and Polynomial Regression

The fitting of curves to the variation of some measurement as a function of time or distance is common to all areas of science. These curves extract major trends in variation and differentiate small-scale fluctuations and random aberrations. Polynomial curves relate the measurement variable to a weighted summation of powers of the distance or time scale, and are fitted by the principle of least squares as a regression analysis.

The shale profile of the reference gamma-ray trace (Figs. 2 and 5) was analyzed by polynomial regression to compute major trends in shale variation. A first-order polynomial is a straight line with the equation

$$\hat{S} = a_0 + a_1 d$$

where $\hat{S}$ is the regression estimate of the shale proportion at depth d, predicted by a line with intercept a_0 and slope a_1. The unknown coefficients are solved from the simple matrix algebra relationship

$$\begin{bmatrix} n & \Sigma d \\ \Sigma d & \Sigma d^2 \end{bmatrix} \begin{bmatrix} a_0 \\ a_1 \end{bmatrix} = \begin{bmatrix} \Sigma S \\ \Sigma S d \end{bmatrix}$$

The fitted first-order polynomial is shown on Figure 8 and defines a basic trend of increasing shale content with depth.

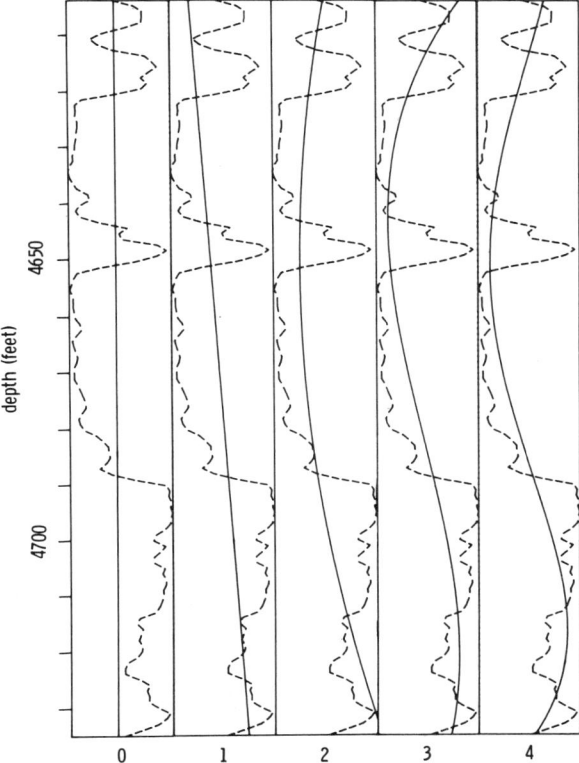

FIGURE 8. Zero- through fourth order polynomial regression trends of the reference gamma-ray log profile.

A quadratic (or second-order) polynomial function describes a curve with a single maximum or minimum and its application to the shale profile is given by the equation

$$\hat{S} = a_0 + a_1 d + a_2 d^2$$

The matrix algebra solution of the unknown coefficients is

$$\begin{bmatrix} n & \Sigma d & \Sigma d^2 \\ \Sigma d & \Sigma d^2 & \Sigma d^3 \\ \Sigma d^2 & \Sigma d^3 & \Sigma d^4 \end{bmatrix} \begin{bmatrix} a_0 \\ a_1 \\ a_2 \end{bmatrix} = \begin{bmatrix} \Sigma S \\ \Sigma Sd \\ \Sigma Sd^2 \end{bmatrix}$$

It should be noted that all information concerning the variation of shale with depth is contained in the vector on the right-hand side of the equation. Further, when divided by the quantity (ΣS), the three vector values are unity and the first two moments of the shale distribution. It follows that moments *define* a polynomial function. In this case, the mean shale content, center of gravity, and standard deviation collectively specify some unique quadratic curve of shale proportion as a function of depth.

In reviewing the matrix algebra equation for the first order polynomial, it can be seen that the straight line relationship is fixed by the mean shale content and the center of gravity. In addition, a "zeroth-order" polynomial would be solved by the equation

$$na_0 = \Sigma S$$

which sets a constant value of shale proportion at all levels within a section and is matched by the mean shale content.

The addition of extra polynomial terms at higher orders specify curves of increasing complexity. However, any polynomial order is fitted to raw data by the same generalized matrix algorithm. For an mth-order polynomial, the relationship is

$$\begin{bmatrix} n & \Sigma d & \Sigma d^2 & \Sigma d^m \\ \Sigma d & \Sigma d^2 & & \\ & & & \\ \cdot & & \cdot & \\ \cdot & & \cdot & \\ \cdot & & & \\ \Sigma d^m & & \Sigma d^{2m} \end{bmatrix} \begin{bmatrix} a_0 \\ a_1 \\ a_2 \\ \cdot \\ \cdot \\ \cdot \\ a_m \end{bmatrix} = \begin{bmatrix} \Sigma S \\ \Sigma Sd \\ \Sigma Sd^2 \\ \cdot \\ \cdot \\ \cdot \\ \Sigma Sd^m \end{bmatrix}$$

and is uniquely determined by the mean and first m moments. The fitted curves matched with polynomials up to fourth order are shown on Figure 8.

There is the interesting corollary that if polynomial curves are defined by distribution moments, then the vertical variability that can be interpreted from a set of moment maps is predetermined. From the combination of a mean shale content and a center of gravity map, only a linear trend can be inferred at any location. The addition of a standard deviation map allows variation to reach either a maximum or minimum within the section, but precludes interpretation of more complex variation.

The dualism that links moments and polynomial curves can be turned to advantage in the prediction of the number of moment maps that are necessary to characterize adequately the vertical variability of a stratigraphic unit. Each polynomial function fitted to data satisfies a proportion of the total variation of the measurement about its mean value. At higher orders, the increasing complexity of the curves results in progressively closer fits until the data are matched perfectly. This characteristic is shown on Figure 8 where a polynomial regression sequence is fitted to the reference log, and the higher-order curves are more realistic expressions of the trend in vertical shale variation.

In a plot of regression fit against order, a polynomial can often be isolated that appears to satisfy a systematic major trend, while the residual variation is compounded from markedly finer-scaled fluctuations. The graph of fit against polynomial order for the reference log (Fig. 9) indicates that the third-order polynomial is the appropriate choice for the Simpson Group in this area, if the well section is typical. Furthermore, the apparent "natural" fit of the cubic polynomial to major shale variation implies that the Simpson Group is comprised of two major subdivisions which match the two extremes allowed by the cubic equation. This mathematical conclusion matches the visual evidence of the gamma-ray log (Fig. 2) and the informal stratigraphy of the area used by explorationists who distinguish an upper "Simpson Sand" from a lower "Simpson Shale." The extension of formal correlations from the Simpson Group of Oklahoma is a subject of minor controversy, but the consensus of opinion from studies by Schramm (1964), Ireland (1965), and Statler (1965) favors the conclusion that these two units are correlated with the Bromide and McLish formations

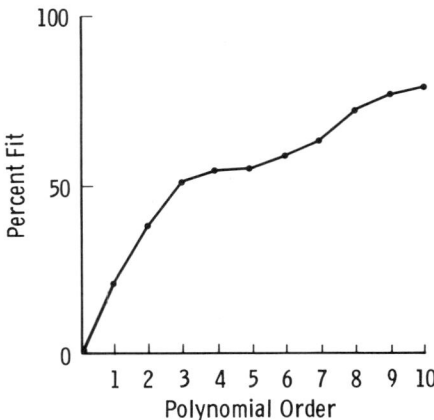

FIGURE 9. Percentage fits of regression polynomials fitted to reference gamma-ray log graphed against polynomial order.

of the Oklahoma subsurface. If a cubic trend is the most appropriate descriptor of trends in shale vertical variability, then at least three moment maps are necessary to trace adequately the regional change of the Simpson Group in three dimensions.

Three-Dimensional Slice Mapping

Conventional moment maps represent surfaces of variation generated by interpolation of moments computed from logs at well control. The reliability of predictions of moment values at undrilled locations is obviously dependent to a large degree on the relative proximity and density of nearby well control. Since the lower-order moments are indices of major depth changes in shale content, it is likely that they can be mapped with about the same degree of confidence as an isopach or structure map of normal stratigraphic units. This condition is an implicit assumption of the method of moment mapping pioneered by Krumbein and Libby (1957).

When interpolated estimates of moments are inserted into the matrix algebra solution for a polynomial regression, the solved coefficients define a curve of vertical variability as a projection of trends from wells in the immediate vicinity. The basic matrix formulation can be summarized as

$$\mathbf{DA} = \mathbf{S}$$

with the solution as

$$\mathbf{A} = \mathbf{D}^{-1}\mathbf{S}$$

where $\mathbf{D}$ is the "depth matrix" whose elements are sums of powers of the depth increments in the digitized record, $\mathbf{D}^{-1}$ is the inverse of this matrix, $\mathbf{A}$ is the "trend coefficient vector," and $\mathbf{S}$ is the "moment vector" which contains moment estimate numerators.

The values of the depth matrix will vary as determined by the thickness of the section, either measured at a well or from interpolation between wells. However, there is a great computational advantage that accrues from standardizing all hypothetical sections to a fixed thickness of unity. The result will be a constant depth matrix regardless of location, so that the matrix and its inverse require only one computation sequence in any application. Secondly, the transformation of depth units to a fixed low range guarantees that matrix elements with high powers will not cause an ill-conditioned matrix with consequent major problems in computer round-off errors. The resultant polynomial regression coefficients will be scaled to the fixed standard, but are easily transformed back to real depth measurements by rescaling to true thickness.

In summary, the consequence of this procedure is that polynomial curve coefficients can be estimated at any location in an area by a matrix algebra operation on moments interpolated from well control. When these coefficients are combined with the estimated thickness of the stratigraphic section, a polynomial curve may be "played back" and is a trend representation of vertical variability. Since polynomials are continuous functions and it is possible to specify an infinite variety of geographic locations, there is no theoretical limitation to the prediction of a shale trend value at any combination of geographic coordinates or depths.

The major obstacle to the application of the full potential of this method is the difficulty of displaying variation across three dimensions on a two-dimensional medium such as paper or a computer terminal screen. As a practical accommodation, variation may be mapped on two-dimensional slices and the separate results combined to give a framework illusion of a three-dimensional model. As an example of the general procedure involved in mapping trend variation on a slice, an east–west cross section was used for illustrative purposes (line A–A' on Figs. 3, 4, 6, 7). A series of alternative cross-section profiles were generated in order to demonstrate the influence of the separate moments in perceptions of regional shale variation (Fig. 10). In each case, a grid was defined in which columns specified vertical prisms whose width matched a geographic cell, while rows corresponded with 10-foot increments of depth. Every column was defined by a unique geographic position on the cross-section trace. Following interpolation of moments from adjacent control, polynomial trend coefficients were solved at the location matched with each column. The cell contents of the columns were generated by solving the polynomial equations for shale trend estimated at the incremental depths registered by each row. The resultant grid of shale trend values were contoured in the same manner as a conventional map. By repeating the same basic procedure, but using a progressively more comprehensive set of moments, a series of more complex profiles were created (Fig. 10).

The most simple profile is the "zeroth order" which is determined by the mean shale content at any location and shows vertical contours of shale variation, because no vertical variability can be perceived. The first order profile was generated by transformation of interpolated mean shale content and center of gravity, which specify a linear trend at any single location and a monotonic character to lateral variation. The higher order profiles incorporate successively the second, third, and fourth moments. These show marked improvements in the resolution of distinctive shapes and gradational changes which reflect vertical and lateral facies changes within the Simpson Group. However, differences between the third and fourth order profile are relatively minor. This observation supports the earlier conclusion drawn from regression analyses of the reference log that a cubic polynomial appears to absorb the major trend in variation. If the reference log is considered fairly typical, then the regression fits of its polynomial trends indicate that the third- and fourth-order profiles are expressions of about 50% of the total variation. The unaccounted residual variation is compounded from relatively fine-scale features, many of which would have only local persistence and of doubtful validity when interpolated laterally.

Polynomial functions have additional simple properties that provide useful parameters for mapping. If a polynomial equation of the form

$$\hat{s} = a_0 + a_1 d + a_2 d^2 + \cdots + a_m d^m$$

is differentiated, the result is

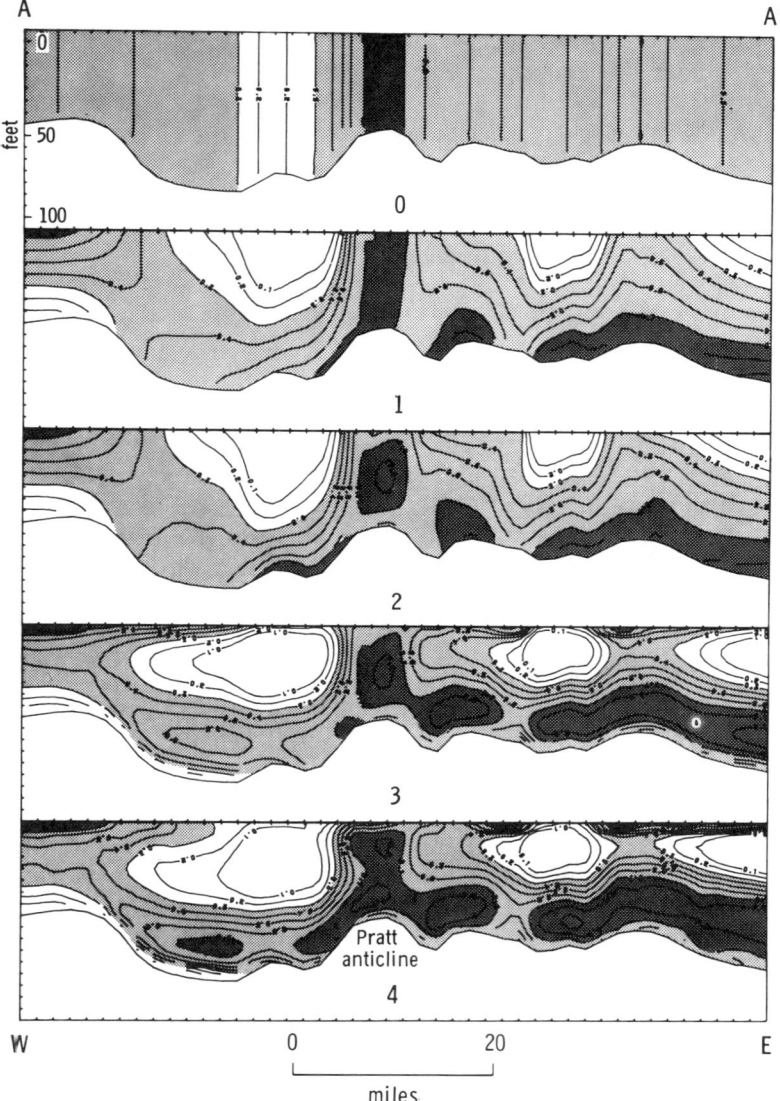

FIGURE 10. Zero through fourth order shale proportion trend profiles of the Simpson Group on east–west cross section (line A–A' on Figures 4, 6, and 7).

$$\frac{d\hat{s}}{dd} = a_1 + 2a_2d + \cdots + ma_m d^{m-1}$$

which gives the slope of the polynomial at any depth. The location of the peaks and troughs of the polynomial can be extracted when the slope equation is set to zero and solved for depth. For the cubic function fitted to the shale profile of the reference log, the slope equation is a quadratic and yields two depth values as solutions. The depths correspond to the trend estimate of the maximum and minimum shale development within the section (Fig. 11).

A further differentiation of the generalized slope equation gives

$$\frac{d^2\hat{s}}{dd^2} = 2a_2 + \cdots + m(m-1)a_m d^{m-2}$$

which is an expression of the rate of change in slope. The rate of change is zero at curve inflection points, which mark the boundaries between peak and trough features. On the reference log, there is a single solution for inflection point depth (Fig. 11), which locates the cubic trend estimate of the contact between the lower

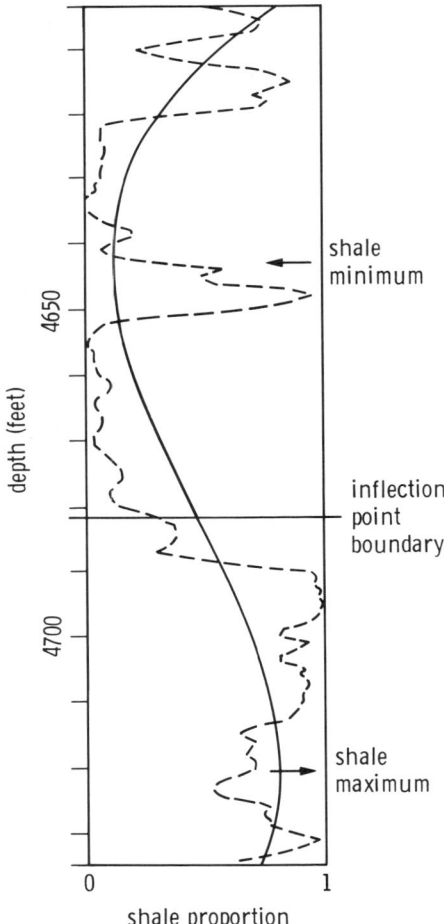

FIGURE 11. Reference gamma-ray log shale profile and fitted cubic trend indexed with maximum, minimum, and inflection-point depths.

shale subdivision ("McLish Formation") and the upper sandstone unit ("Bromide Formation").

These same parameters can be traced on slice profiles, because the solution of the location of maxima, minima, and inflection points depends solely on the polynomial coefficients and section thickness. In particular, the solution of inflection points at successive grid columns can be linked to sketch out subdivision boundaries. The result is a method of "polynomial stratigraphy" and is a mathematical solution of lithostratigraphy based on the trend differentiation of major subdivisions.

The fourth order slice profile of the east–west cross section was used to illustrate the application of this concept. A quartic polynomial has two possible solutions for inflection points and these were solved at all columns of the slice. Only one solution could be traced as a laterally continuous boundary across the Simpson Group section (Fig. 12). In many locations, the second solution proved to be irrational, while at others the additional inflection point picked out localized facies changes of a basal sand in the east and an upper shale division in the west. The isolation of only one regionally continuous boundary is further confirmation that a cubic polynomial (defined by a unique inflection point) is the most appropriate descriptor of major shale facies changes in the Simpson Group.

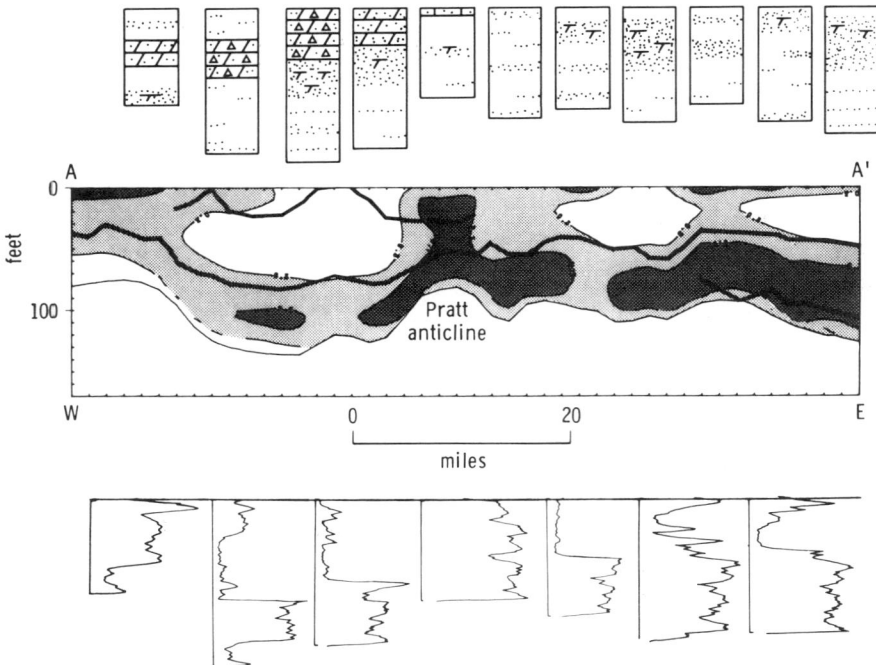

FIGURE 12. Simplified quartic trend shale ratio profile of the Simpson Group on a east–west cross section (line A–A' on Figs. 4, 6, and 7). Internal solid curves trace inflection-point boundaries within the Simpson Group. Cuttings and gamma-ray logs are drawn from wells close to the line of the cross section.

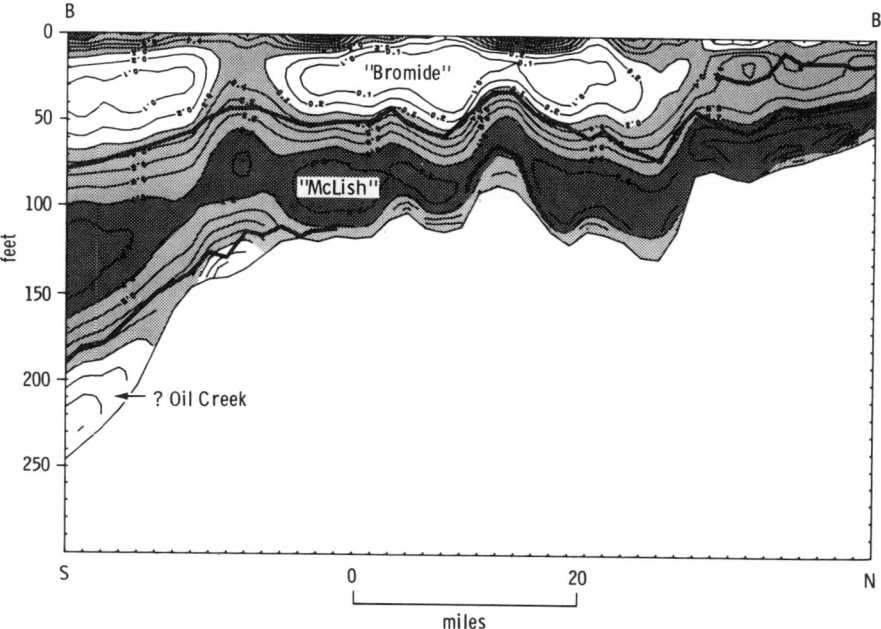

FIGURE 13. Quartic shale proportion trend profile of the Simpson Group on a north–south cross section (line *B–B'* on Figs. 4, 6, and 7).

The profile of Figure 12 is a simplified version of the fourth order profile of Figure 10, where shale isolines have been reduced in number to accentuate the geometry of lithofacies with the Simpson Group. Since the slice is derived totally from a mathematical treatment of gamma-ray logs, which are sensitive only to shale variation, more substantive information is required for detailed interpretation. Cuttings logs of the Simpson Group provide the necessary "ground truth" for validation purposes and a source of more specific lithology data. Representative cuttings logs taken from wells close to the cross-section slice (Fig. 12) indicate that the low-shale feature in the west is a complex of dolomitic sandstones grading upward to sandy dolomites and limestones. Its steep eastern boundary coincides with the western edge of the Pratt Anticline and may be caused by a fault margin rather than an abrupt lateral change in lithofacies. The low-shale anomalies to the east are sandstone bodies overlain locally by thin sandy dolomites and separated by a shale from the overlying Viola Limestone.

A second example of a cross-section slice is shown for a north–south profile (line *B–B'* on Figs. 3, 4, 6, 7), which is oriented at right angles to the regional depositional strike (Fig. 13). This slice was generated from a fourth order polynomial solution. As was the case with the east–west slice, only one regionally continuous inflection point could be traced through the Simpson Group, although a second inflection point indicated localized facies units. The most pronounced secondary facies is marked by a basal sandstone unit that occurs immediately south of the thickness hinge line. The presence of this localized sand is confirmed by cuttings logs from the area. It represents either a basal transgressive sand facies of the McLish Formation or the extreme northern edge of the Oil Creek Formation. The low-shale features of the upper "Bromide" unit are sandstone bodies which contain small, localized oil fields, whose pay zones are generally thin and located toward the top of the sandstone.

The transformation of moment maps to three-dimensional representations is not confined to the generation of vertical slices matched with cross sections. Horizontal slices can be produced which portray trend estimates of shale proportion at any specified depth. The trend projection of the shale proportion at the top of the Simpson Group is particularly easy to compute, since this level corresponds to a depth of zero. Insertion of this null value for depth in any polynomial in the form

$$\hat{s} = a_0 + a_1 d + a_2 d^2 + \cdots + a_m d^m$$

results in an estimate of the shale proportion at the top as simply

$$\hat{s} = a_0$$

The trend estimate of the shale content at the base of the Simpson Group is similarly straightforward. The efficient algorithm for polynomial coefficient solution uses a standardized scale of depth in which the thickness of any Simpson Group is one unit and the solution for the shale content at this level is therefore

$$\hat{s} = \sum_{i=1}^{m} a_i$$

The trend estimate of the Simpson Group basal facies was mapped by calculating estimates of the polynomial coefficients on a gridded areal network and contouring their sum (Fig. 14). The map is particularly interesting in that it defines the lateral extent of the basal transgressive sand. The sand appears to have migrated as a two-pronged advance from the south, and its absence in the center of the area may reflect the influence of the Pratt Anticline as a distinctive syndepositional bathymetric high. The sand development at the northern margin may represent strand-line accumulations that flanked the ancient Central Kansas Arch.

It is possible to compute other trend maps of shale content linked either to proportional levels within the Simpson Group or at depths measured in feet or meters. In fact, there is no intrinsic theoretical objection to the computation of the shale variation of the entire Simpson Group wedge. This would be realized as a three-dimensional cell structure in which each cell would be defined by geographic coordinates, depth level, and an associated shale proportion trend value. As discussed before, the essential problem lies in the design of graphic presentations that transcend their two-dimensional limitations in the portrayal of a three-dimensional object.

Simple Lithofacies Mapping from a Single Log

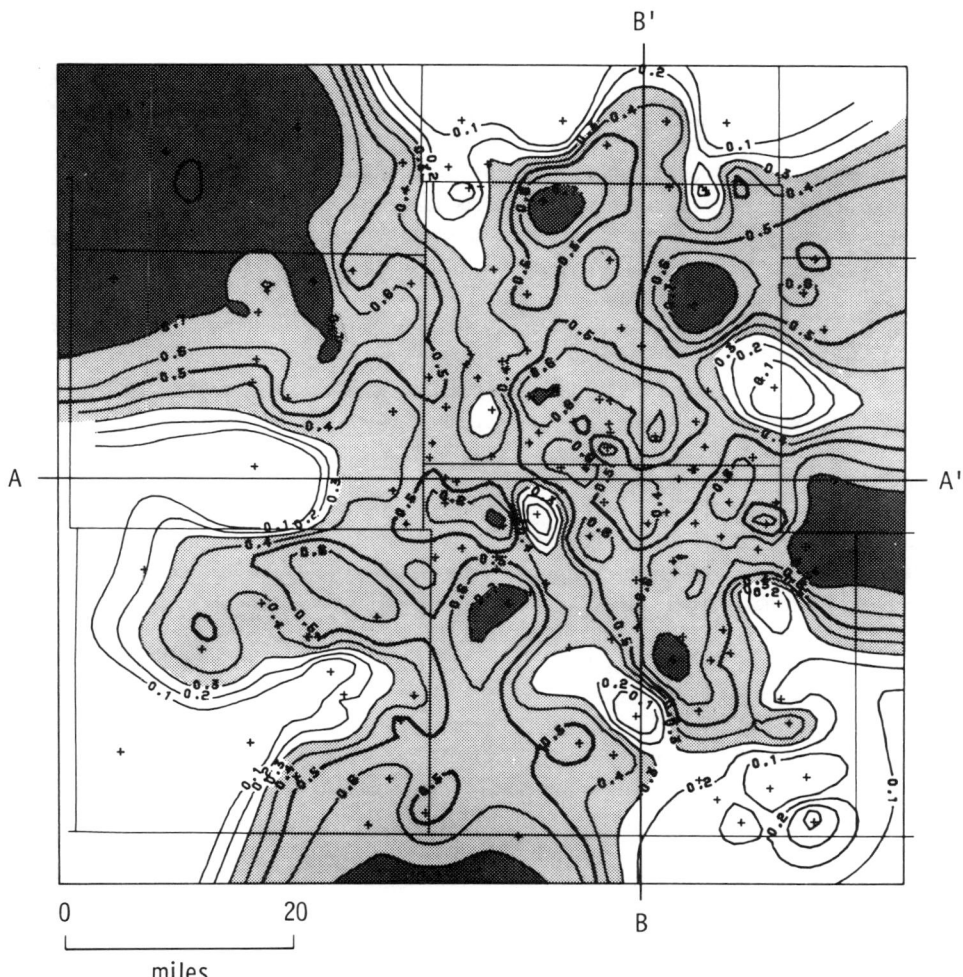

FIGURE 14. Quartic trend shale proportion of the base of the Simpson Group. Proportions greater than 0.3 are shaded; proportions greater than 0.7 are heavily shaded.

Further Applications

Three-dimensional trend mapping need not be restricted to the processing of gamma-ray logs. So, for example, three-dimensional characterizations of trends in porosity within reservoir units may be produced using the same methodology applied to neutron, density, or sonic logs. The major assumption would be that the range of lateral continuity in porosity trends exceeds the spacing of control wells.

An interesting consequence of three-dimensional mapping of sonic logs on cross-section slices is that they should show strong similarities with synthetic velocity profiles generated from the deconvolution of seismic records. This assertion is suggested by the fact that seismic information represents a convolution of a subsurface impedance sequence by a relatively broad energy wavelet. This wavelet behaves as a low-pass filter of velocity variation in an analogous (but not mathematically equivalent) manner in which polynomial trends filter frequency content. This intruiging parallelism suggests that the method may find useful applications in seismic modeling through studies which integrate processed seismic records and regional three-dimensional velocity trends based on sonic logs.

In using three-dimensional mapping of moments, the following considerations should be kept in mind. First, logs should be selected which measure a property whose trends are likely to be adequately sampled by available well control; otherwise it may not be possible to interpolate between the control wells in a meaningful manner. Second, the vertical extent of the analyzed stratigraphic interval should be chosen taking into account both vertical and lateral scales of variability. If the stratigraphic interval is too thick, the low-order trends may account for only an insignificant proportion of the variation in composition shown on the log traces. Although these trends can be mapped laterally with the greatest confidence, the three-dimensional model will give only the most general indications of vertical variation. At the other extreme, selection of too short a vertical interval may result in the definition of features of localized importance which cannot be interpolated between the control wells.

These two considerations express the practical limitations imposed by the scale of the model and the complexity of vertical variability and the lateral spacing between logged wells. The possible polynomial equations can be considered as a set of filters in which low-order polynomials are low-pass filters. At progressively higher orders, the polynomials pass increasingly higher frequencies. In general, low-frequency information shows greater lateral continuity than do high frequencies, which often represent variation caused by thin, impersistent beds. However, the ultimate controls are the thickness of the analytical section, its vertical variability, and the spacing distance between control wells.

The design of a useful application can be aided by preliminary polynomial trend analysis of selected logs that appear to typify vertical variation. The statistics of fitted trends reveal if there is any "natural" division which differentiates large-scale components from less systematic small-scale variation. The numerical values of the fits are estimates of the fraction of total information that will be presented on the three-dimensional model of the associated polynomial trend.

COMPLEX LITHOFACIES MAPPING FROM MULTIPLE LOGS

The use of a single log for lithofacies mapping has obvious limitations. Only two complementary components can be resolved by a log reading. These will either be the systems shale/nonshale or porosity/matrix. In practice, the situation is better than these remarks suggest. Core and cuttings are the source of information which puts flesh on the bones of variability mapped from a log. Bodies that are mapped with low-shale contents can be assigned specific identities as either sandstones, limestones, dolomites, or other low-radioactive lithologies. These distinctions are not recognized explicitly by a single log. Mapped variations in porosity can be interpreted in terms of distinctive lithofacies. Once again, it is the analysis of core and cuttings that is the supplementary key which provides geological meaning.

More detailed assessments of compositional variation may be made if several logs are studied in combination. Useful mathematical techniques for this purpose were discussed in Chapter 6. In those applications, mineral composition was estimated from log combinations for successions of depth zones in a single well. The same mathematics can be applied for mapping purposes. The basic problem involves a change in the descriptive framework from a one-dimensional string of observations to a two-dimensional representation in geographic space. The operation requires a computer contouring program that can handle grid-to-grid manipulations in addition to surface estimation and map production. The theory and practical considerations of this approach are best illustrated in the case study application described by Bornemann and Doveton (1983) which is summarized in the following sections.

The study area is the same four-county region in southern Kansas which was used for the map analysis of the Simpson Group (Fig. 1). The stratigraphic unit selected for compositional analysis is the Viola Limestone (Middle to Upper Ordovician) which immediately overlies the Simpson Group. The Viola Limestone has already been used on several occasions in this book to demonstrate a variety of log analysis techniques. The logged Viola section which was examined in Chapter 7 is of particular interest, since it was taken from a well located in the southwestern quadrant of the study area. The Viola Limestone shows complex patterns in variation between limestones, cherty dolomitic limestones, and cherty dolomites. The Viola subcrops against the Central Kansas Uplift a short distance to the north and, in the northern part of the area, has a variable thickness of residual chert in the uppermost part of the unit.

The mapping of changes in lithology across an area requires an extension of the methodology developed for individual well profiles to techniques suitable for lateral interpolation. In theory, it would be possible to compute computational profiles of the Viola at each well location, calculate average compositions, and use the results as input for a mapping procedure. However, the Viola data set has a severe deficiency which would be a common limitation in other studies of this type. Only two wells in the total sample have a complete neutron–density–sonic log combination. Since only one porosity log was run in the majority of wells, a new approach to the problem must be devised.

If a porosity log suite in a stratigraphic unit can be transformed into a compositional profile, the mean composition of the unit is readily calculated by averaging the total vertical variation. An identical result is obtained by directly transforming the average log responses of the unit. This equivalence implies that in order to compute the unit mean composition at any location, the basic input requirement is merely the average log responses of the unit rather than a complete suite of log traces.

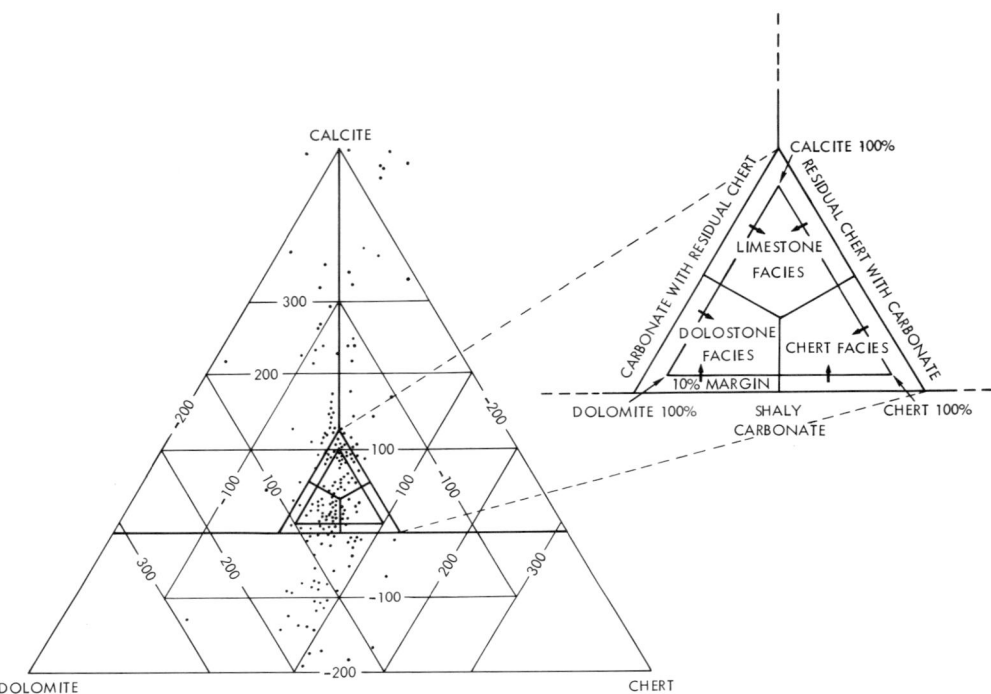

FIGURE 15. Plot of grid element computed compositions of the Viola referenced to a calcite–dolomite–chert triangle.

Since each average log response of a major stratigraphic unit will fluctuate only gradually when traced laterally between wells, these variations will sketch out continuous hypothetical surfaces across the area. Such surfaces are generally approximated by computer mapping packages as grids of numerical values. Grid rows and columns specify geographic position, while the grid node values are estimated by interpolation from data in available well control. On this basis, three surfaces of average Viola neutron, density, and sonic log responses may be computed independently as grids drawn from the separate sources of log control. The three grids can be set to cover the study area and to be geographically congruent so that an estimated average response of each porosity log is available at all geographic locations specified by the grid nodes.

Normalization of logs run in the Viola was made using the trend surface method described in Chapter 8. The normalized Viola average log responses were used as input to an automated contouring program in the generation of grids for neutron, density, and sonic values. The matrix algebra method described for the analysis of vertical variation was extended to the transformation of these log response grids to compositonal grids of calcite, dolomite, chert, and porosity proportions. Rather than convert a sequence of depth zone log responses to compositional profiles, the procedure was modified as a grid-to-grid operation in which corresponding grid nodes were successively transformed.

Grid node proportions were plotted with reference to a calcite–dolomite–chert composition triangle (Fig. 15). This ternary diagram is highly unconventional since many of the nodes are associated with negative component fractions. These anomalous points have a simple explanation when comparison is made with the M–N plot of depth zone log readings from one of the wells (Fig. 16). Points that lie above the calcite endmember indicate systematic coarse-grade porosity which is generally associated with residual chert and weathered carbonate. Points below the dolomite–chert line suggest the influence of minor amounts of shale as an additional component. Following this line of interpretation, the diagram was subdivided into six lithofacies types. The fundamental calcite–dolomite–chert triangle was allocated between three facies of limestone, dolomite, and chert as determined by the dominance of one of the three mineral endmembers. The remainder of the diagram was divided into three lithofacies matched with residual chert with carbonate, carbonate with residual chert, and shaly carbonate.

The lithofacies subdivision of the diagram allowed the four grids of compositional variation to be fused in a single lithofacies grid presentation, where each set of grid nodes could be allocated to one or other of the six lithofacies. The result is shown in Figure 17, where the size of the grid cells conforms approximately to the density of well control in the area.

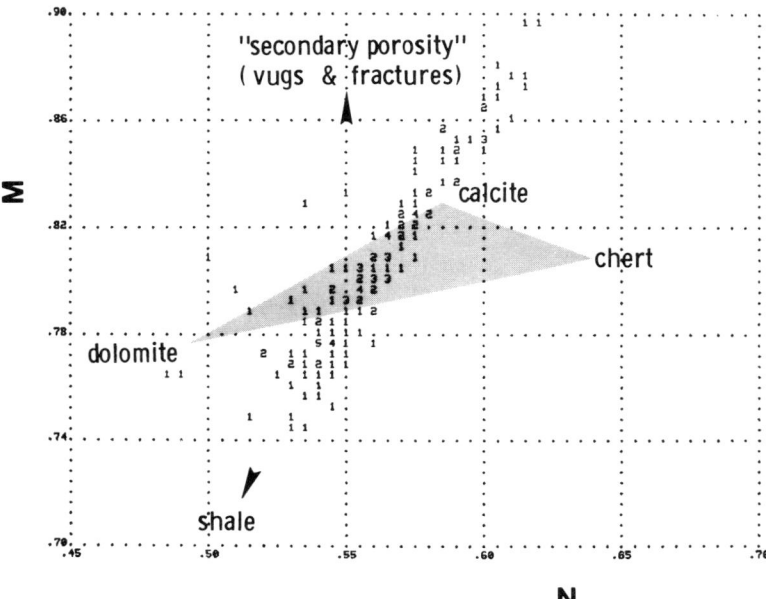

FIGURE 16. M–N plot of the Viola Limestone in the Belcher A-1 well.

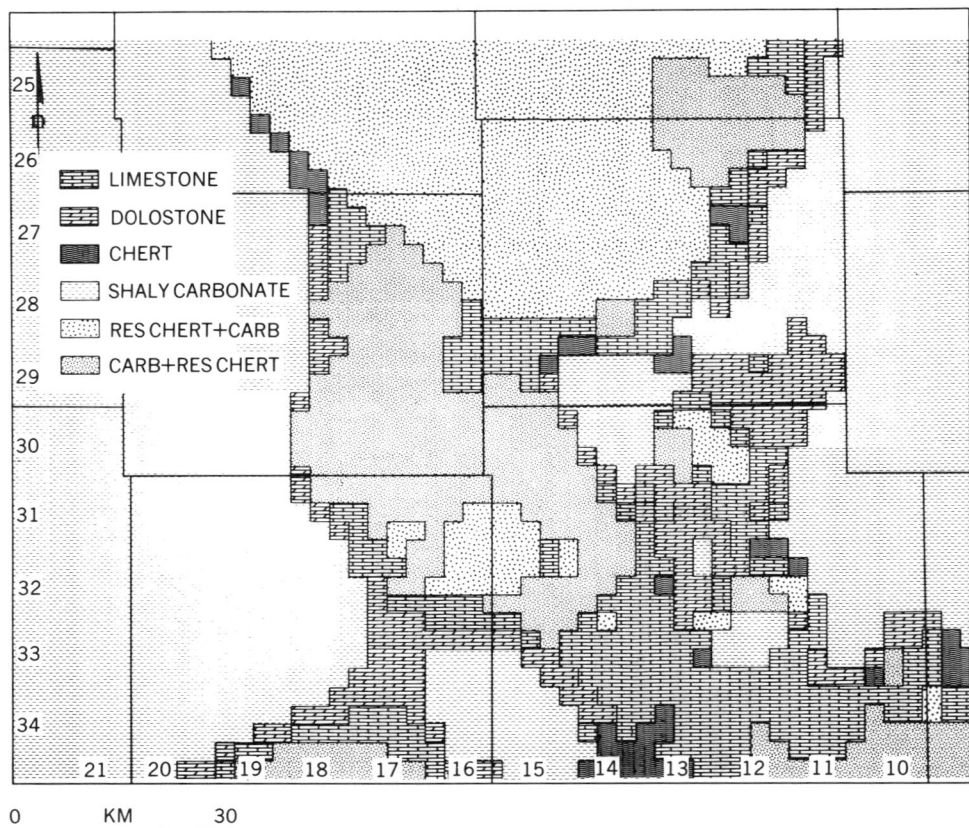

FIGURE 17. Petrophysical lithofacies map of the Viola.

Interpretation and Validation of the Wireline Lithofacies Map

The structural map of the top of the Viola (Fig. 18) shows details of the major features illustrated on the regional map of Figure 1. In the north of the area, the southern limit of the Central Kansas Uplift is linked southward with the Pratt Anticline which marks a boundary between the Hugoton Embayment to the west and the Sedgwick Basin to the east. The structure contours in the eastern half of the area show a marked NNE grain of localized features which persist to the Precambrian basement. This character was observed by Rich (1930) who was "at once struck by the pronounced rectangular pattern of the structure and . . . that the area seems to be divided into blocks bounded by nearly straight sides, or fracture lines." In a discussion on seismic anomalies on the Pratt Anticline, Brewer (1959) suspected faulting and differential erosion of Lower Paleozoic rocks.

The isopach map of the Viola (Fig. 19) shows a regional trend of increasing thickness to the west which is confounded by significant local variations. It is probable that the depositional thickness of the Viola increased in a southwesterly direction away from the Central Kansas Uplift, and would mirror a similar trend in the Viola in the Salina Basin to the north of the Arch (Taylor, 1947). Regional uplift that preceded deposition of the Late Devonian Chattanooga Shale was probably responsible for the westerly reorientation of gross thickness trends as a differential erosion modification to the earlier depositional trend.

The second major tectonic event began its major cycle of activity in the Late Mississippian and culminated in the Early Pennsylvanian (Lee, 1956; Merriam, 1963). Its effect on the Viola was categorically different from the earlier tectonic uplift. While the former caused differential erosion of the Viola with determination of the formation's regional thickness variation, the second event finalized the structural configuration of the area. The present disposition of the Pratt Anticline was established and fault-bounded blocks were activated in the east so that Pennsylvanian erosion reached the Viola Limestone and caused additonal thinning over these features with marked topographic relief.

The influence of these tectonic events is readily seen on the Viola lithofacies map based on well logs. Facies interpreted as "residual chert" forms a belt which

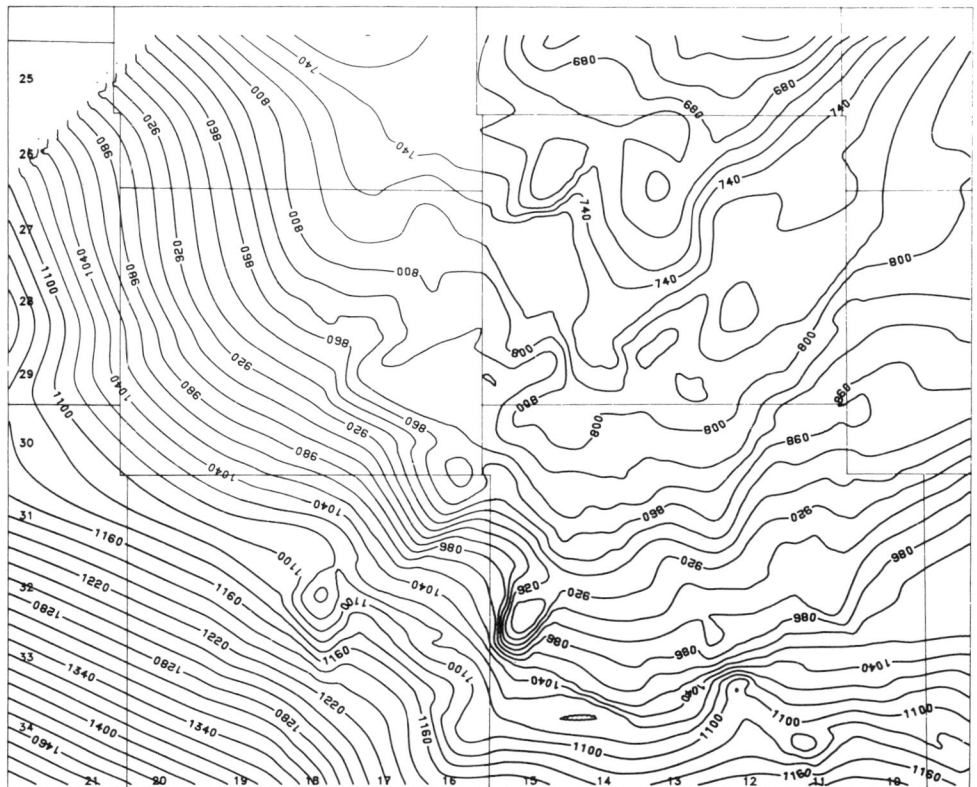

FIGURE 18. Structural map of the top of the Viola (meters subsea).

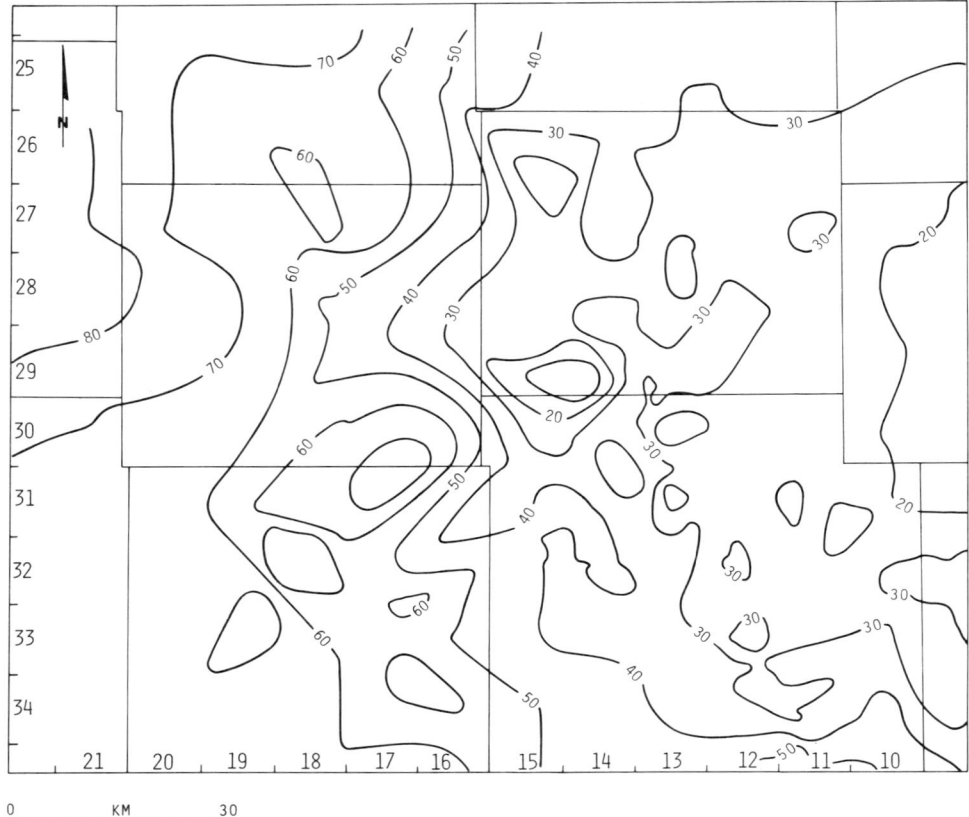

FIGURE 19. Isopach map of the Viola (meters).

borders the Central Kansas Uplift to the north and parallels the subcrop of the Viola. The extension of this facies to the south together with isolated outliers picks up the trend of the Pratt Anticline feature. This petrophysical interpretation shows good concordance with a map of residual chert thickness determined from cutting logs by Adkison (1972) (Fig. 20). The chert regolith resulted from extensive weathering in the Late Devonian, Late Mississippian, and Early Pennsylvanian.

The facies designated as "carbonate with subsidiary residual chert" is located mainly in the center of the area in an intimate association with the "residual chert facies" and has an interesting interpretation. Several sample logs in this facies area report thin developments of shale and sandstone which are very similar to those of underlying Kinderhook rocks. One well in the center of the area shows 46 feet (14 meters) of "intra-Viola detritus" as a jumble of Viola chert and dolomite in clay and shale, which is overlain by shale and cherty clay. Adkison (1972) has suggested that "some of the non-carbonate clastic deposits are probably misidentified cavern fillings of post-Viola age." If this interpretation is correct, then the two residual facies of "residual chert with carbonate" and "carbonate with residual chert" are paleogeomorphic, the former representing a detritus regolith and the latter karstic solution-weathered carbonates.

A map of representative cuttings from the area is shown in Figure 21. These were selected from open file reports of a single geologist (J. D. Davies) in order to exclude interpretative differences in style. The disposition of rock type profiles is in essential agreement with the petrophysical facies patterns and picks up the central limestone facies which grades outward to dolomite. The "shaly carbonate" facies can now be explicitly identified as cherty dolomite. The slightly "argillaceous" character of dolomites in these areas accounts for its shaly aspect as perceived by the logs, which are particularly sensitive to shale contents.

The combined effects of depositional facies patterns and subsequent erosional bevelling of the Viola (summarized in Fig. 22) appears to be the key that explains the regional disposition of the "dolomite" and "limestone" facies of the log analysis map. A central limestone facies trending NW–SE is flanked by dolomite to the northeast and southwest. As a composite analysis of the entire Viola section, the petrophysical facies will tend to reflect the relative dominance of the subdivisions. To the west, all four units are present and, since the dolomitized units are thicker than the limestone units, a composite facies of dolomite is determined. In the central region, the upper dolomite is absent so that the combination of both limestone units with the lower dolomitized unit results in a "limestone facies." To the east, only the thinner basal limestone and thicker lower dolomitized unit are present so that the net effect is a "dolomite facies."

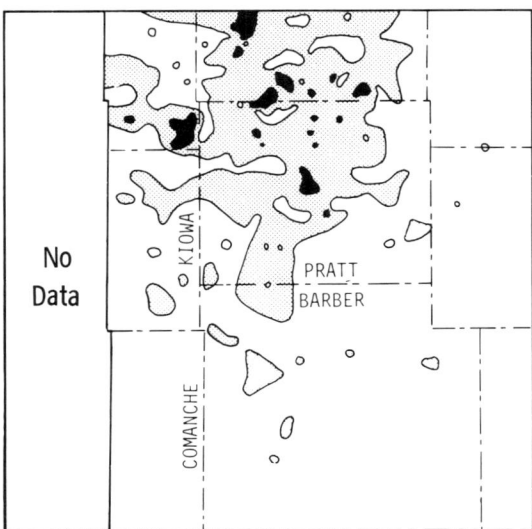

FIGURE 20. Viola residual chert thickness map of the study area, modified after Adkison (1972). Key: dark, residual chert 50–100 feet thick; stipple, 0–50 feet thick; blank, absent.

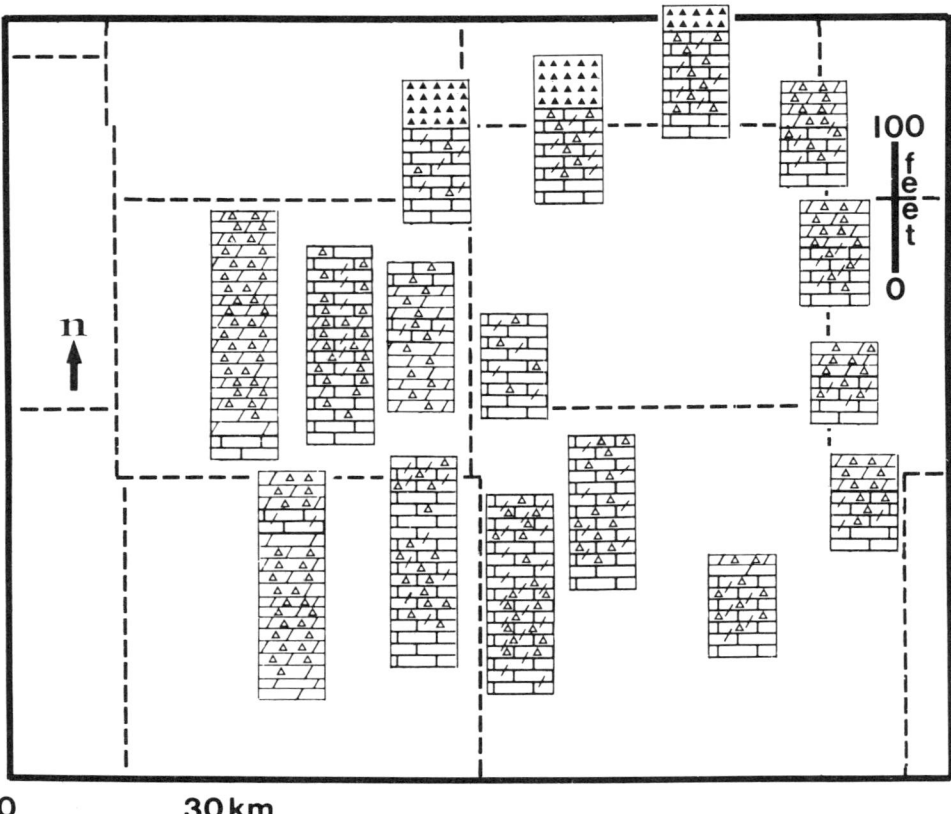

FIGURE 21. Representative cuttings logs of the Viola. The location of each associated well is at the center of the strip.

Of particular interest on the petrophysical facies map is the marked grain, whose lineaments strike approximately NNE–SSW and NW–SE. This characteristic is most pronounced on contour maps of individual components. A contour map generated from the grid of chert proportion (bedded chert facies, rather than residual chert) is shown in Figure 23. The obvious lineaments can be related to features on both structure and isopach maps, and can be traced to the Precambrian basement. The history of these structural elements has been described earlier and they appear to represent ancient zones of weakness which have been reactivated at various periods in the stratigraphic record.

One explanation for the localized increases of chert in the vicinity of these tectonic features is through generation by hydrothermal solutions in fracture systems. This genetic form has been designated "T-chert" by Dunbar and Rodgers (1957, p. 248), and the localized concentrations would contrast with the more normal "S-chert" bedded and nodular forms which can be attributed to more regional mechanisms of diagenesis. Alternatively, the uplift block features associated with these anomalies may have caused extended residence times of the local Viola in phreatic mixing zones, which enhanced chert formation to concentrations greater than normal levels.

Another confirmation for this lineament pattern is provided by the recent aeromagnetic survey of Kansas (Yarger, 1981). The relative total intensity magnetic field map of the area (Fig. 24) is singularly bland. However, a second vertical derivative transformation of this data is sensitive to abrupt changes in magnetic susceptibility (Vacquier et al., 1951). As a measure of the rate of change in slope in magnetic field intensity, a second derivative map of the area (Fig. 25) shows a distinctive lineament grid which has good concordance with the grain perceived on both structural and petrophysical facies maps. These lineaments are even more striking on second derivative maps of the entire state (Fig. 26), since they are

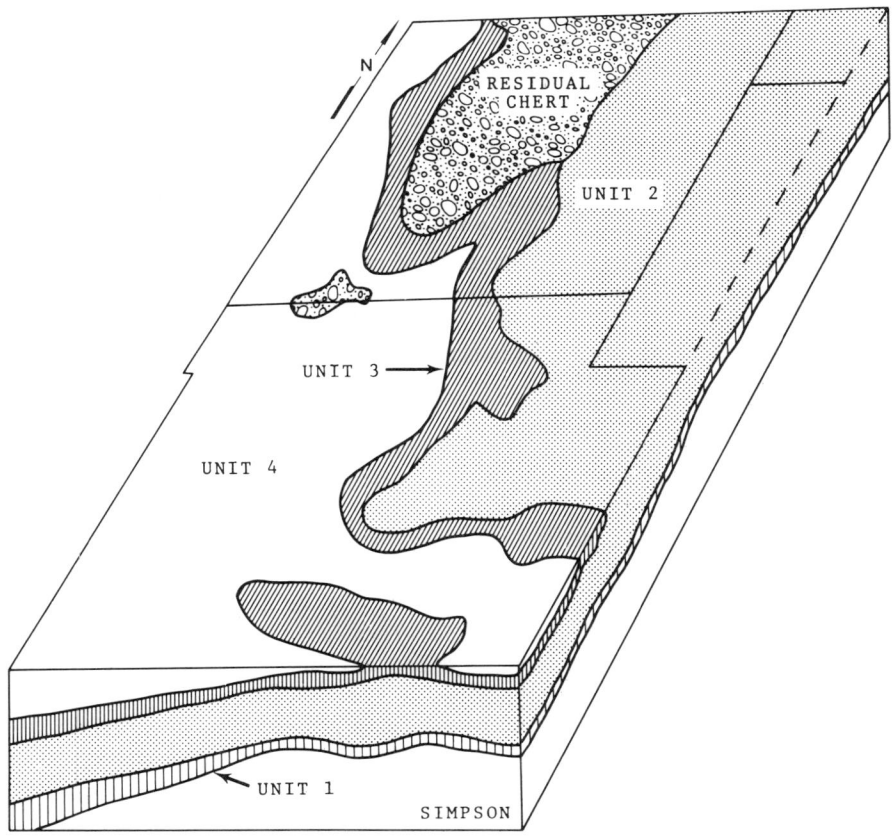

FIGURE 22. Block diagram of the eastern half of the study area, showing the present distribution of the Viola subdivisions. Basal limestone = 1; lower cherty dolomite limestone = 2; upper limestone = 3; upper cherty dolomite limestone = 4. After St. Clair (1981).

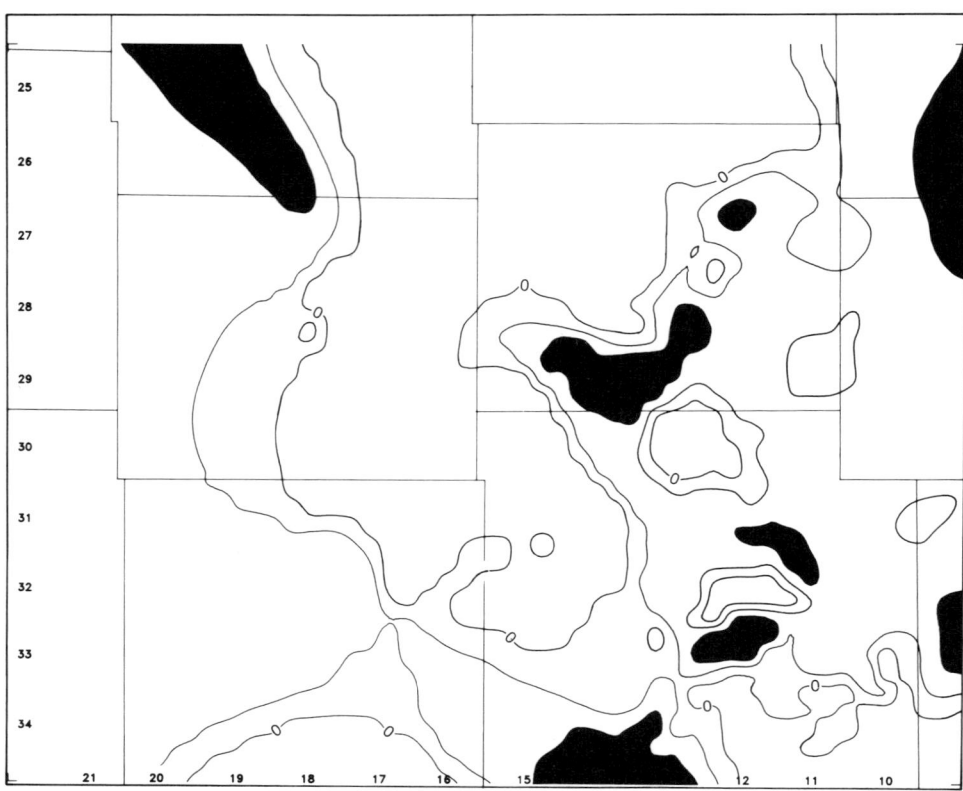

FIGURE 23. Contour map of Viola chert proportion (internal chert facies) drawn from log analysis. Proportions greater than 0.2 are shaded.

features of regional significance. The NNE–SSW trends in the eastern half of the area appear to be southerly continuations of lineaments linked with the gravity/magnetic anomaly associated with the Proterozoic Central North American Rift System (CNARS). In the north of Kansas and beyond, the CNARS is marked by a pronounced gravity anomaly that probably reflects an aborted "oceanic" rift in Pre-Cambrian times, which is floored by Keweenawan mafic rocks (King and Zietz, 1971). The southern Kansas extension does not seem to have moved beyond an early period of block faulting and possible dike intrusion (Yarger, ibid). In summary, it appears that the study area straddles the western margin of an ancient rift system which has been intermittently active throughout geological time and whose effects can be perceived on lithofacies patterns derived from well logs. Indeed, these zones

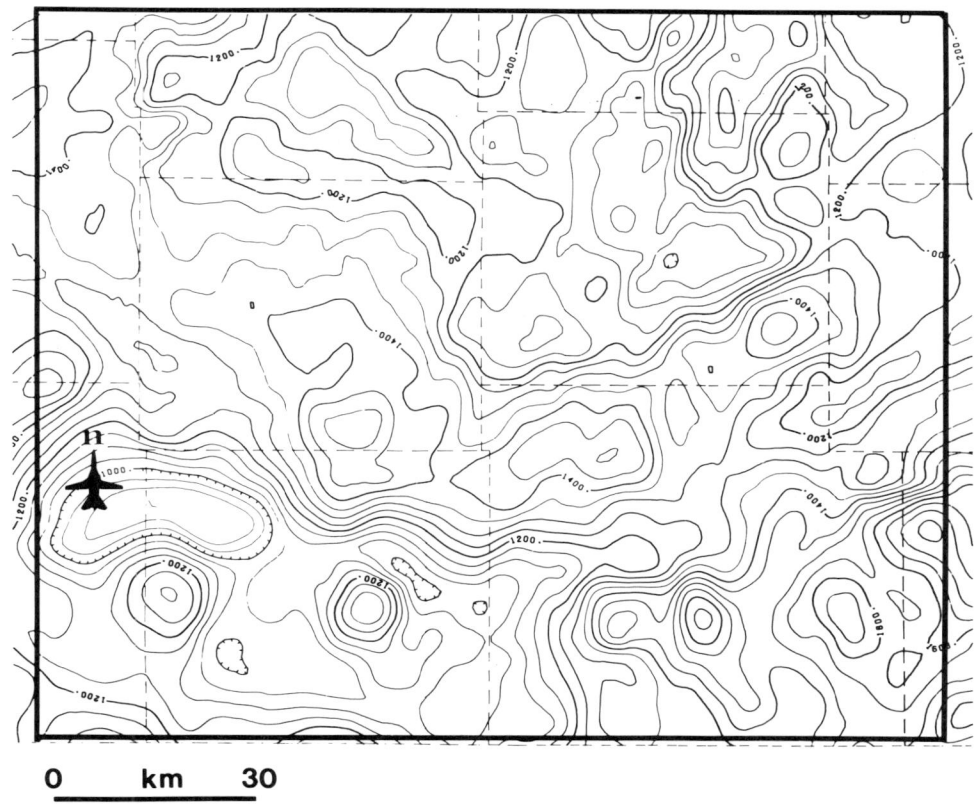

FIGURE 24. Relative total intensity magnetic field map of study area. After Yarger (1981).

of weakness are still active, since four microearthquakes were recorded in the southeast of the area over a recent three-year period (Steeples, 1982).

The location of Viola oil fields is shown imposed on a regional porosity map contoured from the grid of pore volume component, generated as an intermediate step in the production of the lithofacies map (Fig. 27). Northeasterly trends can be seen which are influenced by the tectonic pattern and its history of preferential uplift, and control of depositional facies, diagenesis, and erosion. Structural highs associated with porosities in excess of 7% are favorable conditions for oil accumulation within the Viola. In areas where the Viola is thin and residual chert is absent, reservoirs are rare, probably due to erosional stripping to relatively tight, well-cemented calcite grainstones at the base. Systematic exploration in the area would

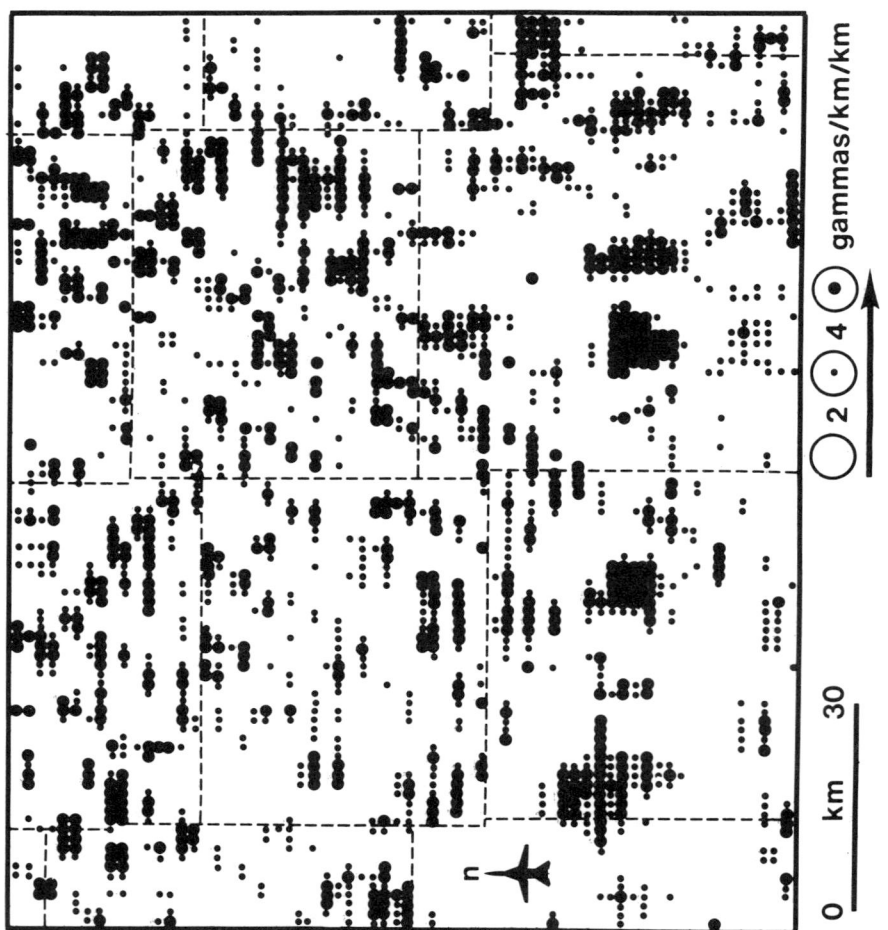

FIGURE 25. Second vertical derivative map of relative magnetic field intensity in study area. After Yarger (1981).

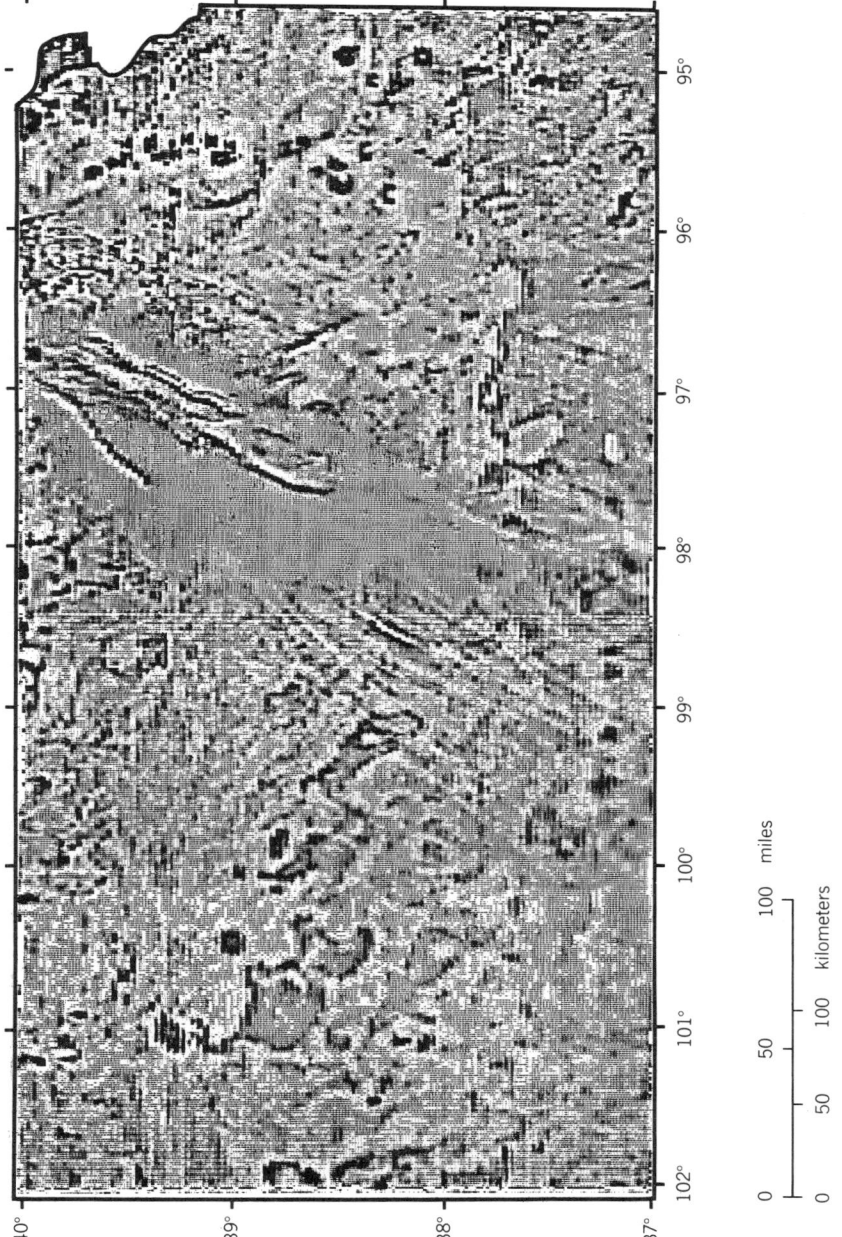

FIGURE 26. Second derivative map of relative magnetic field intensity in Kansas. From Yarger (1983).

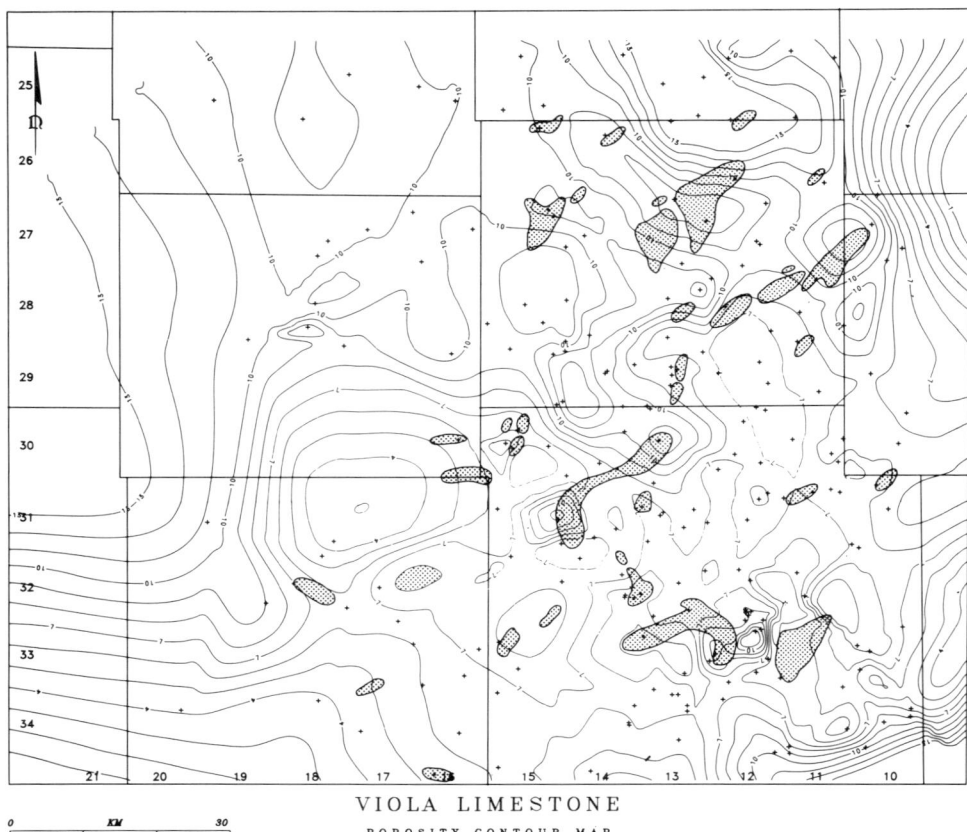

FIGURE 27. Map of the average porosity of the Viola in percent, indexed with Viola oil- and gas-field outlines.

obviously be aided by lithofacies and component maps from well logs in conjunction with more conventional structure and isopach maps.

REFERENCES

Adkison, W. L., 1972, Stratigraphy and Structure of Middle and Upper Ordovician Rocks in the Sedgwick Basin and Adjacent Areas, South-Central Kansas, U.S. Geological Survey Prof. Paper 702, 33 pp.

Bornemann, E., and Doveton, J. H., 1983, Lithofacies mapping of Viola limestone in south-central Kansas, based on wireline logs, *Am. Assoc. Petrol. Geolog. Bull.*, Vol. 67, No. 4, pp. 609–623.

Brewer, J. E., 1959, Geophysical problems on Pratt Anticline, *Kansas Geol. Survey Bull.*, Vol. 137, pp. 275–280.

Busch, D. A., 1950, Subsurface techniques, in *Applied Sedimentation* (P. D. Trask, Ed.) John Wiley & Sons, New York, pp. 559–78.

References

Doveton, J. H., Zhu Ke-an, and Davis, J. C., 1984, Three-dimensional trend mapping using gamma-ray well logs: Simpson Group, south-central Kansas, *Am. Abstr. Petrol. Geolog. Bull.*, Vol 68, No. 6, pp. 690–703.

Dunbar, C. O., and Rodgers, J. 1957, *Principles of Stratigraphy*, John Wiley & Sons, New York, 356 pp.

Forgotson, J. M., Jr., 1954, Regional Stratigraphic analysis of Cotton Valley Group of Upper Gulf Coastal Plain, *Am. Assoc. Petrol. Geolog. Bull.*, Vol. 38, No. 12, pp. 2476–99.

Ireland, H. A., 1965, Regional depositional basin and correlations of the Simpson Group, *Tulsa Geol. Soc. Digest*, Vol. 33, pp. 74–89.

King, E. R., and Zietz, I., 1971, Aeromagnetic study of the Midcontinent Gravity High of central United States, *Geol. Soc. Amer. Bull.*, Vol. 82, No. 8, pp. 2187–2208.

Krumbein, W. C., and Libby, W. G., 1957, Application of moments to vertical variability maps of stratigraphic units, *Am. Assoc. Petrol. Geolog. Bull.*, Vol. 41, No. 2, pp. 197–211.

Krumbein, W. C., and Sloss, L. L., 1963, *Stratigraphy and Sedimentation* (2nd ed.), W. H. Freeman, San Francisco, 660 pp.

Lee, W., 1956, Stratigraphy and structural development of the Salina Basin area, *Kansas Geol. Survey Bull.*, Vol. 121, 167 pp.

Merriam, D. F., 1963, The geologic history of Kansas, *Kansas Geol. Survey Bull.*, Vol. 162, 217 pp.

Rich, J. R., 1930, Fault-block nature of Kansas structures suggested by elimination of regional dip, *Am. Assoc. Petrol. Geolog. Bull.*, Vol. 19, No. 10, pp. 1540–1543.

St. Clair, P. N., 1981, Depositional History and Diagenesis of the Viola Limestone in South-Central Kansas, Unpublished M.S. Thesis, University of Kansas, 66 pp.

Schramm, M. W., J., 1964, Paleogeologic and quantitative lithofacies analysis, Simpson Group, Oklahoma, *Am. Assoc. Petrol. Geolog. Bull.*, Vol. 48, No. 7, pp. 1164–95.

Statler, A. T., 1965, Stratigraphy of the Simpson Group in Oklahoma, *Tulsa Geol. Soc. Digest*, Vol. 33, pp. 162–211.

Steeples, D. W., 1982, Structure of the Salina-Forest City interbasin boundary from seismic studies, *University of Missouri-Rolla Journal*, No. 3, pp. 55–81.

Taylor, H., 1947, Middle Ordovician limestones in central Kansas, *Am. Assoc. Petrol. Geolog. Bull.*, Vol. 31, No. 7, pp. 1242–1282.

Vacquier, V., Steenland, N. C., Henderson, R. G., and Zictz, I., 1951, Interpretation of aeromagnetic maps, *Mem. Geol. Soc. Amer.*, No. 47, pp. 38–45.

Yarger, H. L., 1981, Aeromagnetic survey of Kansas, *EOS*, Vol. 62, No. 17, pp. 173–178.

APPENDIX ONE

KIWI—A COMPUTER PROGRAM FOR COMPOSITIONAL ANALYSIS FROM LOGS

KIWI is a computer program designed to transform log responses from depth zones into estimates of mineral proportions and fluid volumes. The logs used most typically for this purpose are the gamma-ray, sonic, density, and neutron logs. However, the spontaneous potential, resistivity, spectral gamma-ray, or other logs may be used. The choice of logs is restricted obviously to those available in the well. Some thought must be directed to the physical properties of the endmembers which define the extremes of the compositional system to be solved. In particular, the degree of sensitivity of the various logs in the distinction between components is an important control on the accuracy of compositional estimates.

Environmental effects, measurement errors, and imprecise depth registration are aspects which should be dealt with by log corrections and adjustments. This pretreatment phase precedes entry of the data into the KIWI program. The use of less conventional logs should also be considered carefully. For example, if a resistivity log is used as a contributory variable to compositional analysis of a shale-free, water-saturated section, its variation will be controlled not only by pore volume, but by the tortuosity of the pore network. Since the program solves a linear set of equations, an effectively linear relationship between logs and compositional variation must be ensured. The Archie equation stipulates that porosity is a nonlinear function of resistivity. However, this situation can be rectified through the use of the mth root of the conductivity instead of raw resistivity readings. In the case of other logs, some pretreatment of the responses may also be necessary to linearize their relationship with compositional variables.

Input requirements for the program are detailed on the comment lines which precede the operational coding. These specify the name of the well and section to be analyzed, the number of logs, the names of the logs and components to be solved, and the log coefficients of the components. The number of components must exceed the number of logs by one.

The logging data used for analysis can be either from a digitized source recorded at constant increments of depth, or from zone readings of peak and trough features at irregular intervals. In its present form, the program requires depth values to be integers, but can be modified easily to handle real numbers. The output is designed for line printer production. Information concerning well name, formation, names of logs and components, and log coefficients for components are printed first. This sequence is followed by a numerical listing of zone depths, log readings, and component proportions solved by the program. Finally, the zone compositions are plotted as a line printer graphic profile of the analytical section. The alphabetic symbols used are matched with a key printed in the initial phase of the output. When the input consists of digitized data at one unit depth increment, the graphic profile is a continuous system of letter sequences. For zone readings at irregular intervals, the profile spaces the zones according to their differences in depth, with blanked output in the intervening sections. In either case, continuous traces of compositional variation can be drawn by linking sequences of each composition letter.

The coding of KIWI is written in FORTRAN IV. Machine-dependent features have been kept to an absolute minimum. The important exception is statement 3, namely 'FORMAT (V)'. This is the Honeywell convention for reading data in free-format, rather than fixed fields. Most makes of computer have a similar convention but different coding statement.

Experienced programmers will recognize that the bulk of KIWI is concerned with input and output functions, and that relatively few lines are devoted to mathematical processing. Calculations are limited to an inversion of the log coefficient matrix and the transformation of zone log readings to compositional solutions by matrix multiplication. The current upper limit of six components and two hundred sets of log readings can be expanded easily by changes in the sizes of dimensioned arrays and do-loop integers. KIWI is restricted to the solution of a determined system. However, the program can be modified as a more generalized procedure to solve either underdetermined, determined, or overdetermined systems through incorporation of the matrix algorithms described in Chapter 6. Finally, some representative values of log responses which are typical of common minerals and fluids are listed in Appendix 2. These are not unreasonable values for use as input to KIWI, although allowances should be made for the high variability in properties of some components (particularly the clay minerals) and the nonlinearity of neutron log responses over ranges of porosity.

```
C                       KIWI
C     COMPUTATION OF MINERAL AND FLUID COMPONENTS
C     PROPORTIONS, BASED ON LOG RESPONSES
C         LINE 1 = NAME OF WELL
C         LINE 2 = NAME OF FORMATION OR SECTION
C         LINE 3 = NUMBER OF LOGS (N)
C         LINE 4 = NAME OF FIRST LOG
C         ---------------------------------
C         LINE 4+N = NAME OF NTH LOG
C         LINE 5+N = NAME OF FIRST COMPONENT
C         ---------------------------------
C         LINE 6+2N = NAME OF (N+1)TH COMPONENT
C         LINE 7+2N = FIRST LOG COEFFICIENTS FOR COMPONENTS
C         ---------------------------------
C         LINE 7+3N = LAST LOG COEFFICIENTS FOR COMPONENTS
C         REMAINING LINES = DEPTH AND LOG READINGS FOR
C                   EACH ZONE, ONE LINE PER ZONE
C                                       DOVETON 1981
      DIMENSION TITLE(12),C(6,6),CI(6,6),RES(6),R(6),
     1    ID(200),IOUT(100),IL(6),IR(200,6)
      DATA ISTAR,IBLNK,IL/'*',' ','A','B','C','D','E','F'/
      DATA C,CI,RES,R/84*0.0/
      WRITE(6,76)
   76 FORMAT(1H1,25X,'KIWI PROGRAM',//)
      WRITE(6,77)
   77 FORMAT(//100(1H*)/)
      DO 1 I=1,2
      READ(5,2) TITLE
    2 FORMAT(12A6)
      WRITE(6,2) TITLE
    1 CONTINUE
      WRITE(6,77)
      READ(5,3) N
    3 FORMAT(V)
      WRITE(6,61)
   61 FORMAT(15X,'KEY TO LOGS',/)
      DO 4 I=1,N
      READ(5,2) TITLE
      WRITE(6,62) I,TITLE
   62 FORMAT(20X,'LOG ',I1,'=',12A6)
    4 CONTINUE
      WRITE(6,5)
    5 FORMAT(15X,'KEY TO COMPONENTS',/)
      M=N+1
      DO 6 I=1,M
      READ(5,2) TITLE
      WRITE(6,7) IL(I),TITLE
    7 FORMAT(20X,'COMPONENT ',A1,'=',12A6)
    6 CONTINUE
      WRITE(6,77)
      WRITE(6,11) (IL(I),I=1,M)
   11 FORMAT(10X,'LOG COEFFICIENTS',8X,6A10)
      DO 8 I=1,N
      READ(5,3) (C(I,J),J=1,M)
      WRITE(6,9) I,(C(I,J),J=1,M)
    9 FORMAT(21X,'LOG ',I1,5X,6F10.2)
```

```
    8 CONTINUE
      DO 60 I=1,M
   60 C(M,I)=1.0
      WRITE(6,77)
C     INVERT MATRIX OF LOG COEFFICIENTS
      DO 10 I=1,M
   10 CI(I,I)=1.0
      DET=1.0
      DO 12 I=1,M
      DIV=C(I,I)
      DET=DET*DIV
      DO 13 J=1,M
      C(I,J)=C(I,J)/DIV
      CI(I,J)=CI(I,J)/DIV
   13 CONTINUE
      DO 14 J=1,M
      IF(I-J) 15,14,15
   15 RATIO=C(J,I)
      DO 16 K=1,M
      C(J,K)=C(J,K)-RATIO*C(I,K)
      CI(J,K)=CI(J,K)-RATIO*CI(I,K)
   16 CONTINUE
   14 CONTINUE
   12 CONTINUE
C     READ AND PROCESS LOG RESPONSES
      NZ=0
      RES(M)=1.0
      WRITE(6,31)
   31 FORMAT(10X,'LOG RESPONSES AND',
     1   ' COMPONENT PROPORTIONS',/)
      WRITE(6,32)  (I,I=1,N)
   32 FORMAT(5X,'DEPTH',5I8)
      WRITE(6,81)  (IL(J),J=1,M)
   81 FORMAT(1H+,55X,6A8)
      DO 33 I=1,200
      READ(5,3,END=51) ID(L),(RES(J),J=1,N)
      NZ=NZ+1
      DO 34 I=1,M
      R(I)=0.0
      DO 34 J=1,M
   34 R(I)=R(I)+CI(I,J)*RES(J)
      WRITE(6,17) ID(L),(RES(I),I=1,N)
   17 FORMAT(5X,I4,3X,5F8.2)
      WRITE(6,82)  (R(J),J=1,M)
   82 FORMAT(1H+,52X,6F8.2)
      RC=0.0
      DO 18 I=1,M
      IF(R(I).LT.0.0) R(I)=0.0
   18 RC=RC+R(I)
      DO 19 I=1,M
      R(I)=100.0*R(I)/RC
   19 IR(I,I)=R(I)+0.5
   33 CONTINUE
   51 WRITE(6,77)
C     PLOT COMPONENTS AS GRAPHIC LOG
      WRITE(6,20)
```

```
   20 FORMAT(1H1,'GRAPHIC COMPONENT LOG',//)
      WRITE(6,25)
   25 FORMAT(10X,102(1H*))
      IC=ID(1)
      DO 21 L=1,NZ
      DO 22 J=1,100
   22 IOUT(J)=IBLNK
   71 CONTINUE
      IF(ID(L).EQ.IC) GO TO 72
      WRITE(6,24) IC,IOUT
      IC=IC+1
      GO TO 71
   72 IZ=0
      IC=IC+1
      IOUT(100)=IL(M)
      DO 23 J=1,N
      IF(IR(L,J).EQ.0) GO TO 23
      IZ=IZ+IR(L,J)
      IOUT(IZ)=IL(J)
   23 CONTINUE
      WRITE(6,24) ID(L),IOUT
   24 FORMAT(I10,'*',100A1,'*')
   21 CONTINUE
      WRITE(6,25)
      STOP
      END
```

APPENDIX TWO

PHYSICAL PROPERTIES OF COMMON GEOLOGICAL COMPONENTS

	ρ_{ma}	Δt	CNL, %	SNP/GNT, %
Silicates				
Quartz	2.65	55.1	−2	−1
Garnet	4.31		7	3
Hornblende	3.20	43.8	8	4
Tourmaline	3.02		22	16
Zircon	4.50		−3	−1
Carbonates				
Calcite	2.71	47.5	−1	0
Dolomite	2.87	43.5	1	2
Ankerite	2.86		1	0
Siderite	3.89		12	5
Oxides				
Hematite	5.18	44.2	11	4
Magnetite	5.08	73.8	9	3
Limonite	3.59	56.9	60+	50+
Sulfides				
Sulfur	2.02	122.0	−3	−2
Pyrite	4.99	39.4	−3	−2
Sphalerite	3.85		−3	−3

Physical Properties of Common Geological Components

	ρ_{ma}	Δt	CNL, %	SNP/GNT, %
Sulfides (*continued*)				
Chalcopyrite	4.07		−2	−3
Galena	6.39		−3	−3
Clay Minerals (Wet)				
Kaolinite	2.41		37	34
Chlorite	2.76		52	37
Illite	2.52		30	20
Montmorillonite	2.12		44	40
Micas				
Muscovite	2.82	54.9	20	12
Glauconite	2.55		39	23
Biotite	3.00	55.0	21	11
Feldspars				
Orthoclase	2.52	69.2	−3	−2
Albite	2.59	50.2	−2	−1
Anorthite	2.74	45.4	−2	−1
Evaporites				
Halite	2.04	66.7	−3	−2
Anhydrite	2.98	51.8	−2	−1
Gypsum	2.35	52.5	60+	50+
Trona	2.08	65.0	35	24
Sylvite	1.86	74.0	−3	−2
Carnallite	1.57	80.7	60+	40+
Polyhalite	2.79	57.5	25	14
Barite	4.09		−2	−1
Coals				
Anthracite	1.47	105.0	38	37
Bituminous	1.24	120.0	60+	50+
Lignite	1.19	160.0	52	47
Fluids				
Fresh water	1.00	189.0	100	100
Average oil	0.85	238.0	109	109
Methane	Dependent on temperature and pressure			

CAVEAT! This table was compiled from a variety of sources. Almost all the values have ranges associated with them which are generally narrow for rigorously defined minerals, but highly variable for other components (e.g. clay minerals, coals). The user should view all these quantities as "average" estimates.

INDEX

Abstract log, 149–150
Aeromagnetic survey, 255–259
Analysis of variance, 219
Anhydrite, 167–169, 268
Archie equation, 11, 13–16, 18, 20, 22, 25–26
Archie, G.E., 6, 11, 21, 25
Areas described, see Geographic regions
Arp's formula, 26

Band-pass filtering, 196
Borehole environment, 27

Caliper, 107, 213
Canonical analysis, 211
Capillary tube model, 18
Cation exchange, 26, 85, 123
Cementation factor, 13–24
 and carbonate texture, 21–24
 and fractures, 22
 and particle orientation, 19
 and particle shape, 19
 and permeability, 16–17
 and sandstone texture, 17–21
Cementation factor profile, 17–18, 20, 23–24
Center of gravity, 229, 231
Central North American Rift System, 259
Chert, 10, 23, 163, 186–190, 251–253, 254–255
 residual, 247, 252–253
Chlorite, 164, 268
Clay minerals, 9, 74, 85, 87, 90, 95, 141–146, 164–167, 268
Clays, authigenic and allogenic, 142

Closed system, 186
Cluster analysis, 46, 190, 206
Coal, 77, 109, 112, 145, 268
Compaction, 43–45, 59, 113–117
Compactional drape, 43–45, 59
Compaction gradient, 115–118
Computer programs, 36, 45–46, 153, 160–161, 177–178, 224, 247, 262–266
Conductivity, 29, 65, 205
Conglomerate, 48
Conical projection, 135, 147–149
Convolution, 101, 194–195, 201–202, 246
Core analysis, 215–216
Correlation:
 Pearson, 36, 93–94, 189, 205, 215
 Spearman, 16–17
Crinoidal limestone, 120, 186, 188
Cross-bedding, 17, 46–57, 60–62, 65
Crossplot:
 frequency, 128–132, 144, 167
 gamma ray-N, 145
 gamma ray-P, 136
 MID, 137–138, 147
 M-N, 133–136, 147, 249
 MNZ, 178
 normalization, 214–215
 ROMA-UMA, 138–139
 Z, 132–133
Cycle skips, 105
Cyclic components, 196–199
Cylindrical plot, 57, 61

Deconvolution, 180, 246

Deltas, 51–54, 79
Dendrogram, 190
Density log, 106–109, 124–128, 133–134, 139, 144–145, 159, 161, 164, 167, 186, 205, 209, 215, 218, 220–221, 245, 247
  and porosity estimation, 107–108
  and shales, 108, 113, 114, 144–146
Density tool design, 106
Depositional environment, *see* Sedimentary environments
Depositional environment recognition:
  from dipmeter, 45–47
  from gamma ray log, 88
  from porosity, 119
  from SP shapes, 75–83
Derivative:
  first, 193, 195–196, 239–241
  second, 193, 195–196, 241, 242, 255
Determined system, 173–175
Diagenesis, 17, 118–120
Dipmeter:
  color conventions, 39
  and compactional drape, 43–45, 59
  cylindrical plot, 57, 61
  eigenvector analysis, 63–64
  and fabric orientation, 64–67
  and faults, 41–43
  polar plots, 57–58, 62
  processing methods, 36–39, 45–46
  and sedimentary structures, 45–57
  tadpole (vector) plot, 39, 57, 60
  tool, 35
  and unconformities, 40–41
Discriminant function analysis, 150–151, 207–211
Discriminant score log, 210
Dolomitization, 119–120, 186
Dunes, eolian, 46–47

Eigenvalues, 63–64, 205
Eigenvectors, 63, 93, 205
Electrochemical potential, 71
Electrofacies, 203, 206–207
Electrokinetic potential, 71
Electron microscope, 12, 142
Entropy, 170
Error, absolute sum, 175, 177, 178
Error diagnostics, 177–178
Errors, logging tool, 177, 213–224
Evaporites, 45, 90, 112, 167–169, 268

Fabric orientation analysis, 64–67
Factor analysis, 93–95
Faults, 41–43, 243, 251, 259

Feature extraction, 46
Feldspar, 85, 90, 95, 96, 146, 268
Filter, 180, 194–196, 202
First electrical log, 1–2
Flushed zone, 27, 30–31, 32, 105
Formation factor, 12, 15, 16, 20, 22, 25
Fourier analysis, 196–200
Fractures, 22–23, 96–97, 105, 178, 254
Frequency domain, 197

Gamma ray index, 84–87
Gamma ray log, 20, 57, 83–91, 119, 125–128, 132–133, 143, 145, 149, 161, 164, 186, 205, 209, 226–244
  and environmental recognition, 88
  and shale content, 84–88, 226
  spectral log, 90, 92–98, 146, 166
Gas, 9, 21, 75, 105, 108, 112, 114, 144, 268
Geographic regions:
  Alberta, 44, 45
  Beaufort Basin, 82
  California, 78, 116
  Colorado Plateau, 42
  Gulf Coast, 41, 117–118
  Kansas, 10–11, 17, 20, 23–24, 57–64, 119, 127–129, 132–133, 135, 145, 149–150, 158–164, 167–170, 178–180, 185–203, 205, 208–210, 215–216, 217–222, 227–260
  Kentucky, 90
  LeHave Platform, 41
  Michigan Basin, 45
  Niger Delta, 75, 77, 143–144
  North Dakota, 53
  North Sea, 47, 88, 118, 136, 144
  Oklahoma, 65, 79, 114
  Sacramento Basin, 78
  Texas, 44, 67, 96, 120
  Venezuela, 67, 136
  Wisconsin, 21
  Wyoming, 53
Geothermal reservoirs, 96
Glauconite, 88, 268
Grain densities, 108, 267–268
Grain density, apparent, 137–138
Grain orientation, 19, 64–66
Grain shape, 19
Graywackes, 142
Guided minimum variance, 173, 178
Guidepoint, composition, 173

Halite, 107, 167–169, 268
Helmert transformation, 186
Histograms, 217

Index

Humble equation, 14–15
Hydrocarbon effect, 105, 108, 112
Hydrocarbons, 3, 24–27, 31, 75
Hydrogen index, 112–113

Illite, 95, 113, 164, 268
Induction log, 10, 57
Induction resistivity, 10, 23, 29–30, 72
Information theory, 170
Insoluble residues, 161–163
Invasion, 27, 28, 30

Kaolinite, 74, 164, 268
KIWI, 161, 262–266
Kurtosis, 232

Lagrange multipliers, 175
Larionov chart, 85, 88
Laterolog, 23, 30, 32–33, 215
Least-squares, 175, 191, 217
Lithodensity log, 109, 138–139, 146, 166
Lithofacies mapping, 225–260

Mathematical geology, 184
Matrix:
   algebra, 154–161, 170–177, 235–236, 238, 248
   identity, 156–157
   inverse, 156–157, 238
   transpose, 171
Mean absolute deviation, 176
Mica, 90, 96, 145–146, 164, 268
Micrite, 119, 120
Microresistivity, 23, 30–31, 32, 35, 46
Microscope, electron, 12, 142
MID crossplot, 137–138, 147
M-N crossplot, 133–136, 147, 249
M-N-Z crossplot, 178
Moments, statistical, 228–233, 236–239
Montmorillonite, 74
Movable hydrocarbon saturation, 31
Moving average, 194
Mud filtrate, 27, 31, 72–73

Negative components, 175, 178, 249
Neutron log, 109–113, 119, 124–128, 134, 136, 139, 144–145, 159, 161, 164, 186, 205, 209, 215, 218–219, 245, 247
   and porosity estimation, 110–111
   and shales, 113
Neutron tool design, 109–110
Normalization, 213–224
Normal resistivity tool, 28
Normative and modal solutions, 180–182

Ohm's law, 8
Oil, 9, 20, 21, 25–26, 31, 75, 78, 96, 104, 105, 112, 227, 259, 268
Oolitic limestone, 10, 119, 209
Oomoldic porosity, 23–24
Orientation tensor, 63
Overdetermined system, 175–177
Overlays, log, 124–128, 216
Overpressured zones, 117–118

Pattern recognition, 38, 46, 203–211
Péchelbronn, 1
Percolation theory, 19
Perfect Martini, 154–157
Permeability, 17
Photoelectric absorption, 109, 139, 165–166
Polar plots, 57–58, 62
Polynomial stratigraphy, 242
Pore casts, 12
Porosity:
   carbonate types, 22
   and density log, 107–108
   geological implications, 118–121
   map, 119, 219–222, 259
   and neutron log, 110–111
   "primary," 105
   and resistivity, 10–15
   "secondary," 105, 178, 249
   and sonic log, 104–105
Porosity logs, 100–121
   crossplots, 128–132
   overlays, 124–128
Potassium, 83, 89, 92–98
Potential:
   electrochemical, 71
   electrokinetic, 71
Power spectrum, 198, 199, 201
Principal components, 94, 150–151, 204–206
Projection:
   conical, 135, 147–148
   spiral, 149
Projections of crossplots, 147–151
Proportional variance, 170–171
Pseudostatic potential, 73
Pyrite, 114, 267

Radius of gyration, 229, 232
Radius of investigation, 28, 32–33, 83
Reduced major axis, 187
Reefs, 44–45
Reflection coefficient, 101
Regression, *see* Trend
Residual chart, 247, 251–253
Residuals, trend surface, 221–224

Resistance, 8, 11–12
Resistivity, defined, 9
 and porosity, 10–15
 ellipse, 66–67
 formation water, 9–11, 26
 hydrocarbon-bearing rocks, 24–27
 index, 25
 log, 1, 10, 23, 32–33, 57, 77, 78, 79, 215
 rocks, 9–11
 shales, 9, 26, 67
 tools, 28–33
ROMA-UMA crossplot, 138–139
Rose diagram, 63

SP shapes, defined, 76
 environmental recognition, 75–83
 mapping, 79–83
Sandstones, shaly, 26, 140–146
Saturation exponent, 25
Scattering angle, 64
Schlumberger, Conrad, 1
Schmidt net, 57–58
Search length, 36–38
Sedimentary environments:
 barrier bar, 43, 54–56, 79
 beaches, 55, 66
 braided stream deposits, 47–48, 66
 channel, 17, 43, 62, 79
 delta distributary, 51–53, 79
 eolian dunes, 46–47
 estuarine & tidal, 53–54
 oolite shoals, 119, 209
 point bars, 48–50
 reefs, 44–45
 shallow marine, 21, 56–57, 66, 209
Seismic records, 101
Self-similar model, 19
Sequency, 200–203
Shale, 141–146
 compactional trends, 113–118
 and density log, 108, 113, 114
 and gamma ray log, 84–88, 226
 indicators, 74, 84–88, 142–144
 laminar, structural and dispersed, 141
 and neutron log, 113
 and sonic log, 105, 113, 114, 115–118
 and spontaneous potential, 73, 74
 resistivity, 9, 26, 67
Shale ratio:
 map, 227–228
 mean, 227, 230–231
Shales, organic, 90, 95, 114
Shaly sandstones, 26, 140–146
Short normal, 28, 72

Siderite, 114, 267
Simultaneous equations solution, 160
Skewness, 232
Smectite, 113
Sonagram, 199
Sonic log, 101–105, 128, 133–134, 159, 167,
 186, 205, 215, 218, 220, 245, 247
 and porosity estimation, 104–105
 and shales, 105, 113, 114, 115–118
Sonic tool design, 102–103
Span, 103
Spearman correlation, 16–17
Spectral gamma ray log, 90, 92–98, 146, 166
Spiral projection, 149
Spontaneous potential:
 and hydrocarbon zones, 75
 and shale content, 73–74
 and water resistivity, 72–73
Spontaneous potential log, 57, 71–83, 143, 205
Spontaneous potential log shapes, see SP shapes
Square waves, 200
Standard deviation, relative, 232
Static self-potential, 73
Stationarity, 197
Step length, 36–38
Stratigraphic units:
 Agboda Formation, 144
 Arbuckle Limestone, 23, 161–163,
  215–216
 Austin Chalk, 96
 Bartlesville Sandstone, 57
 Cherokee Group, 57, 145, 164
 Coffeyville interval, 79
 Cogollo Group, 136
 Greenhorn cyclothem, 95
 Hunton Group, 10–11, 127, 128
 Kansas City Group, 23–24, 119, 149
 Kinderhook Shale, 10, 133
 Leduc Reef, 44
 Maquoketa Shale, 10, 127, 132–133
 Mississippian, 44, 208–210
 Rotliegendes Group, 47
 St. Peter Sandstone, 20–21, 128
 Simpson Group, 127, 227–245
 Skinner Sandstone, 17, 145, 164
 Viola Limestone, 127, 128, 129, 135,
  158–160, 178–180, 185–203, 217–222,
  247–260
 Wellington Formation, 167–170
Synthetic seismograms, 101, 109
System of equations:
 determined, 173–175
 overdetermined, 175–177
 underdetermined, 170–173

Tadpole plot, 39, 57, 60
Textural maturity, 17
Texture, and cementation factor, 17–24
Thorium, 83, 89, 92–98, 146
Three–dimensional mapping, 228–246
Three wise monkeys, 4–6
Time domain, 197
Time series analysis, 190–203
Tortuosity, 13–16, 23
Transition zone, 27
Transit times, 103–104, 267–268
Trend:
   linear, 188, 217, 219–220
   polynomial, 188, 190–196, 234–237
   quadratic, 188, 217, 219–220
Trend surface analysis, 217–224
Trend surface residuals, 221–224

Unconformities, 40–41, 116
Underdetermined system, 170–173
Uranium, 83, 89, 92–98

Variance:
   analysis of, 219
   guided minimum, 173
   proportional, 170–171
Vector plot, 39, 57, 60
Vertical resolution, 32–33, 107, 111, 168
Vertical variability maps, 228
Vugs, 22, 23, 105, 178

Wackestone, 119, 186, 209
Walsh functions, 200–203
Water saturation, 25, 26
Wiener, Nobert, 6
Window length, 36–38, 46
Wulff net, 57–58
Wyllie time average equation, 104–105, 157

Z/A ratio, 107
Zones, 33
Z-plot, 132–133

RAYMOND H. FOGLER LIBRARY
DATE DUE

BOOKS ARE SUBJECT TO
RECALL AFTER TWO WEEKS